A User's Manual to the *PMBOK®* *Guide*—Fifth Edition

A User's Manual to the *PMBOK® Guide*—Fifth Edition

Cynthia Stackpole Snyder

This book is printed on acid-free paper. ♾

Published by John Wiley & Sons, Inc., Hoboken, New Jersey
Published simultaneously in Canada

For general information about our other products and services, please contact our Customer Care Department within the United States at (800) 762-2974, outside the United States at (317) 572-3993 or fax (317) 572-4002.

Wiley publishes in a variety of print and electronic formats and by print-on-demand. Some material included with standard print versions of this book may not be included in e-books or in print-on-demand. If this book refers to media such as a CD or DVD that is not included in the version you purchased, you may download this material at http://booksupport.wiley.com. For more information about Wiley products, visit www.wiley.com.

Project Management Institute (www.pmi.org) is the leading advocate for the project management profession globally. Founded in 1969, PMI has more than 650,000 members and credential holders in 185 countries. PMI's Project Management Professional (PMP) credential is globally recognized as the gold standard credential in project management.

Library of Congress Cataloging-in-Publication Data is available upon request.

ISBN: 978-1-118-43107-8; 978-111-8-54628-4 (ebk); 978-111-8-54639-0 (ebk); 978-111-8-54660-4 (ebk)

Printed in the United States of America
10 9 8 7 6 5 4 3 2

Contents

Preface ix

Acknowledgements xi

Chapter 1 Introduction 1
- About This Book 1
- Project Management Process Groups 2
- Project Management Knowledge Areas 4

Chapter 2 Key Concepts 7
- Projects, Programs, and Portfolios 7
- Project Life Cycles 8
- Progressive Elaboration 9
- Tailoring 9
- Enterprise Environmental Factors 9
- Organizational Process Assets 10

Chapter 3 Initiating a Project 13
- Initiating Process Group 13
- Project Sponsor Role 13
- Project Manager Role 14
- Develop Project Charter 15
- Identify Stakeholders 18

Chapter 4 Planning Integration 23
- Planning Process Group 23
- Planning Loops 24
- Project Integration Management 25
- Develop Project Management Plan 26

Chapter 5 Planning Scope 31
- Project Scope Management 31
- Plan Scope Management 32
- Collect Requirements 35
- Define Scope 43
- Create WBS 46

Chapter 6 Planning the Schedule 53
Project Time Management 53
Plan Schedule Management 54
Define Activities 56
Sequence Activities 59
Estimate Activity Resources 63
Estimate Activity Durations 68
Develop Schedule 73

Chapter 7 Planning Cost 85
Project Cost Management 85
Plan Cost Management 85
Estimate Costs 88
Determine Budget 94

Chapter 8 Planning Quality 99
Project Quality Management 99
Plan Quality Management 101

Chapter 9 Planning Human Resources 111
Project Human Resource Management 111
Plan Human Resource Management 112

Chapter 10 Planning Communications 117
Project Communications Management 117
Plan Communications Management 117

Chapter 11 Planning Risk 123
Project Risk Management 123
Plan Risk Management 124
Identify Risks 129
Perform Qualitative Risk Analysis 134
Perform Quantitative Risk Analysis 138
Plan Risk Responses 142

Chapter 12 Planning Procurement 147
Project Procurement Management 147
Plan Procurement Management 148

Chapter 13 Planning Stakeholder Management 157
Project Stakeholder Management 157
Plan Stakeholder Management 157

Chapter 14 Executing the Project 161
Executing Process Group 161
Direct and Manage Project Work 162

Chapter 15 Executing Quality Management 167
Perform Quality Assurance 167

Chapter 16 Executing Human Resource Management 173
Acquire Project Team 173
Develop Project Team 176
Manage Project Team 181

Chapter 17 Executing Communications Management 187
Manage Communications 187

Chapter 18 Executing Procurement Management 191
Conduct Procurements 191

Chapter 19 Executing Stakeholder Management 197
Manage Stakeholder Engagement 197

Chapter 20 Monitoring and Controlling the Project 201
Monitoring and Controlling Process Group 201
Monitor and Control Project Work 202
Perform Integrated Change Control 205

Chapter 21 Monitoring and Controlling Scope 211
Validate Scope 211
Control Scope 213

Chapter 22 Monitoring and Controlling the Schedule 217
Control Schedule 217

Chapter 23 Monitoring and Controlling Cost 221
Control Costs 221

Chapter 24 Monitoring and Controlling Quality 231
Control Quality 231

Chapter 25 Monitoring and Controlling Communications 237
Control Communications 237

Chapter 26 Monitoring and Controlling Risks 241
Control Risks 241

Chapter 27 Monitoring and Controlling Procurements 245
Control Procurements 245

Chapter 28 Monitoring and Controlling Stakeholder Engagement 251
Control Stakeholder Engagement 251

Chapter 29 Closing the Project 255
Closing Process Group 255
Close Project or Phase 255
Close Procurements 258

Appendix 261

Index 289

Preface

This book is designed to help make the *Guide to the Project Management Body of Knowledge (PMBOK® Guide)*—Fifth Edition more accessible to project managers.

It presents information from the *PMBOK® Guide* -Fifth Edition, in easily understandable language, and it describes how to apply the various tools and techniques. In short, it makes the *PMBOK® Guide* easier to understand and helps you implement the practices described in the *PMBOK® Guide*.

The information in this book is based solely on information from the *PMBOK® Guide*—Fifth Edition.[1] Therefore, you will find identical definitions and many of the same tables and figures. Thus, we will not footnote each reference to the *PMBOK® Guide* because, as we have stated, that is the sole source for content.

We have included some forms in Appendix A that show you how to use a form or template to record the information in a specific document. These forms can be found in *The Project Manager's Book of Forms*,[2] published by PMI and Wiley. Again, since this is the sole source for forms, we will not footnote each reference.

To help make this book easier to read, we are using various icons, tables, data flow diagrams, and call-out boxes. For instance, when we use a definition from the *PMBOK® Guide* we have inserted a dictionary icon. At the beginning of each process we describe the process and then show a data flow diagram from the *PMBOK® Guide* so you can see how information flows through the process, where it comes from, and where it goes. Call-out boxes may be used to list elements of a particular document.

---

[1] *A Guide to the Project Management Body of Knowledge (PMBOK® Guide)*—Fifth Edition © 2012 Project Management Institute, 14 Campus Blvd., Newtown Square, PA.

[2] *A Project Manager's Book of Forms: A Companion to the PMBOK® Guide—Fifth Edition* © John Wiley & Sons, Inc., 111 River Street, Hoboken, NJ.

The information is presented by Process Group as opposed to how the *PMBOK® Guide* presents it, by Knowledge Area. Because this book is designed to assist you in managing a project we felt it would be helpful to present information more consistent with how you will apply it on a project. We hope this *User's Manual* helps you in delivering successful projects!

Acknowledgements

This book is a wonderful example of the team work it takes to get something published. First, as always, my thanks to Bob Argentieri for your continual support and promotion of my work. I appreciate the effort you put into making sure everything gets done on time and in the most productive way possible. I always look forward to catching up at congress and celebrating the latest achievements! I appreciate the work Amy Odum and Kerstin Nasdeo do to get everything cleaned up and published on a compressed timeline. I know you had to move a lot of things around to get this book out on time. Thank you!

Thank you to my good friends at PMI Publishing. Donn Greenberg, you are a force of nature. You know so much about publishing and how to make things work at PMI. I so appreciate your ongoing support for my books! Barbara Walsh, you are a gem of a human being and so smart in your field. I really appreciate your insight into how to make this second edition more user friendly. Thank you for your professional support and your friendship. Both mean a lot to me. Roberta Storer, you are an editor extraordinaire and a wonderful person! Thank you for your help in making this a better book.

Finally, thank to the *PMBOK® Guide*—Fifth Edition team. You did a wonderful job updating the standard!

Chapter 1

Introduction

TOPICS COVERED

About This Book

Project Management Process Groups

Project Management Knowledge Areas

About This Book

This book is designed to help make the *A Guide to the Project Management Body of Knowledge* (*PMBOK® Guide*)—Fifth Edition more accessible to project managers. The *PMBOK® Guide* is a standard, therefore it defines **what** is considered to be a good practice on most projects most of the time. Notice it does not define *best* practices, it defines *good* practices. Best practices tend to be industry and organization specific. Because the *PMBOK® Guide* is a standard, it is not descriptive. In other words, it doesn't tell you **how** to implement those practices, it merely identifies them.

The *PMBOK® Guide* also promotes a common vocabulary for project management, thereby enabling effective communication about project management between project managers, their sponsors, and their team members.

Many project managers, PMOs, and organizations mistake the *PMBOK® Guide* as a project management methodology. It is not. A project management methodology is a set of practices, policies, procedures, guidelines, tools, techniques, and so forth that are used to manage projects. This book is not a methodology. This book takes the information in the *PMBOK® Guide* and describes it in easily understandable language and explains how to apply the various tools and techniques. In short, it makes the *PMBOK® Guide* easier to understand and helps you implement the practices described therein.

The information in this book is based solely on information from the *PMBOK® Guide*—Fifth Edition. Therefore you will find identical definitions and some of the same tables and figures.

To help make this book easier to read we are using various features such as definitions, examples, tips, and data flow diagrams. At the beginning of each process we describe the process, show inputs, tools and techniques, and outputs and then show a data flow diagram from the *PMBOK® Guide* so you can see how information flows through the process, where it comes from, and where it goes next. In some instances, we provide a list of elements typically found in a particular document. Sometimes we include references of forms that show how you can record the information in the document. These forms can be found in the Appendix and are available in print and electronic form in *The Project Manager's Book of Forms*, published by PMI and John Wiley & Sons.

Project Management Process Groups

The project management standard is presented as 47 discrete processes. A process is a set of interrelated actions and activities performed to achieve a pre-specified product, result, or service. Processes are comprised of inputs, tools and techniques, and outputs. Therefore, this book will follow that structure of presenting a process and then discussing the individual inputs, tools and techniques, and outputs that comprise the process.

Input. Any item, whether internal or external to the project, that is required by a process before that process proceeds. May be an output from a predecessor process.
Tool. Something tangible, such as a template or software program, used in performing an activity to produce a product or result.
Technique. A defined systematic procedure employed by a human resource to perform an activity to produce a product or result or deliver a service, and that may employ one or more tools.
Output. A product, result, or service generated by a process. May be an input to a successor process.

To facilitate understanding of the processes, PMI has identified five Process Groups. These groups are: Initiating Process Group, Planning Process Group, Executing Process Group, Monitoring and Controlling Process Group, and the Closing Process Group.

Initiating Process Group. Those processes performed to define a new project or new phase of an existing project by obtaining authorization to start the project or phase.
Planning Process Group. Those processes required to establish the scope of the project, refine the objectives, and define the course of action required to attain the objectives that the project was undertaken to achieve.
Executing Process Group. Those processes performed to complete the work defined in the project management plan to satisfy the project specifications.

Monitoring and Controlling Process Group. Those processes required to track, review, and regulate the progress and performance of the project; identify any areas in which changes to the plan are required; and initiate the corresponding changes.

Closing Process Group. Those processes performed to finalize all activities across all Process Groups to formally close the project or phase.

Note in Figure 1-1 how the Process Groups interact with each other in each phase of the project and for the project overall. The processes in the Initiating Process Group are used to identify the high-level definition of the project or phase and obtain authorization to proceed. Once this is accomplished the high-level information can be further elaborated in the Planning Process Group. Of course, we don't only plan at the start of the project. We spend much of the first part of our project planning, but as we get into the Executing Process Group, where we are actually creating and developing the work of the project, we will need to plan in finer levels of detail and re-plan when things do not go as expected. In fact, the Monitoring and Controlling Process Group is used to compare our planned progress to our actual progress. If the two are acceptably consistent, we continue on with the project work. If they are not, we will need to plan corrective or preventive actions to get our performance aligned with our plan. Finally, we will use the Closing Process Group to finalize the work and archive the phase or project information.

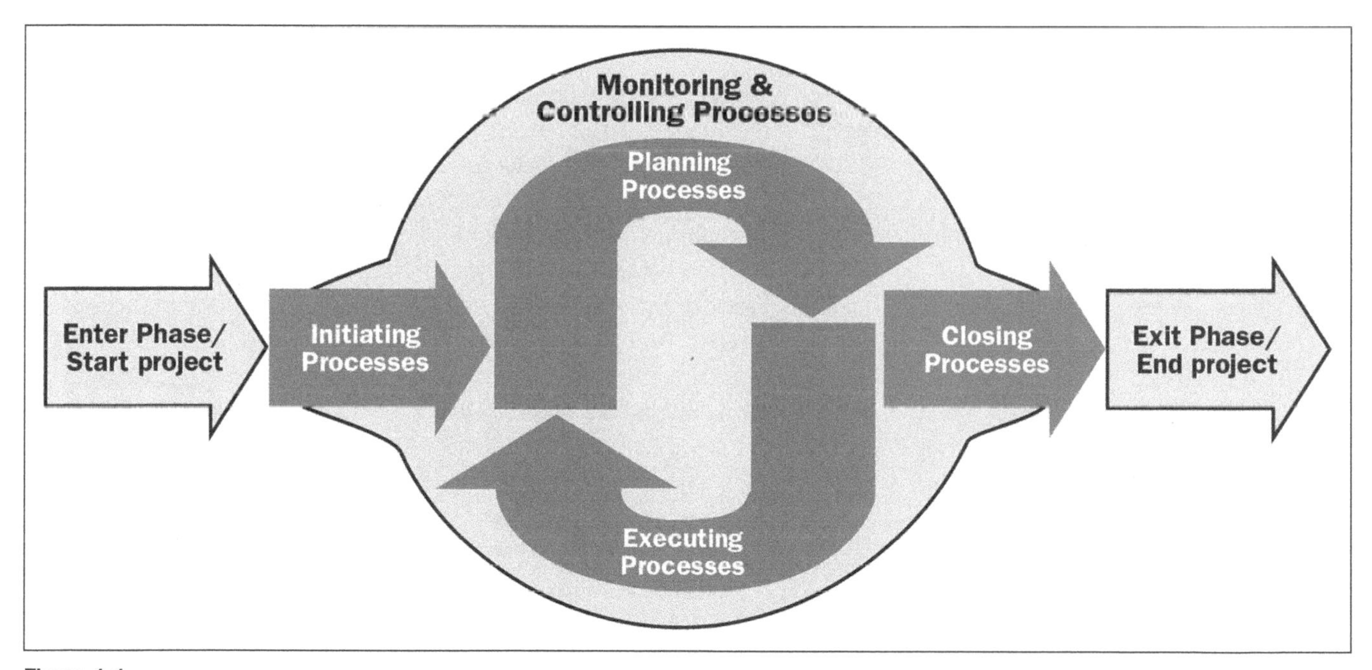

Figure 1-1
Project Management Process Groups
Source: *PMBOK® Guide,* Fifth edition

Project Management Knowledge Areas

Another way to categorize the project management processes is by Knowledge Area. PMI identifies ten Knowledge Areas:

- Project Integration Management
- Project Scope Management
- Project Time Management
- Project Cost Management
- Project Quality Management
- Project Human Resource Management
- Project Communications Management
- Project Risk Management
- Project Procurement Management
- Project Stakeholder Management

Figure 1-2 shows how each of the 47 project management processes aligns with the Project Management Process Groups and the Project Management Knowledge Areas.

This book will use the Process Groups rather than the Knowledge Areas to present information. In Chapter 2 we will review some of the key concepts in project management; in Chapter 3 we will discuss the Initiating Processes. The next several chapters will discuss the Planning Processes. This will be followed by chapters on the Executing Processes, Monitoring and Controlling Processes, and finally, the Closing Processes.

Project Integration Management. Project Integration Management includes the processes and activities needed to identify, define, combine, unify, and coordinate the various processes and project management activities within the Project Management Process Groups.

Project Scope Management. Project Scope Management includes the processes required to ensure that the project includes all the work required, and only the work required, to complete the project successfully.

Project Time Management. Project Time Management includes the processes required to manage the timely completion of the project.

Project Cost Management. Project Cost Management includes the processes involved in estimating, budgeting, funding, managing, and controlling costs so that the project can be completed within the approved budget.

Project Quality Management. Project Quality Management includes the processes and activities of the performing organization that determine quality policies, objectives, and responsibilities so that the project will satisfy the needs for which it was undertaken.

Project Human Resource Management. Project Human Resource Management includes the processes to organize, manage, and lead the project team.

Knowledge Areas	Project Management Process Groups				
	Initiating Process Group	**Planning Process Group**	**Executing Process Group**	**Monitoring and Controlling Process Group**	**Closing Process Group**
4. Project Integration Management	4.1 Develop Project Charter	4.2 Develop Project Management Plan	4.3 Direct and Manage Project Work	4.4 Monitor and Control Project Work 4.5 Perform Integrated Change Control	4.6 Close Project or Phase
5. Project Scope Management		5.1 Plan Scope Management 5.2 Collect Requirements 5.3 Define Scope 5.4 Create WBS		5.5 Validate Scope 5.6 Control Scope	
6. Project Time Management		6.1 Plan Schedule Management 6.2 Define Activities 6.3 Sequence Activities 6.4 Estimate Activity Resources 6.5 Estimate Activity Durations 6.6 Develop Schedule		6.7 Control Schedule	
7. Project Cost Management		7.1 Plan Cost Management 7.2 Estimate Costs 7.3 Determine Budget		7.4 Control Costs	
8. Project Quality Management		8.1 Plan Quality Management	8.2 Perform Quality Assurance	8.3 Control Quality	
9. Project Human Resource Management		9.1 Plan Human Resource Management	9.2 Acquire Project Team 9.3 Develop Project Team 9.4 Manage Project Team		
10. Project Communications Management		10.1 Plan Communications Management	10.2 Manage Communications	10.3 Control Communications	
11. Project Risk Management		11.1 Plan Risk Management 11.2 Identify Risks 11.3 Perform Qualitative Risk Analysis 11.4 Perform Quantitative Risk Analysis 11.5 Plan Risk Responses		11.6 Control Risks	
12. Project Procurement Management		12.1 Plan Procurement Management	12.2 Conduct Procurements	12.3 Control Procurements	12.4 Close Procurements
13. Project Stakeholder Management	13.1 Identify Stakeholders	13.2 Plan Stakeholder Management	13.3 Manage Stakeholder Engagement	13.4 Control Stakeholder Engagement	

Figure 1-2

Project Management Process Groups and Knowledge Areas Mapping

Source: *PMBOK® Guide,* Fifth edition

Project Communications Management. Project Communications Management includes the processes that are required to ensure timely and appropriate planning, collection, creation, distribution, storage, retrieval, management, control, monitoring, and the disposition of project information.

Project Risk Management. Project Risk Management includes the processes of conducting risk management planning, identification, analysis, response planning, and controlling risk on a project.

Project Procurement Management. Project Procurement Management includes the processes to purchase or acquire the products, services, or results needed from outside the project team.

Project Stakeholder Management. Project Stakeholder Management includes the processes required to identify all people or organizations impacted by the project, analyzing stakeholder expectations and impact on the project, and developing appropriate management strategies for effectively engaging stakeholders in project decisions and execution.

Chapter 2

Key Concepts

TOPICS COVERED

Projects, Programs, and Portfolios

Project Life Cycles

Progressive Elaboration

Tailoring

Enterprise Environmental Factors

Organizational Process Assets

Projects, Programs, and Portfolios

The difference between a project and a program can sometimes be fuzzy. And the difference between a program and a portfolio of projects can also be confusing. Let's start by looking at definitions for these words and then explore some additional key concepts in project management.

Project. A temporary endeavor undertaken to create a unique product, service, or result.
Program. A group of related projects, subprograms, and program activities, managed in a coordinated way to obtain benefits not available from managing them individually.
Portfolio. Projects, programs, subportfolios, and operations managed as a group to achieve strategic objectives.

Some people consider a program to be a jumbo-sized project. While this can be the case, it is not always true. For example, the Olympic Games could be considered a very large project with many subprojects. However, because of the size, cost, duration, and the sheer number of projects it takes to produce the Olympic Games, it is more like a collection of projects that is managed in a coordinated fashion—in other words, a program. Many of the projects are construction-related, many are production-related, many are related to press and broadcast, some are technology specific, and still others are about cultural events.

Within the program of the Olympic Games, you could even consider all the construction projects as a portfolio of projects. For the 2012 London Olympic Games, they were grouped together under the Olympic Delivery Authority (ODA) to facilitate effective management. Another portfolio could be considered the projects of the LOCOG. LOCOG is the London Organizing Committee for the Olympic Games. They were responsible for staging the Olympic and Paralympic Games.

Another way to look at the Olympic Games is having a portfolio for ODA, for the Olympics and another for the Paralympics. So you can see that much of the way you organize projects, programs, and portfolios is subjective. You can have programs with projects and portfolios of projects. You can also have portfolios with projects and programs made up of many projects. The main differentiator is that projects are always temporary, while programs and portfolios may have one or more elements that entail ongoing operations.

Project Life Cycles

Most large projects have a defined project life cycle made up of phases.

Project Phase. A collection of logically related project activities that culminates in the completion of one or more deliverables.
Project Life Cycle. The series of phases that a project passes through from its initiation to its closure.

There can be some confusion about the difference between a project life cycle and the Project Management Process Groups. Remember, the Process Groups are: Initiating, Planning, Executing, Monitoring and Controlling, and Closing. While these appear to be sequential, and could be mistaken for phases, they are groups of processes that are applied iteratively and as needed throughout the project. In some cases, the Project Management Process Groups are applied to each phase in a project. For example, a construction project might have three phases: design, procure, construct. An IT project might have phases such as: requirements, planning, design, detail design, build, test, deploy. Each phase is completed sequentially. The needs of the performing organization(s) and the project will determine the number and the names of the phases.

Many organizations use the end of a project phase to review the progress on the project. This gives the project manager, the sponsor, and the customer the opportunity to review the Charter, the progress, and deliverables to determine if the project should continue, if the approach should change, or if the project should be cancelled. There are times when the need for the project is no longer valid.

Circumstances or market forces may have changed, or the duration and cost of the project may no longer justify the expenditure of resources. The end of a phase (sometimes known as a phase gate or kill point) is often the right time to make those decisions.

Progressive Elaboration

One of the key concepts in project management is progressive elaboration.

Progressive Elaboration. The iterative process of increasing the level of detail in a project management plan as greater amounts of information and more accurate estimates become available.

One of the common laments of project managers is that customers and sponsors want accurate estimates in the beginning of a project, before the scope is even fully defined. The concept of progressive elaboration clearly articulates that we can't have detailed estimates until we have detailed and specific information about the project scope. As we progress in the project we can develop more accurate and complete information.

Tailoring

Projects, by their nature, are unique. Therefore, not all projects will use all processes defined in the *PMBOK® Guide*. Tailoring means that the project manager and the project team should carefully determine which processes are appropriate for their project, which outputs are appropriate, and the degree of rigor that should be applied when using the various tools and techniques. Some will use more robust processes, some will use less robust processes. It is up to the project manager and his or her team to determine the appropriate approach for the individual project.

Enterprise Environmental Factors

Enterprise environmental factors (EEF) are conditions, not under the immediate control of the team, that influence, constrain, or direct the project, program, or portfolio. These factors are from any or all of the enterprises involved in the project and include organizational culture and structure, infrastructure, existing resources, commercial databases, market conditions, and project management software. The following is a list from Chapter 2.1.5 in the *PMBOK® Guide*.

- Organizational culture, structure, and processes
- Government or industry standards
- Infrastructure

- Existing human resources
- Personnel administration
- Company work authorization systems
- Marketplace conditions
- Stakeholder risk tolerances
- Political climate
- Organization's established communication channels
- Commercial databases
- Project management information system

As project managers we don't have control over these factors, but overlooking them can lead to problems on our projects. Therefore, we need to be conscious of how these factors influence and constrain our options on projects.

Examples of how enterprise environmental factors can influence your project include:

Organizational Culture. Some organizations operate in an 8 AM–5 PM environment. People don't work at home; they work their regular shift, and then leave work behind. In other organizations, people bring their laptops home, they answer emails on their PDAs at dinner, they check email before bed and when they get up in the morning. They work weekends and stay late. If you are used to working in the latter environment and manage a project in the first environment you will probably need to modify your expectations.

Marketplace Conditions. The availability, price stability, and quality of resources available on the market can have a significant influence on your project schedule and budget.

Project Management Information System. Some organizations have a detailed project management methodology with policies, life cycles, procedures, templates, and software that work together to support effective projects. Other organizations use spreadsheets to manage projects and everyone uses their own version of documents. The rigor and robustness of the project management information system can either help, or hinder, in getting projects done efficiently.

Organizational Process Assets

Organizational process assets (OPA) include plans, policies, procedures, and knowledge bases that are specific to and used by the performing organization. These process assets include formal and informal plans, policies, procedures, and guidelines. The process assets also include the organizations' knowledge bases such as lessons learned and historical information. There are two general categories of organizational process assets: processes and procedures and the corporate knowledge base. The following is a list from Chapter 2.1.4 in the *PMBOK® Guide*.

PROCESSES AND PROCEDURES

- Guidelines and criteria for tailoring the organization's set of standard processes to satisfy the specific needs of the project
- Specific organizational standards such as, policies, product and project life cycles, and quality policies and procedures
- Templates
- Change control procedures, including the steps by which official company standards, policies, plans, and procedures—or any project documents—will be modified, and how any changes will be approved and validated
- Financial controls procedures
- Issue and defect management procedures defining issue and defect controls, issue and defect identification and resolution, and action item tracking
- Organization communication requirements
- Procedures for prioritizing, approving, and issuing work authorizations
- Risk control procedures, including risk categories, probability definition and impact, and probability and impact matrix
- Standardized guidelines, work instructions, proposal evaluation criteria, and performance measurement criteria
- Project closure guidelines or requirements

Some examples of how processes and procedures can be an asset include:

Reporting Templates. Templates can be used to specify the information you and your team need to compile on a regular basis. They document the information in a consistent format so the audience knows what they are looking at and can compare progress over time and across projects.

Risk Management Procedures. Predetermined definitions of impact on scope, schedule, cost, and quality can save the team time in developing the risk management procedures and avoid a hassle in trying to reach agreement on the definitions.

CORPORATE KNOWLEDGE BASE

- Configuration management knowledge bases containing the versions and baselines of all official company standards, policies, procedures, and any project documents
- Financial databases containing information such as labor hours, incurred costs, budgets, and any project cost overruns
- Historical information and lessons learned knowledge bases

- Issue and defect management databases containing issue and defect status, control information, issue and defect resolution, and action item results
- Process measurement databases used to collect and make available measurement data on processes and products
- Project files

> An example of how the corporate knowledge base acts as an asset includes:
>
> **Project Files.** Looking at project files from previous similar projects can provide a good starting point for developing cost and duration estimates and identifying assumptions and risks.

The difference between enterprise environmental factors (EEF) and organizational process assets (OPA) can be confusing. One way of looking at it is that EEFs tend to limit or constrain your options on a project and OPAs tend to assist, or provide guidance. For example, industry standards significantly restrict options in product development. You can develop a product that is not consistent with industry standards, but it will not be very successful. So this EEF is something that the project team must take into consideration in planning, and it is something that will constrain their options.

OPAs that provide assistance, such as templates for the Project Charter, or the WBS, are artifacts that the team does not have to invent for themselves. They provide a shortcut. Lessons learned from prior projects can also provide guidance and assistance when planning a project. In short, EEFs constrain, OPAs assist.

Chapter 3

Initiating a Project

TOPICS COVERED

Initiating Process Group

Project Sponsor Role

Project Manager Role

Develop Project Charter

Identify Stakeholders

Initiating Process Group

The Initiating Process Group consists of those processes performed to define a new project or a new phase of an existing project by obtaining authorization to start the project or phase.

Initiating processes are used to gain understanding of a project at a high level, authorize funding for the project, identify a project manager and authorize him or her to apply organizational resources to the project. At this point the key stakeholders are identified and their characteristics are documented in a Stakeholder Register.

Projects can be authorized by various methods. Sometimes a portfolio steering committee authorizes projects. Some organizations use the Project Management Office (PMO) to prioritize and authorize them. In other organizations, it is the project sponsor who has the authority to initiate a project.

Project Sponsor Role

Most projects have a project sponsor in addition to the project manager. The sponsor is usually someone at a level in the organization who can commit the funds necessary for the project and guide the strategic direction of the project. For interdepartmental projects, the manager or director in charge of the department may be the sponsor.

For organization-wide projects, or projects that have a significant strategic impact on the organization, the sponsor is generally someone at the corporate level.

Below are some of the key responsibilities associated with the sponsor role.

- Provide financial resources
- Champion the project to senior management
- Participate in developing the project charter
- Approve baselines for the project
- Authorize any significant project changes, including changes to the baselines
- Provide strategic direction and oversight to the project manager
- Be the escalation point for project conflict that the project manager is not able to resolve
- Participate in phase end reviews

Project Manager Role

At the heart of the project is the project manager. The project manager is the hub of communication, planning, execution, and control for the project. The project manager is assigned in the Initiating Process Group and may play a role in developing the project charter. If they do not develop it, they generally have input into the charter.

Below are some of the key responsibilities of the project manager.

- Documenting the project objectives, requirements, and other high-level project information in a project charter (if this has not already been done)
- Leading the team in progressively elaborating the project requirements and developing a complete and concise scope baseline
- Leading the team in developing supporting documents for the project management plan, such as the schedule, budget, risk management plan, and so forth
- Identifying the approach for carrying out the project work, including tailoring the project management plan and project documents to meet the needs of the project and the project stakeholders
- Maintaining performance consistent with the project baselines
- Managing change
- Communicating with and managing team members and other project stakeholders
- Reporting project status
- Managing risk

- Enabling the contracting and procurement process
- Closing out the project and collecting lessons learned

We'll start by looking at the first bullet in more detail by describing the Develop Project Charter process.

Develop Project Charter

Develop Project Charter is the process of developing a document that formally authorizes the existence of a project and provides the project manager with the authority to apply organization resources to project activities.

The project charter is the first formal project document. It includes the purpose of the project, a project description, and other high-level information. It is generally signed by the sponsor, the project manager, and the customer. The signed charter signifies an agreement about the work involved and the end product, service, or result.

In multiphase projects the charter is revisited at phase gates. Reviewing the charter at phase gates ensures that the results to date are aligned with the expectations listed in the charter. It also allows the project manager, sponsor, and customer to determine if the project should continue. In some cases, a business case may indicate that a project is a good investment for a specified amount of money, but as the project progresses it becomes obvious that the project cost will surpass that amount. In this case, the decision may be made to cancel the project.

Figure 3-1 and the data flow diagram in Figure 3-2 show the inputs needed to create the charter, the output of the process, and where the output(s) is/are used in future processes. Each process described in this book will have a similar data flow diagram. In future diagrams, the inputs will show discrete inputs as well as the processes from which they come. Since this is the first process in the project, there are no predecessor processes in this diagram.

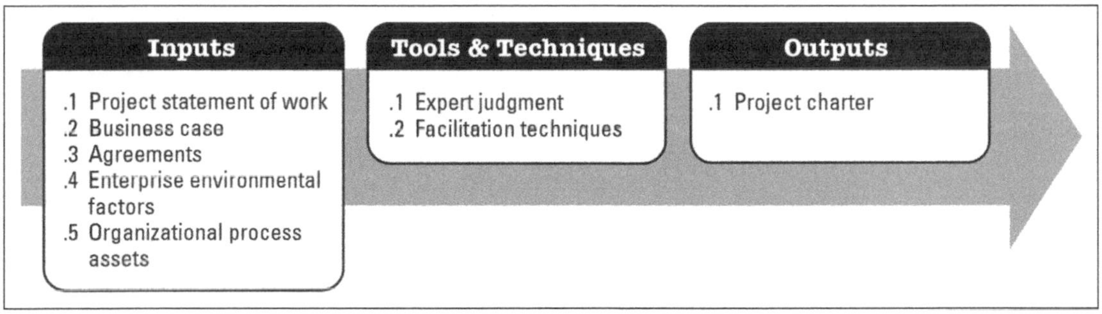

Inputs	Tools & Techniques	Outputs
.1 Project statement of work .2 Business case .3 Agreements .4 Enterprise environmental factors .5 Organizational process assets	.1 Expert judgment .2 Facilitation techniques	.1 Project charter

Figure 3-1
Develop Project Charter: Inputs, Tools and Techniques, and Outputs
Source: *PMBOK® Guide*—Fifth Edition

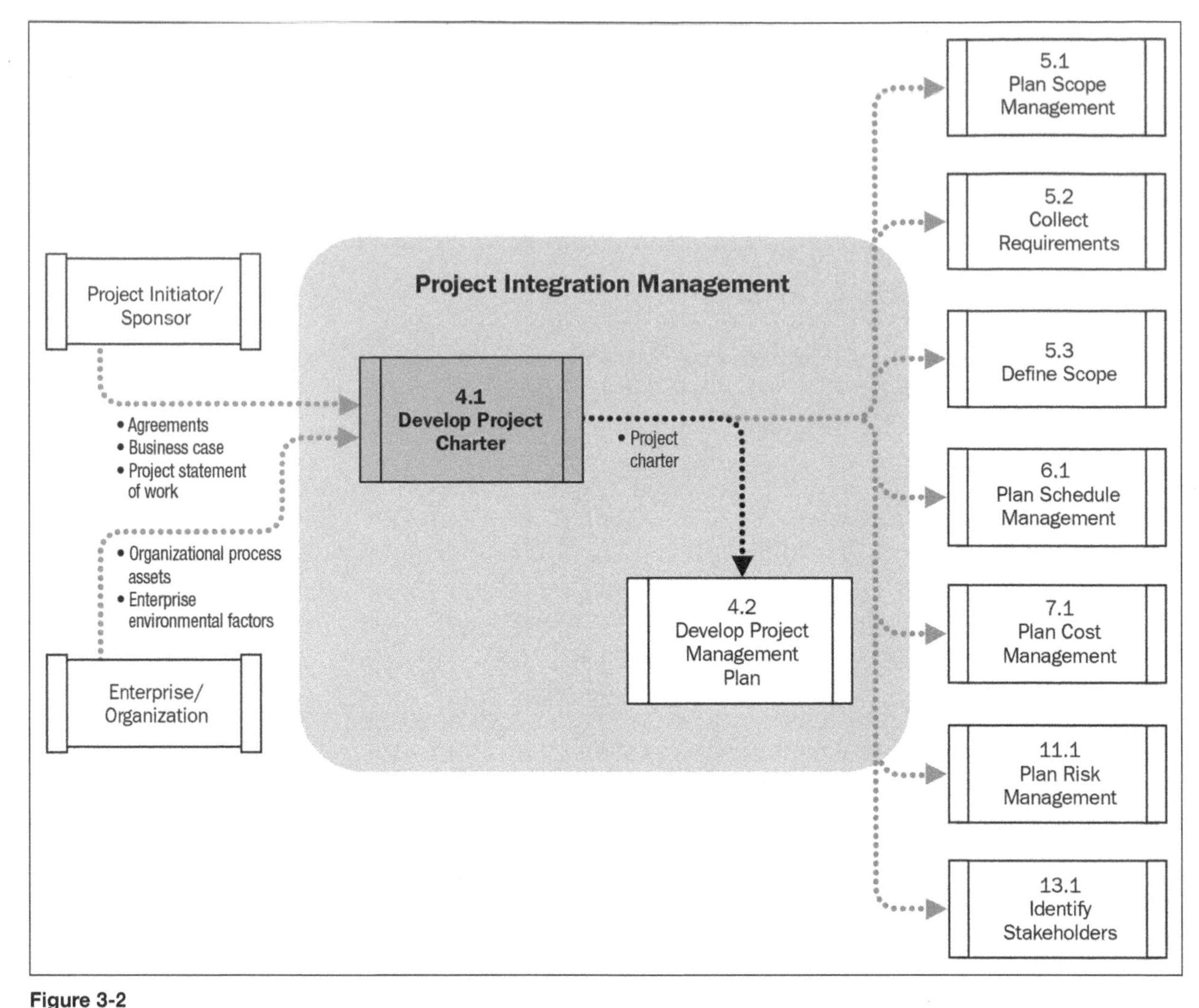

Figure 3-2
Develop Project Charter Data Flow Diagram
Source: *PMBOK® Guide* – Fifth Edition

INPUTS

The project statement of work describes the product that will be produced by the project. If the project is done for an external organization, the statement of work will be part of the *agreement*. If the project is internal to the organization it may be provided by the sponsor along with a justification for the project, such as the business need and how the project aligns with the organization's strategic plan. Often, internally initiated projects are accompanied by a *business case* that justifies the investment. The business case describes why the project is being undertaken. Reasons for undertaking projects include:

- Market demand such as developing products that are more environmentally friendly
- Internal organizational need such as making operations more efficient

- Customer request such as customizing a product to more effectively meet a customer's specific needs
- Technology advance such as upgrading hardware, software, or other equipment
- Legal or regulatory requirements such as meeting new information, privacy, and system security requirements
- Ecological reasons such as needing to meet stricter air quality requirements
- Social needs such as a fundraiser for a local charity

The business case also contains a cost-benefit analysis. Common cost-benefit analysis techniques include

- Return on investment (ROI)
- Payback period
- Net present value (NPV)
- Future value (FV)
- Internal rate of return (IRR)

All these methods attempt to determine the total expected costs against the total expected benefits, using currency as a common denominator. Most techniques recognize the time value of money through a discount rate to ensure a common comparison of alternatives and options.

In situations where the project is initiated because of a change in a legal requirement, the project may not be optional, but a cost-benefit analysis could be used to show the financial implications for various alternatives in implementing the project.

Of course financial investment and return are not the sole determining factors. Social, ecological, community perception, employee morale, and altruistic considerations can also be evaluated when undertaking new projects.

Agreements, such as contracts or purchase orders can also be an input to creating the charter as the product description as well as some of the terms and conditions will provide information for the project charter.

The enterprise environmental factors that are considered when developing a charter can include the marketplace conditions, such as interest rates, availability of skills or materials, and the cost of resources. In addition, any government regulations or industry standards that will impact the project should be identified. Finally, the organization's infrastructure, such as the available equipment, technology, hierarchy, and so forth, will influence the project approach.

Organizational process assets that are helpful include charters from previous projects; lessons learned from previous projects; and policies, procedures, and templates for developing the charter.

TOOLS AND TECHNIQUES

When developing the project charter you may need to talk to people with expert judgment in technical, business, or project management areas to help develop a robust, concise document. This can come from internal sources such as other project managers, the sponsor, technical subject matter experts, or the project management office. Sometimes it is useful to look outside the organization for expertise. For example, you may want to consult professional associations, industry groups, or independent consultants.

Developing a project charter can be challenging, especially if you have numerous stakeholders with conflicting opinions. In these instances you will need to employ facilitation techniques such as brainstorming, conflict resolution, problem solving, and meeting management to reach agreement on the project description and other elements in the project charter.

CHARTER CONTENTS

- Project purpose or justification
- Project description
- High-level requirements
- Measurable project objectives
- Success criteria
- High-level risks
- Summary milestone schedule
- Summary budget
- Stakeholder list
- Approval requirements
- Project manager authority
- Name and authority of authorizer

OUTPUTS

The project charter provides the initial understanding of the needs that the product, service, or result will fulfill. This information is used in later processes to collect requirements, define the scope, identify stakeholders, and provide input to the subsidiary plans that are part of the project management plan.

Depending on the nature, size, and complexity of the project the charter may contain all the listed elements, or it may contain additional information such as the initial project organization, functional areas that are involved, related projects, and the like. You should tailor the project charter to reflect the needs of the project.

Identify Stakeholders

Identify Stakeholders is the process of identifying the people, groups, or organizations that could impact or be impacted by a decision, activity, or outcome of the project; and analyzing and documenting relevant information regarding their interests, involvement, interdependencies, influence, and potential impact on project success.

Figure 3-3 shows the inputs, tools and techniques and outputs for the Identify Stakeholders process. Figure 3-4 shows a data flow diagram for the Identify Stakeholders process.

Stakeholder. An individual, group, or organization who may affect, be affected by, or perceive itself to be affected by a decision, activity, or outcome of a project.

Many of the stakeholders on a project will be obvious, such as the customer, end user, sponsor, team members, and vendors. However, there are times when there are certain stakeholders who may not be readily apparent. In a project to install a stoplight at an

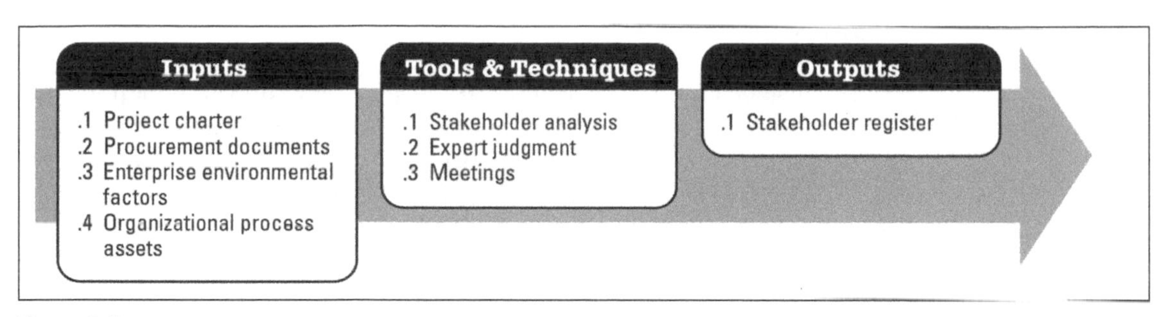

Figure 3-3
Identify Stakeholders: Inputs, Tools and Techniques, and Outputs
Source: *PMBOK® Guide*—Fifth Edition

intersection, a homeowners' group may insert itself into the project. They are affected by the execution of the project, and they may exert influence. At first they may not appear to be an important stakeholder, but failure to include and inform them of the project could have negative repercussions if, late in the project, they have a problem with some aspect of the project.

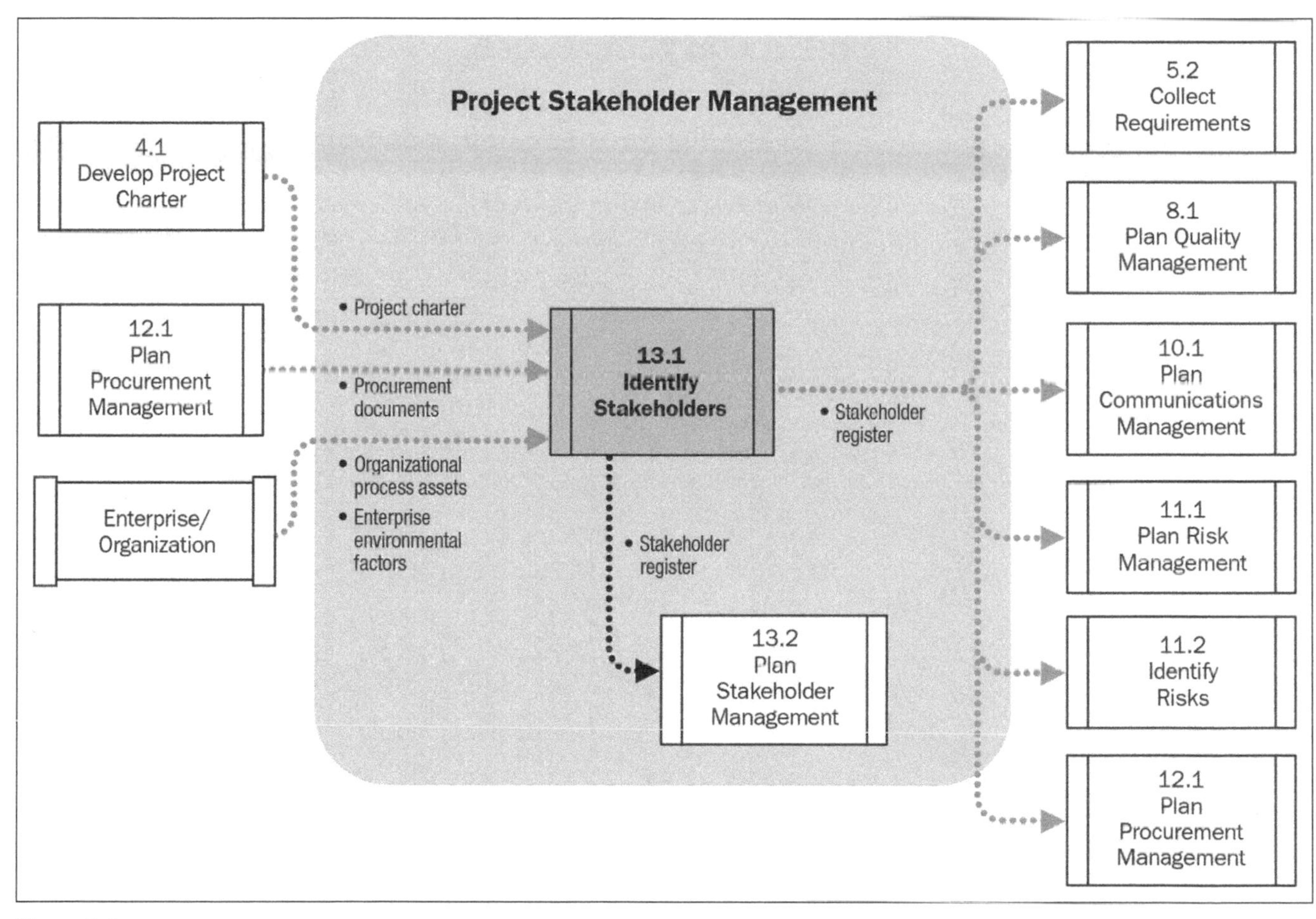

Figure 3-4
Identify Stakeholders Data Flow Diagram
Source: *PMBOK® Guide*—Fifth Edition

INPUTS

The project charter provides the project description which is the starting place to identify stakeholders. You will be able to identify at least the sponsor, customer, and end user by reading the charter. In most cases you can determine which departments, organizations, and regulatory agencies would need to be considered as well.

Procurement documents identify vendors and contractors that are involved in the project and should be considered stakeholders.

Enterprise environmental factors you should take into consideration are any applicable standards, regulatory, or legal entities. Often there are industry standards or government regulatory agencies that influence your project. Those bodies and entities may end up being stakeholders on your project if you need to obtain permits or provide notification about your project. You should also scan the external environment to see if there are any global, regional, or local trends or practices that will influence your project.

Organizational process assets such as lessons learned and stakeholder registers from previous projects, as well as templates for stakeholder registers are helpful in creating the stakeholder register for your current project.

TOOLS AND TECHNIQUES

Conducting a *stakeholder analysis* entails gathering and organizing information on project stakeholders. In addition to the generic contract, role, and department information, it can be useful to look a little deeper. For example, you may want to determine if stakeholders are likely to support your project or not, whether they have significant influence on the outcome of your project, or if they are even

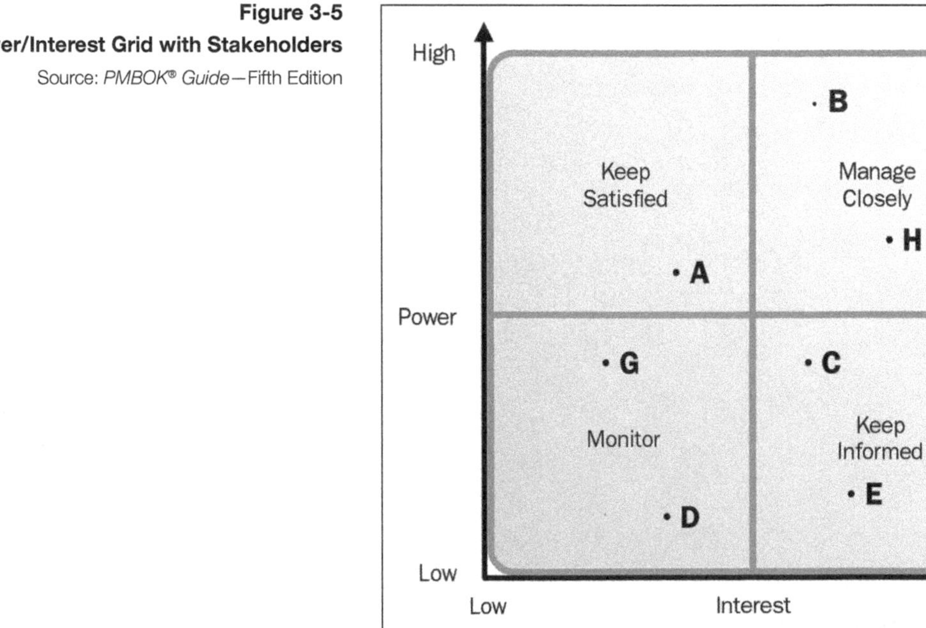

Figure 3-5
Example Power/Interest Grid with Stakeholders
Source: *PMBOK® Guide*—Fifth Edition

interested in its outcome. You can also use this information to create a matrix or a grid that groups common stakeholders together. For example, Figure 3-5 shows a Power/Interest Grid that identifies stakeholders who are interested in your project and the relative power, or lack of power, to influence the outcome of the project.

Meetings to gain expert judgment can be very helpful in identifying and categorizing stakeholders. Talking with project managers who have done a similar type of project, your sponsor, subject matter experts in the field, and previously identified stakeholders will help you identify additional stakeholders and determine how to manage them effectively.

OUTPUTS

The stakeholder register is usually documented in a tabular format. It lists the stakeholders' names and the relevant information you collect when conducting a stakeholder analysis. It can include only the contact and department information, or you can include the information from the stakeholder analysis matrix as well. Tailor the information to meet the needs of your project.

Chapter 4

Planning Integration

TOPICS COVERED

Planning Process Group

Planning Loops

Project Integration Management

Develop Project Management Plan

Planning Process Group

The Planning Process Group consists of those processes required to establish the scope of the project, refine the objectives, and define the course of action required to attain the objectives that the project was undertaken to achieve. Much of what we do on a day-to-day basis is planning. Of course, a good deal of planning happens at the beginning of a project, but planning continues throughout the course of the project as well.

At the start of the project we have a high-level understanding of the product and project. This is documented in the project charter. As we go through the various planning processes we take this information and further refine it. This continues until the project closes. This concept of gathering more detailed information is called *progressive elaboration*. A specific type of progressive elaboration occurs when there is concerted effort to define detail for specified planning horizons. This is called *rolling wave planning*. The intent of rolling wave planning is to document the detail of the work that is happening in the near future and keep the work that is further out in the future (for example, 90 days or more) at a less detailed level. As future work gets closer, it is decomposed into activities, sequenced, and resource loaded in the schedule. Let's look at the *PMBOK® Guide* definition of these two terms.

Progressive Elaboration. The iterative process of increasing the level of detail in a project management plan as greater amounts of information and more accurate estimates become available.

Rolling Wave Planning. The iterative planning technique in which the work to be accomplished in the near term is planned in detail, while the work in the future is planned at a higher level.

These methods of planning allow us to develop more accurate cost and duration estimates as the project progresses. However, it is a mistake to rush through initial project planning with the excuse that you are doing rolling wave planning so you will get to it later. It is also not an excuse for allowing scope creep. With rolling wave planning, the scope of the project does not change; the details of the project are merely refined.

Progressive Elaboration

Let's say you are building a new home for yourself. When you first buy the land, you decide where you want to place the house, how big you want it, the size of the garage, maybe even the materials you will use for the house.

At some point you hire an architect. You work with the architect and refine your ideas. As the plans are drawn you know the number of bedrooms and bathrooms, how big the kitchen and living areas will be, and so forth. You will even know the exact square footage and layout of the house. At this point, you have a good idea about how your house will look, but there is more detail to go.

The next step is to hire a contractor. When you hire a contractor you select the materials you will use, whether the exterior will be wood, stone, or stucco. As the contractor hires the subcontractors, you start purchasing materials and the actual construction begins. At this point, you should have a very detailed bill of materials, such as 18 boxes of copper river slate tiles that are 12" x 12". You may even know where the cuts in the tiles will be made and how they will be laid out.

There is no way you could have defined that level of detail when you were first looking at the plot of land you bought. Or if you did, it would likely change a few times. But the scope did not change; it was merely elaborated.

Planning Loops

Before we begin looking at the planning processes there are a couple of points you should keep in mind:

- Planning processes will overlap and interact throughout the project in many different ways.
- There are certain planning processes that by their nature create planning loops that go through several iterations before they are ready to be baselined.
- Developing a project management plan and the subsidiary plans that are part of it creates a loop that allows for continual update and refinement.

- Planning will continue throughout the project.
- Planning is integrative in nature.

For example, in Chapter 6 we will discuss the schedule planning processes. When you are developing your initial schedule, you will define and sequence your activities, then identify resources and estimate the durations to build your schedule. Inevitably this will not be your final schedule. You will have to determine which resources will be available, if they are available at the times indicated on your schedule, if the cost of the resources is acceptable given your budget, and so forth. You will also need to identify risks associated with the work and the schedule and then circle back and add in actions to reduce risk. You may end up adjusting scope and quality because of cost limitations. This will cause you to adjust your activities, sequencing, resources, durations, and so on. This entire chain of events will continue until you develop a schedule, budget, resource plan, firm scope, and achieve an acceptable level of risk.

At this point you may baseline your scope, schedule, and cost. However, changes will occur on your project that will cause you to re-plan and revise your schedule, cost estimates, risk responses, and so forth. This will entail more planning (some people refer to this as re-planning). It may even require you to re-baseline, depending on the nature of the changes. Figure 4-1 shows the iteration of inputs as the project management plan and other project planning documents are refined.

Project Integration Management

Project Integration Management includes the processes and activities needed to identify, define, combine, unify, and coordinate the various processes and project management activities within the Project Management Process Groups.

While project management is described as a series of processes categorized by Process Group and Knowledge Area, they are not implemented that way. The processes are implemented in an integrated fashion. Project managers don't define scope in a vacuum without considering schedule, cost, quality, risk, and so forth. We don't plan our human resource needs without an understanding

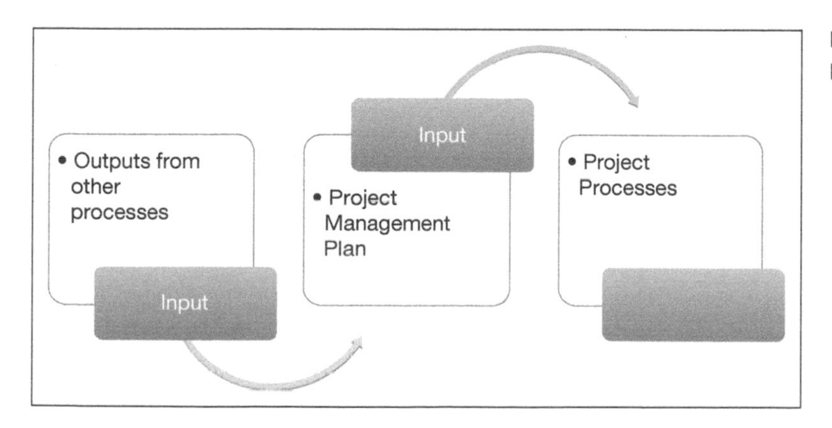

Figure 4-1
Project Management Plan Refinement Loop

of scope, schedule, cost, quality, or communication. It is easier to describe project management as a series of processes. However, the real art takes place in integrating, combining, unifying, making tradeoffs, and consolidating, while managing stakeholder expectations. Projects are successful to the degree we can balance and integrate all the project management processes.

Four of the processes in the Project Integration Management Knowledge Area have discrete outputs, the other two, Direct and Manage Project Execution and Monitor and Control Project Work, are more about the day-to-day effort required to deliver a successful project. We have already discussed the first integration process, Develop Project Charter. Now we will focus on the Develop Project Management Plan process.

Develop Project Management Plan

Develop Project Management Plan is the process of defining, preparing, and coordinating all subsidiary plans and integrating them into a comprehensive project management plan. The project management plan is the central document for executing, monitoring, controlling, and closing the project. A good part of the project management plan is comprised of the subsidiary management plans that are outputs from the various planning processes. The balance of the project management plan contains the project baselines and additional information on how the project will be executed, monitored, controlled, and closed.

Developing the project management plan is an iterative and ongoing process. Initially you may start with a framework for the project management plan that includes the project lifecycle, key management review points, and the high-level milestones and budget from the project charter. This framework is an input to the first planning[1] process in each of the other knowledge areas.

Figure 4-2 shows the inputs, tools and techniques and outputs for the Develop Project Management Plan process. Figure 4-3

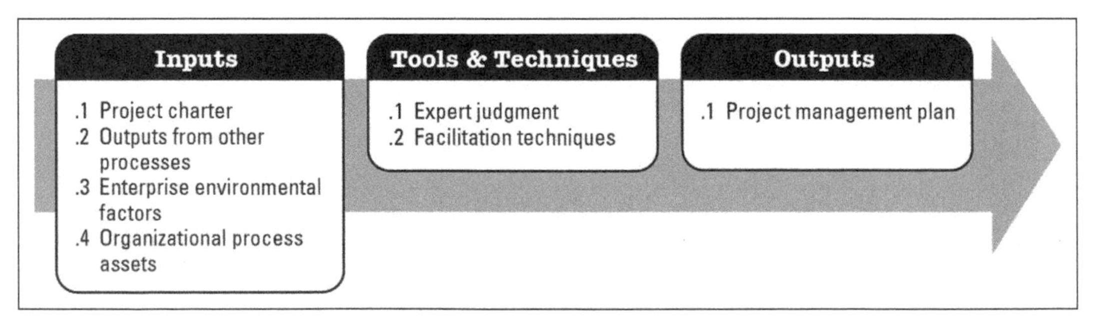

Figure 4-2

Develop Project Management: Inputs, Tools and Techniques, and Outputs

Source: *PMBOK® Guide*—Fifth Edition

[1]*Identify Stakeholders is the first process in the Project Stakeholder Management Knowledge Area. This is an Initiating process. Therefore the Charter is the key input for this process. The project management plan is an input to the first planning process in Project Stakeholder Management, which is Plan Stakeholder Management.*

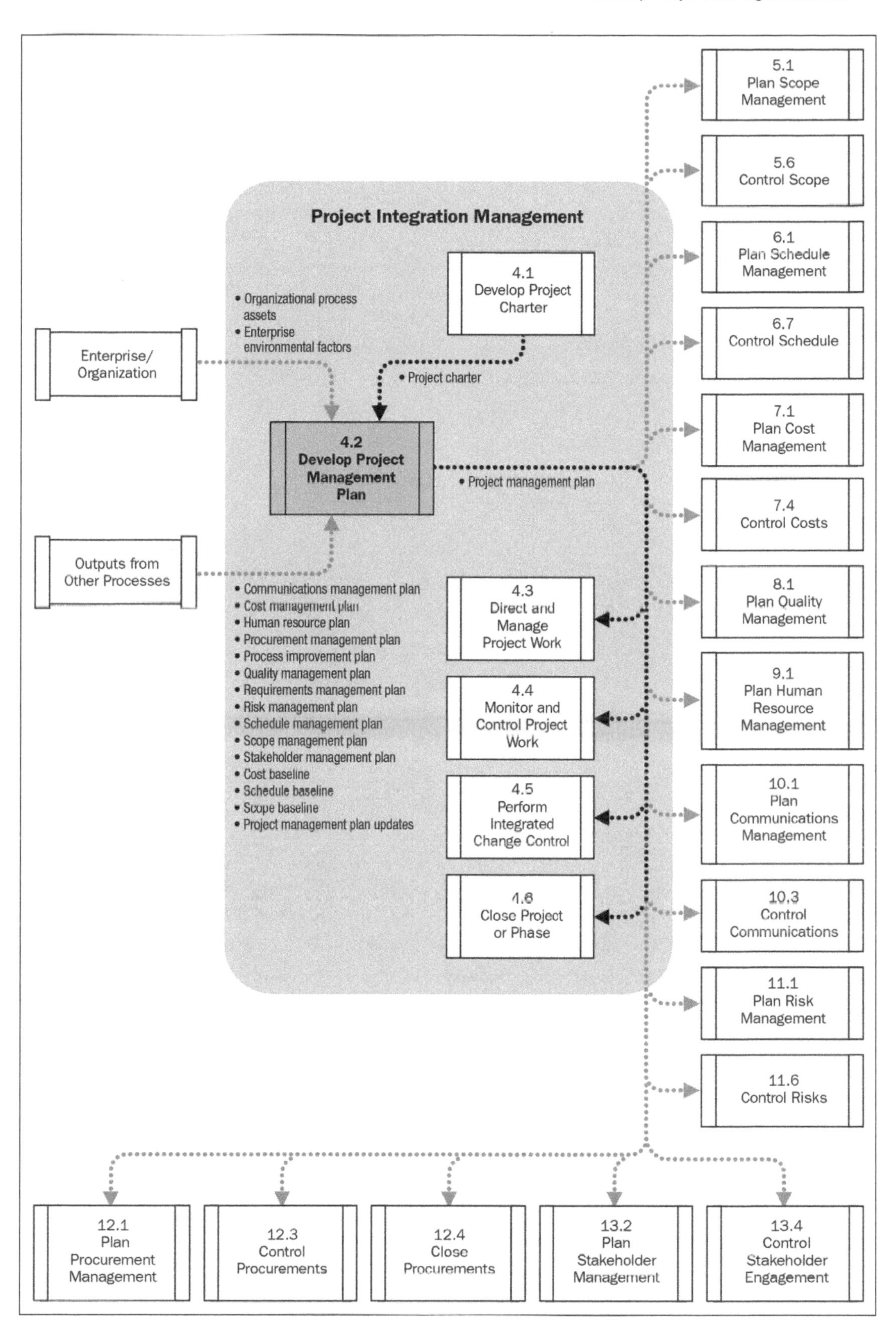

Figure 4-3
Develop Project Management Plan Data Flow Diagram
Source: *PMBOK® Guide*—Fifth Edition

shows a data flow diagram for the Develop Project Management Plan process.

INPUTS

The project charter has the project justification, project description, objectives, success criteria, and the project manager authority level. Most of the other information in the charter eventually will be progressively elaborated in subsidiary plans and baselines.

The outputs from planning processes includes at least:

- Requirements documentation
- Scope baseline
- Schedule baseline
- Cost baseline
- Scope management plan
- Requirements management plan
- Schedule management plan
- Cost management plan
- Quality management plan
- Human resources management plan
- Communications management plan
- Risk management plan
- Procurement management plan
- Stakeholder management plan

Enterprise environmental factors that influence this process include the organization's information management systems, industry standards and regulations, and organizational culture and structure. The information management systems may allow for automated document sharing and storage, along with controls on who can view, edit, or print information. They also determine specific software use for various components of the project management plan contents, such as the schedule, budget, and change control.

Industry standards and regulations determine the need for additional components to the plan such as a safety plan, a security plan, system engineering plan, training plan, and the like. The organization culture and structure will determine the amount of autonomy and authority the project manager has to direct his or her project.

Organizational process assets include policies, procedures, and templates. Specific policies that are of interest in this process include:

- Change management and control
- Configuration management and control
- Variance thresholds

- Tailoring guidelines
- Work authorization procedures
- Performance measurement guidelines

Project files, historical information, and project management plans from similar projects are also very helpful in developing the project management plan.

TOOLS AND TECHNIQUES

Expert judgment can include technical and project management expertise. It is used to help identify the documents that should make up the project management plan, the degree of tailoring to use, and the level of rigor for each document.

One of the areas that should be closely examined is the degree of change control and configuration control that makes sense for the project. Being overly rigid about project changes, or overly rigorous with change control procedures can reduce stakeholder satisfaction in some cases. In many more cases, not being rigorous enough can lead to schedule and cost overruns.

Identifying, labeling, managing, and auditing parts and documents via configuration control can make a big impact in helping a project run smoothly, but going overboard can waste time. Therefore, careful consideration should be given to these project policies.

Facilitation techniques, like those used in developing the project charter, will be useful in developing the project management plan as well. Meeting management, conflict resolution, problem solving, and brainstorming are excellent techniques to use to develop a project management plan that everyone buys into.

OUTPUTS

The project management plan is the overarching document that describes how the project will be managed. In addition to the subsidiary plans and project baselines it will also include at least the following information:

- Project life cycle
- Tailoring decisions for each process
- Approach to developing the deliverables (development methodology)
- Description of special tools and techniques that will be used in the project
- Key project interfaces (with other projects or organizations)
- Variance thresholds for scope, schedule, cost, and quality
- Approach for baseline management for scope, schedule and cost
- Number, timing, and description of project reviews

- Change control processes
- Configuration management processes

The project management plan is the master document for the project. It should be finalized (baselined) toward the end of the planning or elaboration phase. Though it will most likely be updated throughout the project, it is one of the documents that will require strict change and configuration control.

Sometimes there is confusion about what is an element of the project management plan and what isn't. There are a number of documents that help you manage a project that are not part of the project management plan. Collectively these are called *project documents*. Figure 4-4 lists elements from the project management plan and other types of project documents. The contents of the project management plan and the project documents are by no means complete. Every project will have unique project documents and the project management plan will be tailored to meet the needs of the individual project. However, Figure 4-4 provides a foundation for comparison of the project management plan and project documents.

Project Management Plan	Project Documents	
Change management plan	Activity attributes	Project staff assignments
Communications management plan	Activity cost estimates	Project statement of work
Configuration management plan	Activity duration estimates	Quality checklists
Cost baseline	Activity list	Quality control measurements
Cost management plan	Activity resource requirements	Quality metrics
Human resource management plan	Agreements	Requirements documentation
Process improvement plan	Basis of estimates	Requirements traceability matrix
Procurement management plan	Change log	Resource breakdown structure
Scope baseline • Project scope statement • WBS • WBS dictionary	Change requests	Resource calendars
Quality management plan	Forecasts • Cost forecast • Schedule forecast	Risk register
Requirements management plan	Issue log	Schedule data
Risk management plan	Milestone list	Seller proposals
Schedule baseline	Procurement documents	Source selection criteria
Schedule management plan	Procurement statement of work	Stakeholder register
Scope management plan	Project calendars	Team performance assessments
Stakeholder management plan	Project charter Project funding requirements Project schedule Project schedule network diagrams	Work performance data Work performance information Work performance reports

Figure 4-4

Differentiation between the Project Management Plan and the Project Documents

Source: *PMBOK® Guide*—Fifth Edition

Chapter 5

Planning Scope

TOPICS COVERED

Project Scope Management

Plan Scope Management

Collect Requirements

Define Scope

Create WBS

Project Scope Management

Project Scope Management includes the processes required to ensure that the project includes all the work required, and only the work required, to complete the project successfully.

There are two types of scope: product scope and project scope.

Product Scope. The features and functions that characterize a product, service, or result.

Project Scope. The work performed to deliver a product, service, or result with the specified features and functions.

For example, if your project is to develop a project management training curriculum for your organization, the product scope would be the actual course content. The project scope would include all project management documents such as the schedule, budget, and resource pool, as well as editing the work, doing a trial run of the course, and so forth. The project scope makes it possible to deliver the product scope, but it is not the actual end product.

Everything starts with scope. Once the scope is understood, via the requirements, scope statement, and work breakdown structure (WBS), you can start to define activities and resources, cost estimates, risks, procurement needs, and the like. When scope changes, usually one or more project processes will undergo a change or update as well.

Plan Scope Management

Plan Scope Management is the process of creating a scope management plan that documents how the project scope will be defined, validated, and controlled. Scope creep and losing control of requirements are two of the main reasons that projects get out of control. Spending time to determine how you will manage the project and product scope and the requirements will alleviate much of the confusion and conflict during project planning and execution. Figure 5-1 shows the inputs, tools and techniques and outputs for the Plan Scope Management process. Figure 5-2 shows a data flow diagram for the Plan Scope Management Process.

The scope management plan may identify the tools and techniques and approaches that will be used to define the scope statement, such as systems engineering, value engineering, and so forth. It can also identify if there are templates for the WBS, and the techniques that will be used to verify scope. The scope management plan should address the difference between progressive elaboration, scope creep, a scope change, and balancing scope needs to the budget and schedule. If there is not a separate requirements management plan, requirements planning information should be incorporated into the scope management plan.

As mentioned in Chapter 4, the project management plan is an input to the first planning process in every knowledge area. Then the resulting subsidiary management plan is an input to the project management plan. This results in a refinement and progressive elaboration of the project management plan and all its subsidiary documents. Figure 5-3 shows how the project management plan is an input to develop the scope management plan, which in turn is an input to refine the project management plan.

INPUTS

The project charter provides the project description and the high-level requirements that provide the initial information needed about

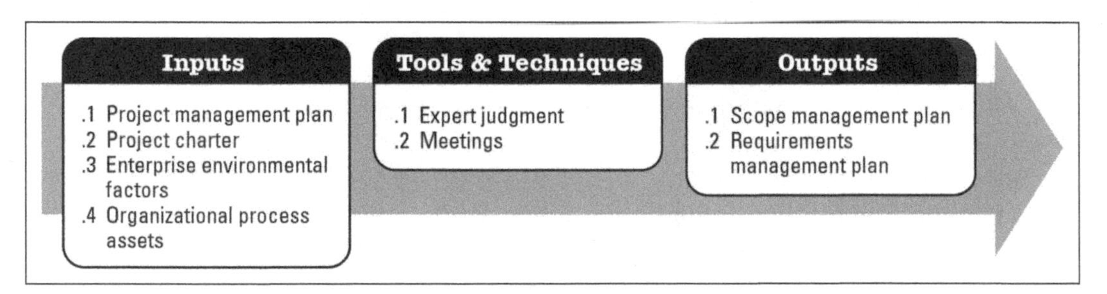

Figure 5-1
Plan Scope Management: Inputs, Tools and Techniques, Outputs
Source: *PMBOK® Guide*—Fifth Edition

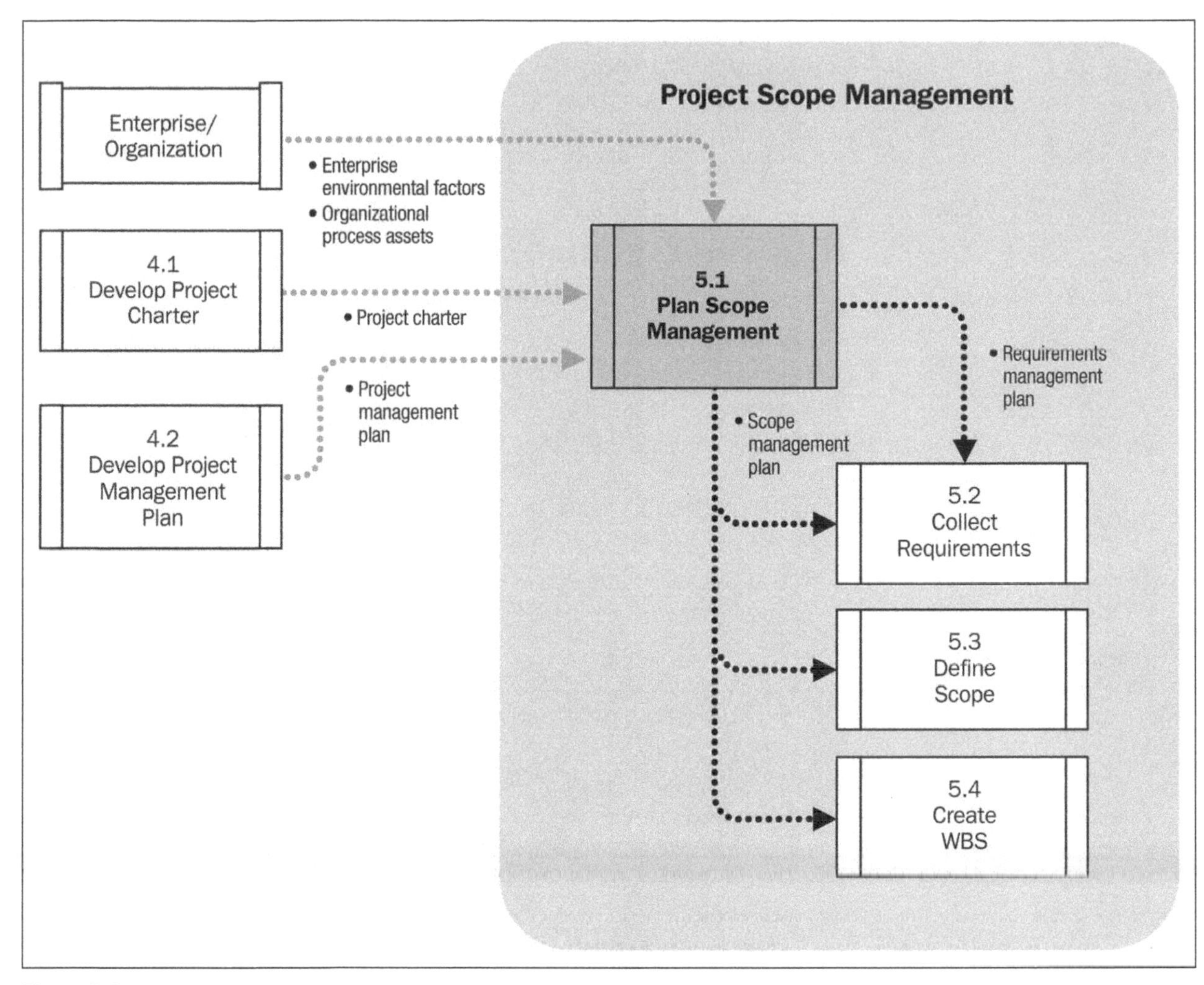

Figure 5-2
Plan Scope Management Data Flow Diagram
Source: PMBOK® Guide—Fifth Edition

the project scope in order to determine how best to manage the scope processes. In the initial phase of the project the project management plan will be more of a framework and contain high-level information. However, having it as an input provides a consistent method of developing subsidiary management plans.

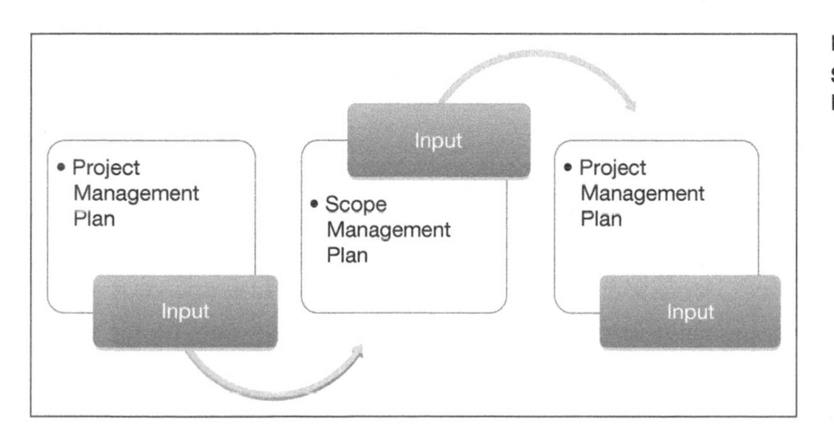

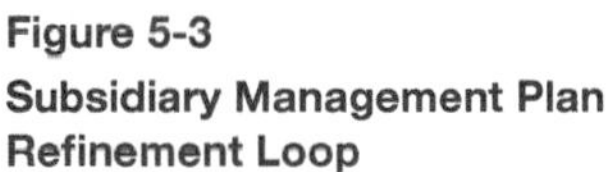
Figure 5-3
Subsidiary Management Plan Refinement Loop

The enterprise environmental factors that influence this process include the company culture and infrastructure as well as the marketplace conditions. The organizational process assets include information from previous, similar projects, templates, policies, and procedures.

TOOLS AND TECHNIQUES

Meetings with subject matter experts and other stakeholders who can lend their expert judgment are useful to discuss the best methods to elaborate and organize the scope. Additionally you should discuss how to manage scope creep, progressive elaboration, and scope changes so there are clear guidelines once project execution begins.

In a separate series of meetings you can pull together the team members who will be collecting and managing requirements to discuss how to ensure that all the requirements are captured, and that they are maintained and controlled in an organized fashion.

OUTPUTS

The scope management plan is a component of the project management plan. It describes the process for managing the scope processes including:

- The approach for developing the project scope statement, including methods for product analysis and the use of any facilitated workshops that will be used to clarify the project scope
- Guidelines for creating the WBS and the WBS dictionary, including the information that will be included in the WBS dictionary
- Methods used for deliverable verification and validation
- The approach for managing scope changes and methods for avoiding scope creep

Scope Management Plan. A component of the project or program management plan that describes how the project scope will be defined, developed, monitored, controlled, and verified.

Requirements Management Plan. A component of the project or program management plan that describes how the project requirements will be analyzed, documented, and managed.

The requirements management plan provides direction for maintaining control over requirements. The plan can document the various techniques that will be used to manage requirements including:

- Methods used to collect requirements
- Configuration management techniques used to control requirements
- How requirements will be categorized
- Requirements prioritization
- Information on how the requirements traceability structure will be set up, including the fields that will be documented, tracked, and traced
- Steps necessary to request changes to requirements and the authorization levels needed to approve changes

REQUIREMENTS MANAGEMENT PLAN CONTENTS

Requirements collection methods
Configuration management
Requirements categories
Prioritization
Traceability
Verification
Change management

Collect Requirements

Collect Requirements is the process of determining, documenting, and managing stakeholder needs and requirements to meet project objectives. A project requirement is a need or expectation for the project or product. Missing requirements, changing requirements, and unclear or undefined requirements are the main reasons projects are challenged or fail. Requirements can be progressively elaborated throughout the project, especially for complex projects, or projects that deal with new and evolving technology. Figure 5-4 shows the inputs, tools and techniques and outputs for the Collect Requirements process. Figure 5-5 shows a data flow diagram for the Collect Requirements Process.

There are project and product requirements. Product requirements can be decomposed into functional and nonfunctional requirements. Nonfunctional requirements can be further decomposed into reliability, security, safety, performance, and so forth. Figure 5-4

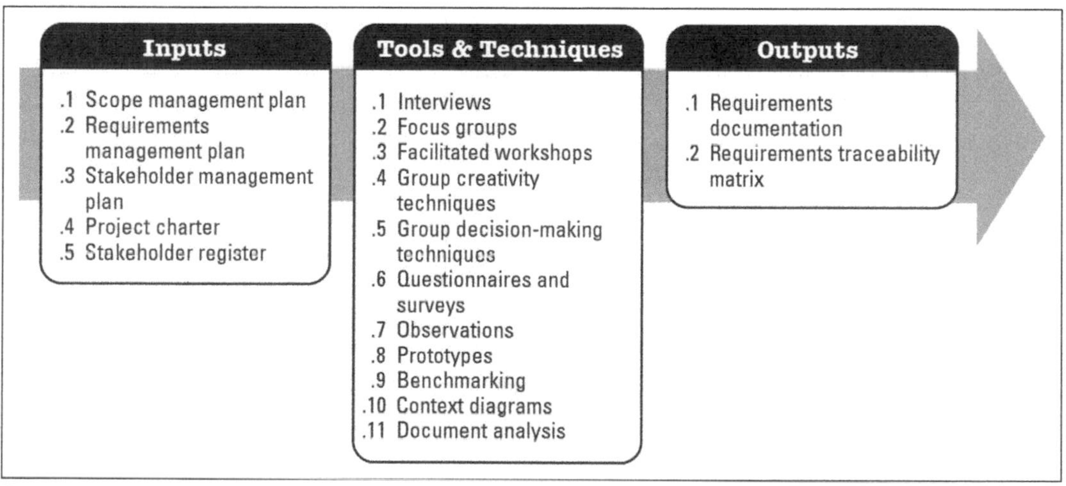

Figure 5-4
Collect Requirements: Inputs, Tools and Techniques, Outputs
Source: PMBOK® Guide—Fifth Edition

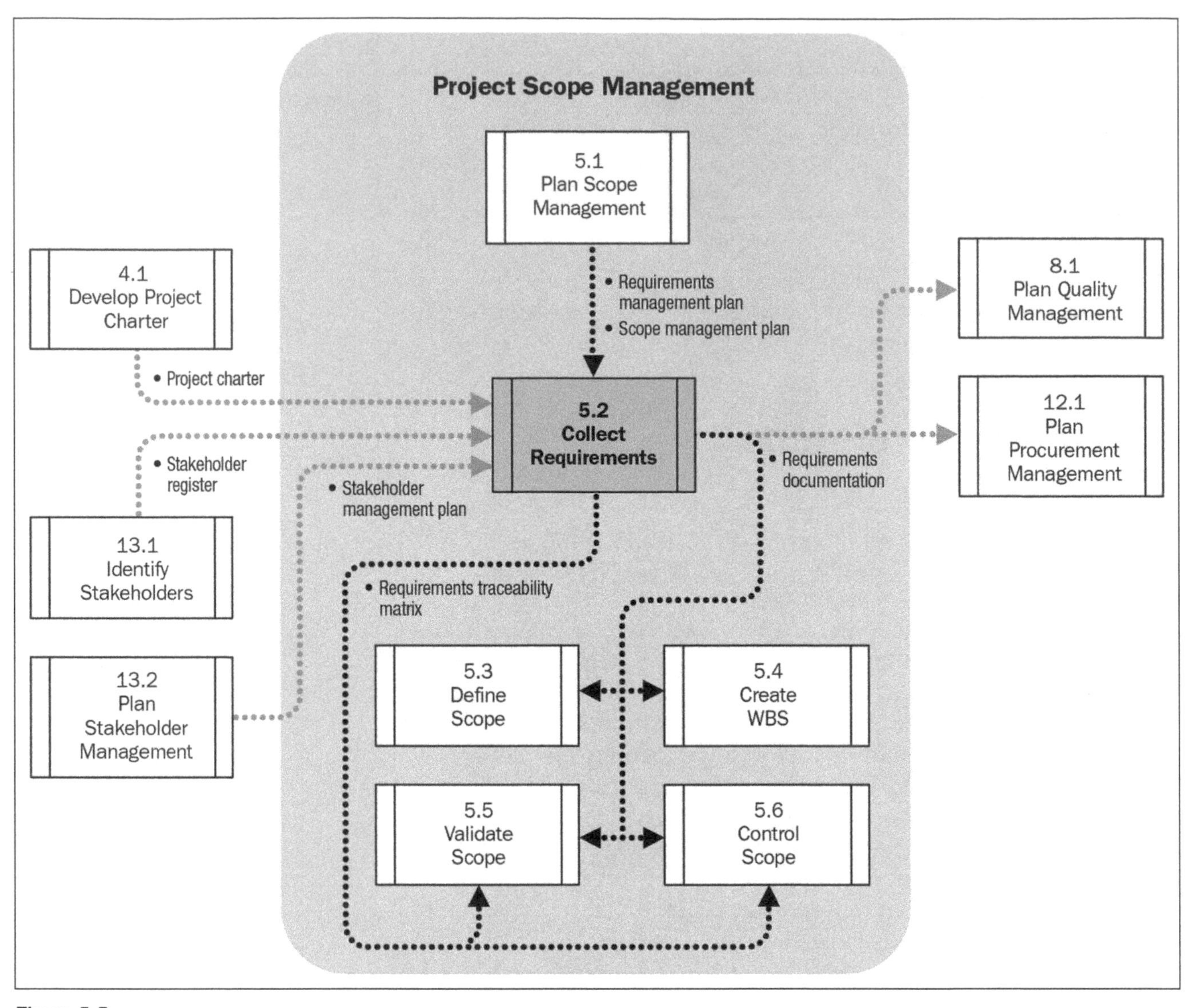

Figure 5-5
Collect Requirements Data Flow Diagram
Source: PMBOK® Guide—Fifth Edition.

shows how requirements can be decomposed. Decomposing requirements into categories as shown in Figure 5-6, helps to identify, manage, and control requirements.

INPUTS

There are several subsidiary management plans that you should reference when collecting requirements. The scope management plan and requirements management plan provide information on how project and product scope and the detailed requirements will be identified, defined, and managed throughout the project life cycle.

The stakeholder management plan identifies interests that categories of stakeholders might have and the type and degree of engagement in the requirements collection process. For example, end users may have functional requirements, while departments within the

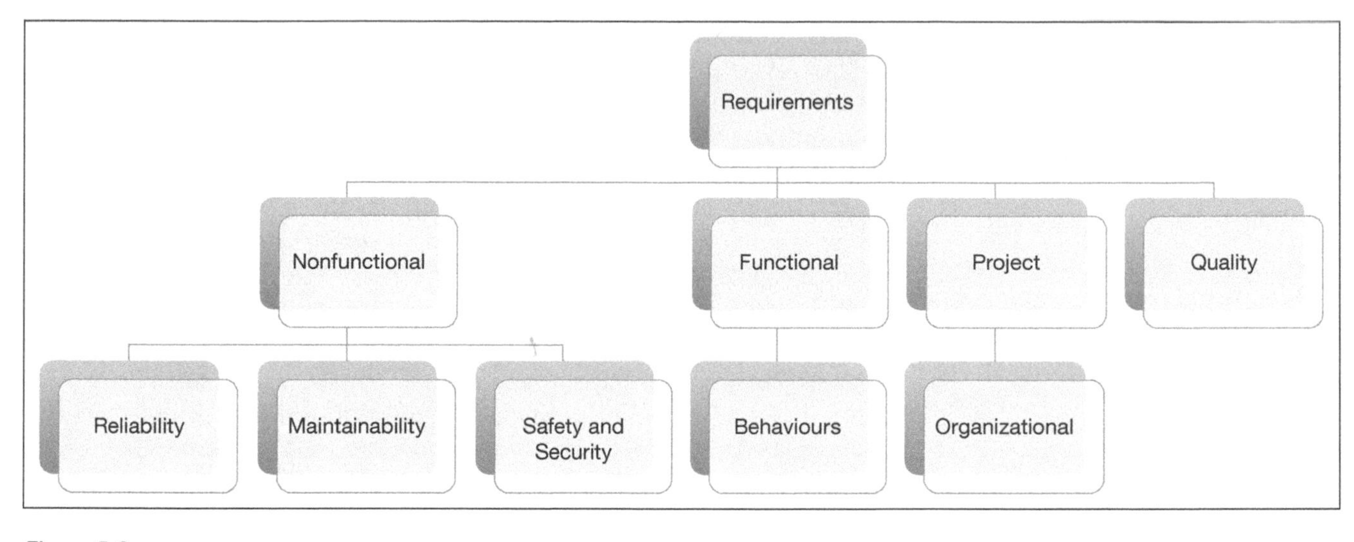

Figure 5-6
Types of Requirements

organization may have nonfunctional requirements for security, or maintenance and so forth.

The project charter contains the high-level information you need to start identifying and documenting requirements. You may find both project and product requirements in the charter. Project requirements can include information on the project approach, fixed delivery dates, pre-identified vendors or technology, and so forth. Product requirements can include information about performance, security, size, reliability, technical information, and the like.

The stakeholder register identifies the specific stakeholders that should be considered for collecting requirements. All requirements come from a stakeholder want or need.

TOOLS AND TECHNIQUES

There are a variety of techniques available to collect requirements from stakeholders. Some are very structured and some are informal.

The most common way to collect requirements is to interview the customer or end user of the product, service, or result. Interviews can be formal or informal one-on-one interviews, or they can include multiple interviewers and interviewees.

Focus groups are like interviews but they include prequalified or prescreened participants, groups of stakeholders, or subject matter experts. Focus groups are generally led by a trained facilitator. The facilitator usually has some specific questions they want feedback on as well as some questions that are designed to elicit interaction and discussion. A focus group may result in more qualitative information regarding attitudes, expectations, and thoughts about the end product, service, or result.

Facilitated workshops are structured workshops or meetings led by trained facilitators to bring numerous stakeholders together to define, prioritize, and agree on requirements. Two common types of facilitated workshops are Quality Function Deployment (QFD) and Joint Application Development (JAD) workshops.

QFD is used in new product development. The intent is to consider the "voice of the customer" (VOC) in the design process. Workshops will generally include stakeholders from multiple market segments and the engineers and designers of the product. The intended outcome is a product that captures the customer wants and needs and accurately reflects them in the engineering, design, and manufacturing processes.

JAD sessions are seen in the Software Development Life Cycle. They include "knowledge workers" who will use the system and the IT specialists who will build and maintain the system. The intent is for the IT specialists and the knowledge workers to collaborate and work through any misunderstandings and conflicts early in the process. Facilitated workshops can take from a few days up to many weeks for complex projects.

Many of the group creativity techniques are informal, the exception being a Delphi technique. When using the Delphi technique to gather requirements, a facilitator poses a series of questions to a group of anonymous stakeholders. Their answers are combined and sent back to the group. The group then revises their input based on seeing the collective information. This continues for several rounds until there is some degree of consensus from the group about the requirements.

The more informal group creativity techniques include brainstorming, nominal group technique, affinity diagrams, mind mapping and multicriteria decision analysis. Most people are familiar with brainstorming, so we won't discuss that here.

- The nominal group technique is used to prioritize the information generated from brainstorming. The participants vote on or rate the ideas in order of importance according to their personal perspective. Then the votes are tallied and ranked.
- Affinity diagrams take the information from brainstorming and lay out the natural groupings of requirements that are then used for further review and analysis. Figure 5-7 shows an affinity diagram for activities at a new community childcare center. The brainstorming ideas ended up in four categories.

Figure 5-7
Affinity Diagram for a Childcare Center Activities

Arts	Curriculum	Playground	Indoor Activities
▪ Music	▪ Letters	•Open areas	•Games
o Listening	▪ Numbers	•Sandbox	•Group activities
o Creating	▪ Stories	•Monkey bars	•Individual activities
▪ Visual		•Swing set	
o Painting		•Slide	
o Coloring		•Teeter totter	
▪ Crafts			

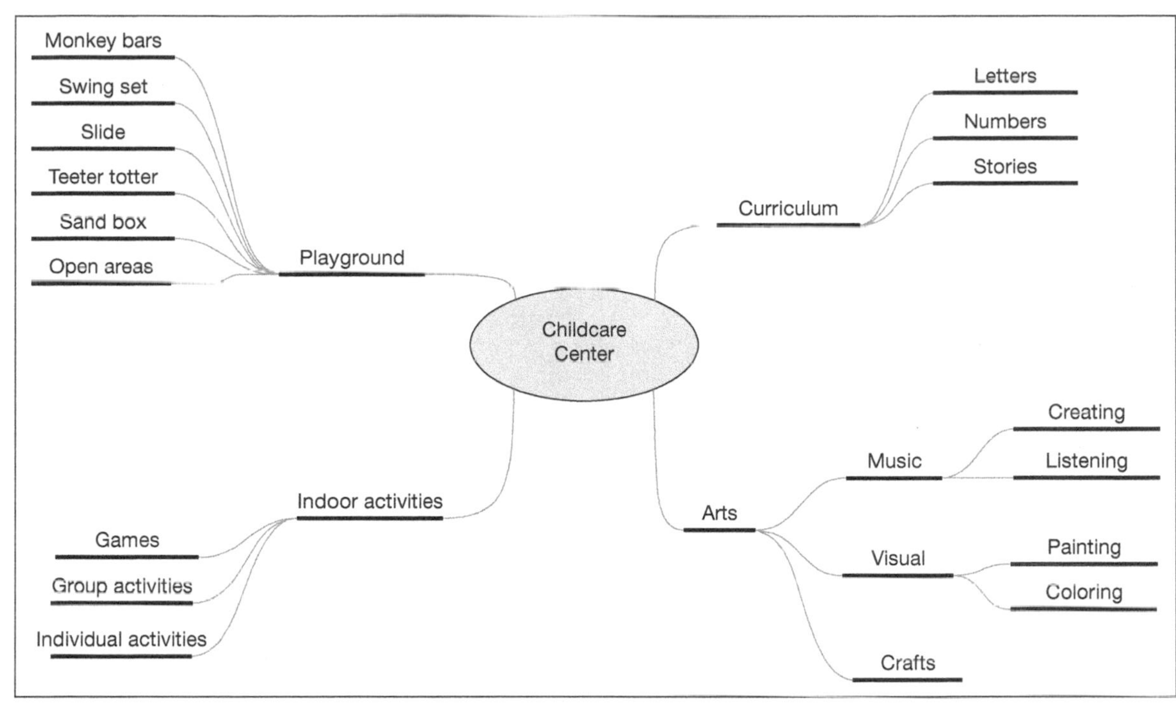

Figure 5-8
Mind Map for Childcare Center Activities

- Mind mapping takes ideas from the individual contributors and draws a map that shows the connectedness and the differences in requirements based on the individual needs. The mind map in Figure 5-8 presents the same information for the community childcare center, but in a mind map.
- Multicriteria decision analysis is used to weight several criteria to provide a quantitative method for prioritizing or including requirements. Tables 5-1 and 5-2 show a multicriteria evaluation matrix for a purchasing curriculum for the childcare center. There are three criteria that stakeholders are evaluating: consumer rating information on a scale of 1 to 10, cost, and product life.
  1. Identify the criteria you will use to rate the various options.
  2. Determine the relative weighting of the criteria so the total equals 100 percent.
  3. Define the rating scale. This one uses 1 to 5, with 5 being rated the highest.
  4. Define the method for assigning each rating.
  5. Create a matrix with each criteria and the weight on it.
  6. Evaluate each alternative using the rating system.
  7. Sum the totals of each alternative to determine the highest rated option.

Table 5-1 shows how stakeholders would evaluate options for meeting a requirement of having a curriculum package for the childcare center.

Table 5-1 Multicriteria Decision Analysis Weighting Criteria

	Consumer Rating	Cost	Product Life
5	9 to 10	Lowest cost	> 7 years
4	7 to 8	Within 5% of lowest cost	6 to 7 year
3	5 to 6	Within 10% of lowest cost	5 to 6 years
2	3 to 4	Within 20% of lowest cost	4 to 5 years
1	1 to 2	Greater than 20% of lowest cost	< 4 years

Table 5-2 Rated Multicriteria Decision Analysis

	Weight	Wiz Kids	Learning Kit	Scholastic Olympics
Consumer rating	40%	3	4	5
Cost	35%	4	5	4
Product life	25%	2	3	5
Totals	100%	3.1	4.1	**4.65**

Table 5-2 shows how three curriculum packages were evaluated against the criteria. In this example, the team would choose the Scholastic Olympics because it scored highest.

In some situations, the information gathered from the group creativity techniques, workshops, and interviews are compiled and then voted on to identify the requirements that will be included in the project. The group decision-making techniques employed can include a unanimous vote (unanimity), majority rules, plurality (where the largest block decides), or one individual making the decision for the group (dictatorship).

Questionnaires and surveys can be used for quantitative or qualitative data collection. They are very useful when collecting quantitative data from a large number of respondents. Automated survey tools that are available today make it very easy to ask yes/no questions, or questions that ask people to rate things on a numeric scale. The answers are available immediately and can help prioritize the project requirements.

Observation can be used to identify how people perform a job or a process. This can be helpful when mapping an "as-is" process for process improvement projects or to see where errors or redundancies can occur in a six sigma type of project.

Prototypes are used to help stakeholders see how their requirements will manifest in an end product. Some examples of prototypes, ranked from the simple to complex, include:

- A model of an apartment building
- A 1:10 model of a refinery

- A limited functionality product, such as a website that has some of the functions and the look and feel of the desired end product
- A single, fully functional unit that can be reviewed and tested before going into production

Benchmarking is identifying industry best practices for processes and operations and comparing the organization's current practices to the best practices. The comparison can be inside the organization or outside the organization. Benchmarking can be used to establish the desired end result of a process or quality requirements.

Context diagrams show the relationship between the users, data, and a process or a system. The person or system that provides or receives input to the process or system is called an "actor." Figure 5-9 shows an enrollment system for the community childcare center. The parents provide demographic information about the children. The county provides funding, and in return receives demographic information. The staff receives information on the age and number of children. The administrators receive management reports.

Document analysis is a comparatively simple way to gather requirements when compared with many of the previous techniques. It simply entails reviewing all current information from the charter, agreements, technical specifications, proposals, business plans, and any other relevant information. The data collected from the analysis is used to develop requirements for the project.

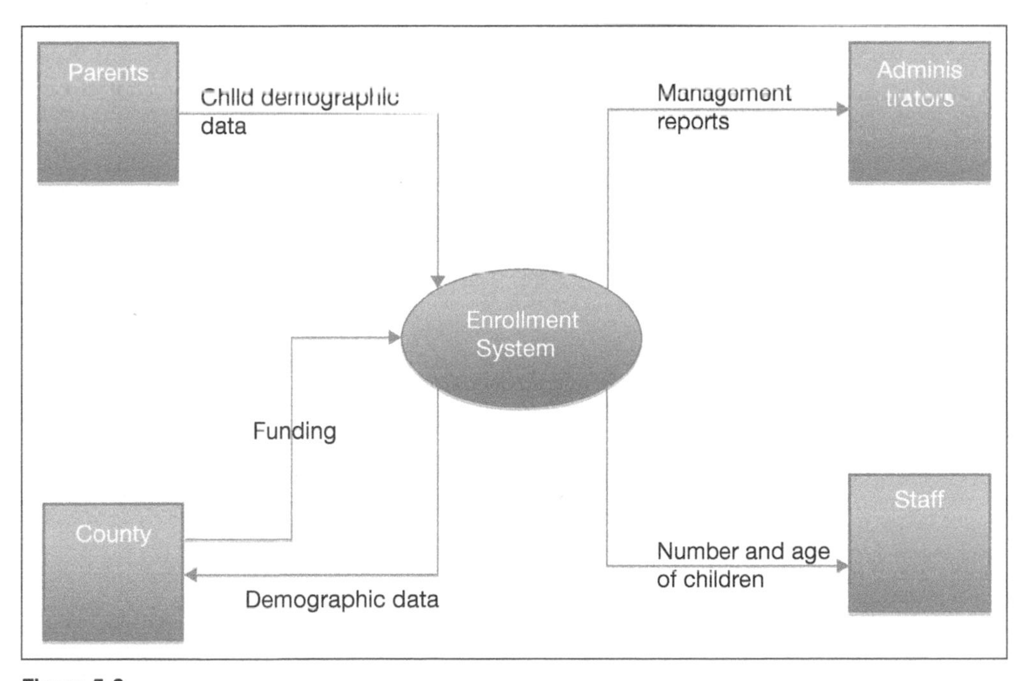

Figure 5-9
Context Diagram for Childcare Center

OUTPUTS

The amount of information needed in the requirements documentation will depend on the level of complexity of the project. Projects that are complex with many interfaces and interactions with multiple systems require more complete documentation. Projects that are less complex, or where the product is similar to previous products, will require less documentation. Requirements can be recorded in a document, a spreadsheet, or a database format. Regardless of the amount of documentation and the vehicle used to document, all requirements need to be clear, unambiguous, have a method to prove they are present, and be internally consistent.

A requirements traceability matrix is used to track and trace various attributes of requirements through the life cycle. For example, you may want to trace user requirements to technical requirements. If your project was to produce a conference you might have a business requirement of online registration. The technical requirements associated with that might be something along the lines of:

- A database that collects user information
- A way to take credit cards online
- A secure environment that protects personal and financial data
- A method to send an automated message confirming registration

These requirements might further trace to detailed specifications that contain requirements for each field, such as whether the field takes numbers or letters, the number of characters allowed in a field, whether information has to be entered in a specific format, and so forth.

The requirements traceability matrix can contain more detailed attributes about each requirement such as:

- Identification number
- Source
- Priority
- Version
- Status
- Complexity
- Acceptance criteria

The benefit of a requirements traceability matrix is that when someone wants to change a specification or a requirement, it is relatively easy to trace the impact of the change to the technical and business requirements to determine the impact. You can also use the matrix to determine when a requirement has been met, and to track the progress on a requirement. Other information that can be traced includes how requirements relate to project objectives and goals, and how they relate to WBS elements and to testing scenarios.

REQUIREMENTS DOCUMENTATION CONTENTS

Stakeholders and their requirements
Project requirements
Product requirements
Functional requirements
Nonfunctional requirements
Category
Priority
Acceptance criteria

See Appendix A for an example of requirements documentation and a requirements traceability matrix.

Define Scope

Define Scope is the process of developing a detailed description of the project and product. Once you have the charter, stakeholder information, and requirements documentation you can start to more fully define the project scope. The document used to do this is a scope statement. This document helps you answer questions like:

- What components and interim deliverables do I need to create the end product?
- What are my options for delivering the end product?
- What will provide the most value to the customer, for the least investment?
- Do my stakeholders have ideas that I should incorporate to further develop the product?
- If this product is part of a bigger program or system, how will it fit? Do I have to worry about system engineering?
- How good is good enough? What are the acceptance criteria for each component and the product as a whole?
- What are the assumptions, constraints, or exclusions?

Figure 5-10 shows the inputs, tools and techniques and outputs for the Define Scope process. Figure 5-11 shows a data flow diagram for the Define Scope process.

The time and effort needed to develop a really good scope statement can be significant, but it is well worth it. This is a key document that, coupled with the WBS and WBS dictionary, forms the scope baseline for the project. Let's take a look at how we get there.

INPUTS

The scope statement outlines the work needed to produce the product and the project. The scope management plan provides guidance on some of the techniques that will be used to define the detailed

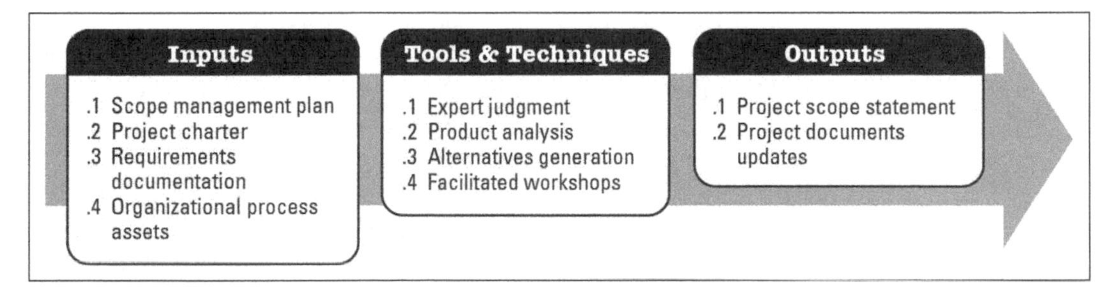

Figure 5-10
Define Scope: Inputs, Tools and Techniques, Outputs
Source: PMBOK® Guide—Fifth Edition

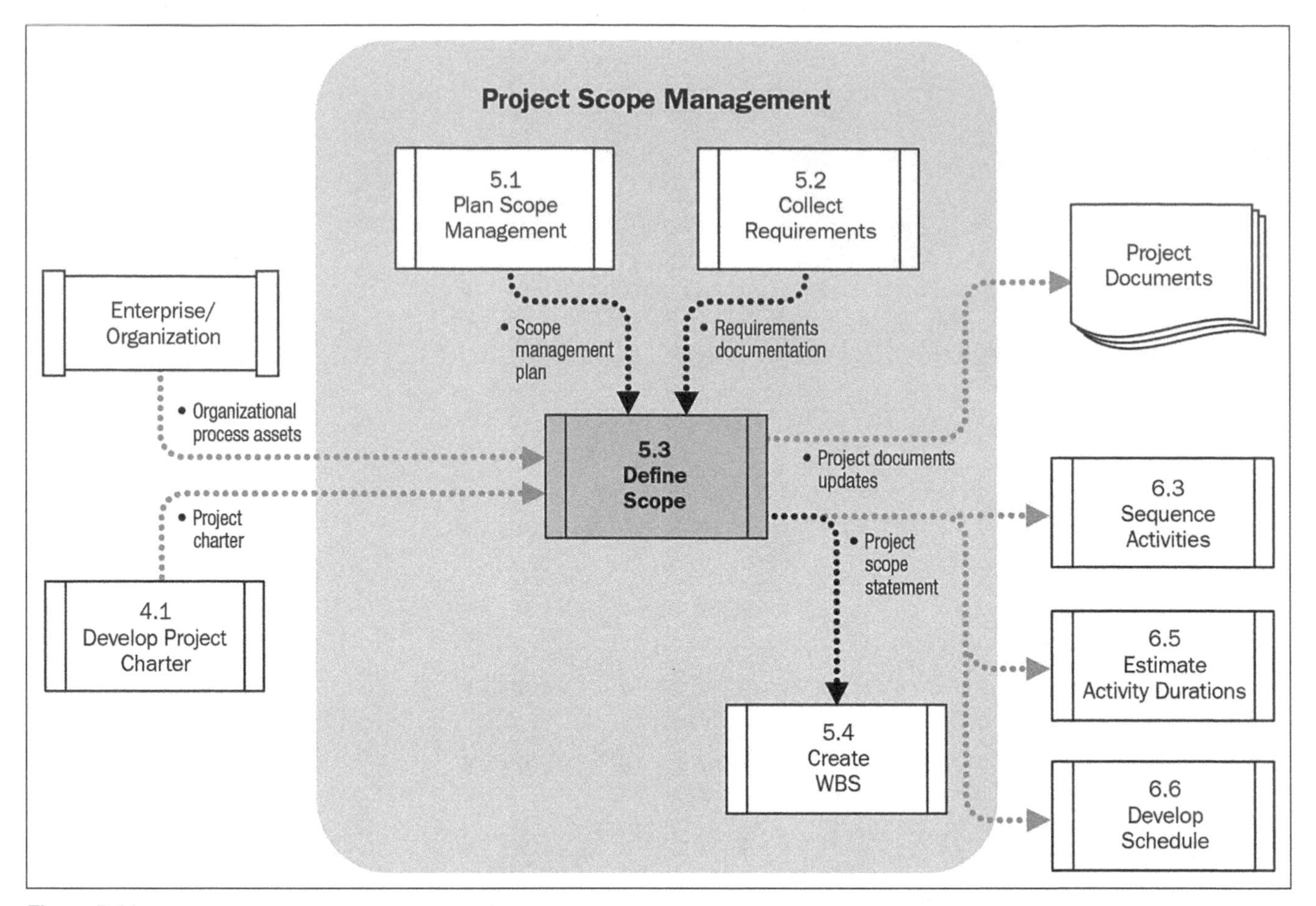

Figure 5-11
Define Scope Data Flow Diagram

product and project scope. The information from the project charter provides the high-level project description and objectives. This is the starting point for progressively elaborating the information into a scope statement. You start with the high-level project description and identify each interim deliverable or component needed to produce the product. As you identify the deliverables you will have a greater understanding of the project work involved such as training, contracting, documentation, and so forth.

The requirements documentation has better defined project requirements than the project charter. You can start to delve into the technical requirements, the business requirements, security and safety requirements, and various stakeholder requirements. As you understand all these needs, you and your team can start to come up with options and alternatives that will meet the needs of the stakeholders and meet the objectives in the charter.

Organizational process assets such as information from previous, similar projects, forms, templates, and procedures can give you a head start. It is perfectly acceptable to customize information from previous projects, and it can save you a lot of time. Don't forget to look at the lessons learned as well. There is nothing as embarrassing as repeating mistakes!

TOOLS AND TECHNIQUES

In order to define the scope in more detail you need input from numerous people. Some examples include:

- Project team members
- Subject matter experts in a particular field
- Consultants
- End users
- Customers
- Legal personnel

This is called expert judgment. The experts can be the same people who gave you the requirements, or they can be different. You will use your experts and their judgment to conduct a product analysis.

Some common types of analysis include decomposing the product down into its component parts, also known as a product breakdown structure. For a project that entails building a system, this process is called systems analysis or systems engineering. Many times, product development techniques, such as value engineering and value analysis, are employed to determine how to provide the most value for the least cost or effort. Regardless of what it is called, performing a product analysis will help you understand the components that make up the final product.

> **Tailoring**
>
> Your product analysis can be very brief with three to five components, or, for large and complex projects, you may end up with many levels of a product breakdown or a substantial system engineering document.

Many times the product analysis is combined with alternatives generation techniques. This can include brainstorming, lateral thinking, and other methods of generating alternatives. Sometimes facilitated workshops, such as brainstorming sessions, joint application development (JAD) sessions, or other techniques described in the Collect Requirements process are used.

OUTPUTS

The project scope statement is one of the key documents necessary to effectively manage the project scope. It provides detail for the product and the component parts of the product. It also defines the acceptance criteria for the component parts (when appropriate) and the product as a whole.

Project deliverables can be defined in this document as well. This may include testing and validation documentation, training materials, project management documents, and so on.

SCOPE STATEMENT CONTENTS

Product scope description
Product acceptance criteria
Project deliverables
Project exclusions
Project constraints
Project assumptions

Assumptions are considered to be true, real, or certain, but without proof or validation. It is a good practice to begin documenting assumptions in the scope statement, but often project managers start a separate assumption log. On large projects, the assumptions proliferate and can run as long as several pages. In this case, you would note that a log is being used and refer the reader to that log.

It is a good practice to continually update your assumption log. You can determine if assumptions are still valid, if they have been proved and can therefore be closed, or if they are not valid. If the assumptions are not valid or true, they may be moved to an issue or risk log. You can keep an assumption log as a spreadsheet or a table. Document the assumption, who should validate it, by when it should be validated, the status (open, pending, closed), and any additional information you feel is appropriate. See Appendix A for an example of an Assumption Log.

The project exclusions and constraints put boundaries around the project. It is really important to document exclusions in the scope statement. You can be certain that if you do not explicitly exclude something, one of your stakeholders will be sure that it is in scope. So save yourself some heartache up front and list what is not included in the project.

Remember, constraints are those criteria or requirements that limit your options. For example, a fixed budget; compliance with specific codes, regulations, and policies; a set end date, or the like.

It makes sense to document these so your team members understand the boundaries that constrain them. It is okay to have an overlap in constraints and requirements. Sometimes requirements are constraints. It is up to you whether you want to document them in both places.

Creating the project scope statement can lead to project document updates, such as requirements documentation, a requirements traceability matrix, and the stakeholder register.

Create WBS

Create WBS is the process of subdividing project deliverables and project work into smaller, more manageable components. The work breakdown structure is the cornerstone of project scope. It takes the information from the scope statement and fully defines and organizes the scope. It is similar to an outline or an "org" chart. The top level represents the project. The next level down (or indented) starts to break the project into levels of detail until all the work has been decomposed into discrete deliverables.

The WBS is an input into several other planning processes. It will also be used to manage and control the project work. The scope statement, along with the WBS and its attendant WBS dictionary, form the scope baseline for the project. Figure 5-12 shows the inputs, tools and techniques and outputs for the Create WBS process. Figure 5-13 shows a data flow diagram for the Create WBS process.

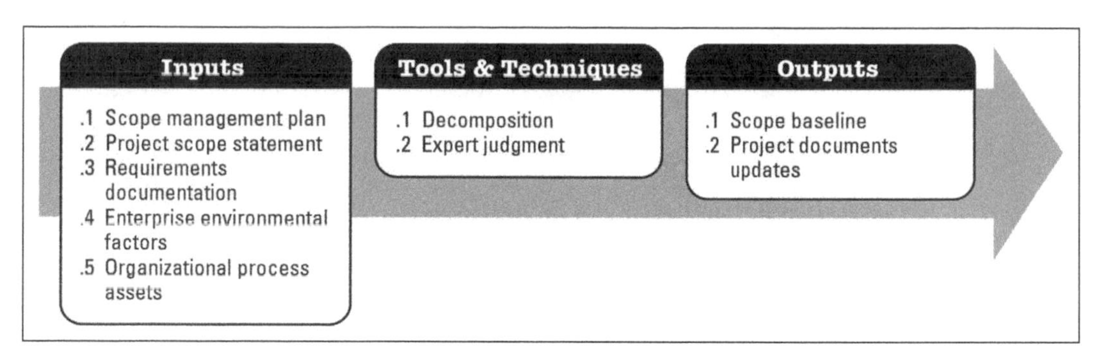

Figure 5-12
Create WBS: Inputs, Tools and Techniques, Outputs
Source: *PMBOK® Guide*—Fifth Edition

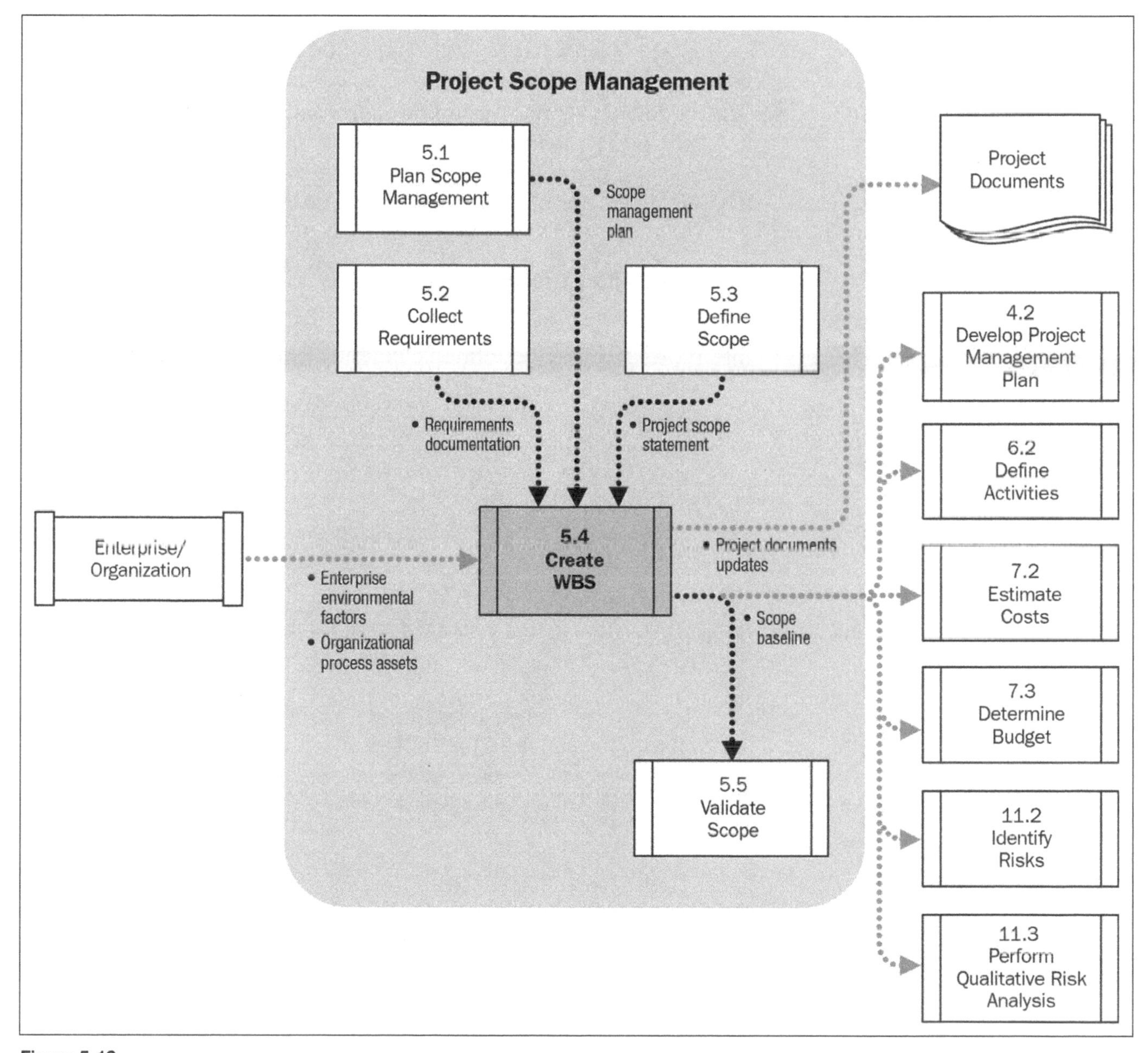

Figure 5-13
Create WBS Data Flow Diagram
Source: *PMBOK® Guide*—Fifth Edition

INPUTS

The scope management plan contains the guidelines for creating the WBS. This can include direction to use specified WBS templates or numbering systems. The scope management plan may also provide direction on the information that should be included in the WBS dictionary.

The project scope description and the product deliverables found in the project scope statement provide a good starting place for developing the WBS. Many people start with the product deliverables and decompose them into their component parts.

The requirements documentation will provide information that will help flush out the lower levels of the WBS and also provide information needed to create the WBS dictionary.

Enterprise environmental factors that influence the WBS include industry standards, such as specific development methodologies or lifecycles. Organizations that do similar types of projects often have WBS templates as part of their organizational process assets. Using a WBS from a previous project is also a good way to shorten the process. You can start with a WBS from a similar project and tailor it to meet the needs of your current project.

TOOLS AND TECHNIQUES

Decomposition is subdividing each of the higher-level elements into more detailed and refined elements. Decomposition stops when the lowest-level deliverable is reached. This may mean that some levels are broken down into five levels or more, and others stay up at level 3.

In many instances, the lowest-level deliverables are not understood when the WBS is first being developed. In this case, you should decompose as low as you can and then come back later and add the lower level of details using rolling wave planning.

WBS

If you are implementing a new computer system that will take many months of work, you may have a WBS that has a level 2 element called "Testing." This is fine, since testing is far in the future. As the time for testing gets closer, you may want to break it into different types of testing such as:

1. Testing
    a. Unit test
    b. System test
    c. Integration test

As you get even closer to the testing phase, you may further decompose "unit test" by identifying the specific units you will be testing.

Deliverables Only!

The WBS is only used to define the deliverables or work products of the project. The activities necessary to produce the work products are identified in the schedule!

One of the more challenging aspects of developing a WBS is figuring out how to structure it. Most people start at the top and decompose down. However, some teams find it easier to start with the deliverables at the lowest level and create the WBS from the bottom up, similar to an affinity diagram. We mentioned in the Inputs section that you can structure the WBS by using the deliverables mentioned in the scope statement. There are, of course, other options. Some of the more common ways include:

- By life cycle phase
- By subproject, particularly when major components of the project are conducted by subcontractors
- By geography, when there are multiple sites involved
- By type of system or deliverable

Using the expert judgment of your team members and subject matter experts can assist in identifying all the deliverables in the WBS and determining the best way to organize the WBS.

OUTPUTS

The scope baseline is the approved version of the scope statement, WBS, and WBS dictionary. The scope baseline is what you will use to manage and control the project scope. You will use these documents to determine if something is in scope or out of scope. Sometimes what one person considers progressive elaboration another person considers scope creep. Therefore, it is wise to be as detailed and descriptive as possible with the documents in the scope baseline.

The WBS organizes and defines the entire scope of the project, therefore, it is important to make sure it has project work as well as product work. It is common to have "Project Management" at level 2 of a WBS. You can keep it there and not decompose it any further, or you can put deliverables such as the schedule, budget, status reports, and the like as level 3. This is the only section of the WBS where you may need to list activity-oriented elements. Some common elements under project management include:

- Risk management
- Stakeholder management
- Communication

- Status reporting
- Conflict management
- Meetings
- Negotiating
- Planning
- Monitoring and controlling
- Management

Before the WBS is finalized you will also need to identify those levels of the WBS where you will collect and report status. For a large project you will not report status for each deliverable at the lowest level of the WBS (called a work package). You will find some level above the lowest level where you can maintain control and visibility, but not micromanage the people accountable for the work. This level is called a control account.

Control Account. A management control point where scope, budget, actual cost, and schedule are integrated and compared to earned value for performance measurement.

Work Package. The work defined at the lowest level of the work breakdown structure for which cost and duration can be estimated and managed.

It is a good practice to establish a numeric coding system for the WBS. You can see an example in Figure 5-14, which is taken from

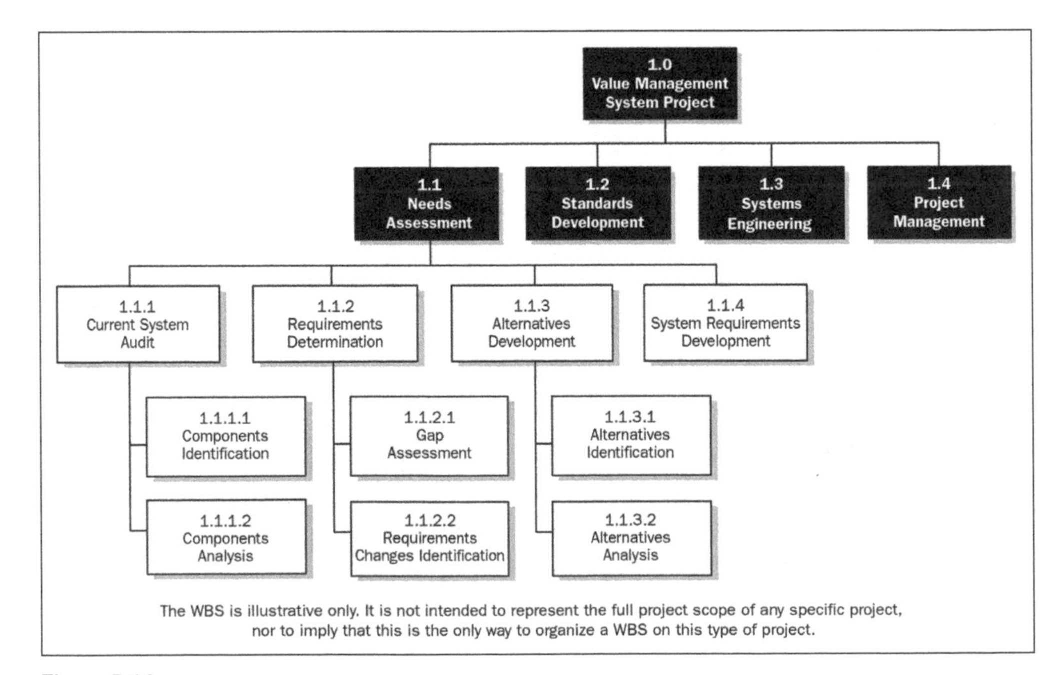

Figure 5-14
Sample WBS Decomposed Down Through Work Packages
Source: PMBOK® Guide—Fifth Edition

the *PMBOK® Guide*. It shows deliverables and work packages and a numbering system.

A WBS dictionary is a document that supports the WBS by providing a detailed description of each element in the WBS. It will need to be progressively elaborated because, in the beginning, the only information is a description of the WBS elements. But as more information is known, it will include resources, cost estimates, predecessor and successor information, and additional data. You can tailor the information in the WBS dictionary to suit the needs of your project.

Project document updates may include updates to the requirements documentation or the scope statement. Additionally, you may find that based on the work completed in developing the WBS, you identified items that you will need to procure from outside the organization. This will lead you to planning your procurements.

See Appendix A for an example of a WBS dictionary.

WBS DICTIONARY CONTENTS

- Code of account identifier
- Description of work
- Organization or person accountable for the work
- List of schedule milestones
- Activities associated with the milestones
- Required resources
- Cost estimates
- Quality requirements
- Acceptance criteria
- Technical references
- Agreement information

Chapter 6

Planning the Schedule

TOPICS COVERED

Project Time Management

Plan Schedule Management

Define Activities

Sequence Activities

Estimate Activity Resources

Estimate Activity Durations

Develop Schedule

Project Time Management

Project Time Management includes the processes required to manage timely completion of the project.

The project schedule is one of the most important documents a project manager develops. One of the most common questions project managers hear is, "When will it be done?" The schedule can help answer that question. But in order to answer it realistically, the schedule has to be accurate and relevant. Taking the time to gather the information necessary to develop a good schedule will pay off throughout the project. For smaller projects, the planning processes in time management can be linked together and combined into one process. For projects that have approximately 75 to 100 activities or less, don't worry about going through each step individually. However, as projects get larger, you will need to pay attention to each individual process to make sure your information is relevant and accurate.

The phrase "garbage in, garbage out" could have been written specifically for schedule development. In fact, professional schedulers call the schedule a schedule model. It is a model that describes what is possible if all the data in the scheduling tool is accurate. Don't mistake the model for reality!

Schedule Components

The scheduling tool describes the manual or automated tool used to develop a schedule. You can use a sticky pad as a tool, a word processor application, a spreadsheet, or software designed specifically for scheduling.

Schedule data refers to the information regarding the activities, sequences, resources, duration estimates, calendars, constraints, milestones, and so forth.

A scheduling methodology is a set of rules and assumptions used in scheduling—for example, critical path or critical chain methodology.

Combining the scheduling methodology, scheduling tool, and schedule data gives you a schedule model. However, remember, the model is only as good as the data that built it!

Plan Schedule Management

Plan Schedule Management is the process of establishing the policies, procedures, and documentation for planning, developing, managing, executing, and controlling the project schedule. When planning how to manage the schedule the team will take the time to consider the scheduling methodology and scheduling tools they will use to manage the project along with the techniques they will use to build and manage the schedule. Figure 6-1 shows the inputs, tools and techniques and outputs for the Plan Schedule Management process. Figure 6-2 shows a data flow diagram for the Plan Schedule Management Process.

INPUTS

At the start of the project the project management plan will be more of a framework and contain high-level information. However, having it as an input provides a consistent method of developing subsidiary management plans. As the project progresses the project management plan will contain the scope baseline and other more robust and detailed information that can be used to iteratively refine the schedule management plan. The project charter

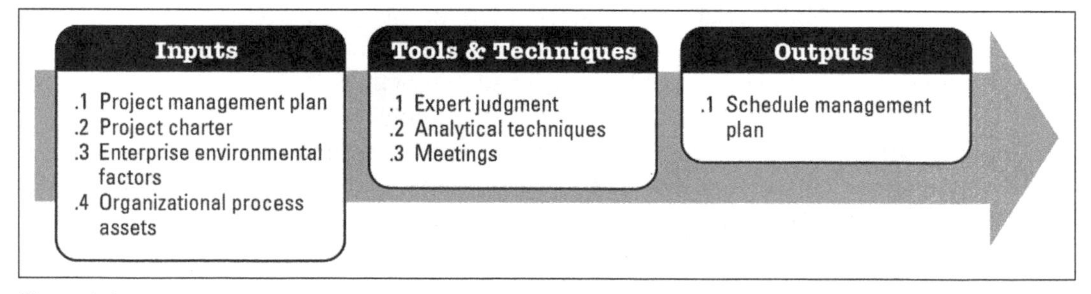

Figure 6-1

Plan Schedule Management: Inputs, Tools and Techniques, Outputs

Source: PMBOK® Guide—Fifth Edition

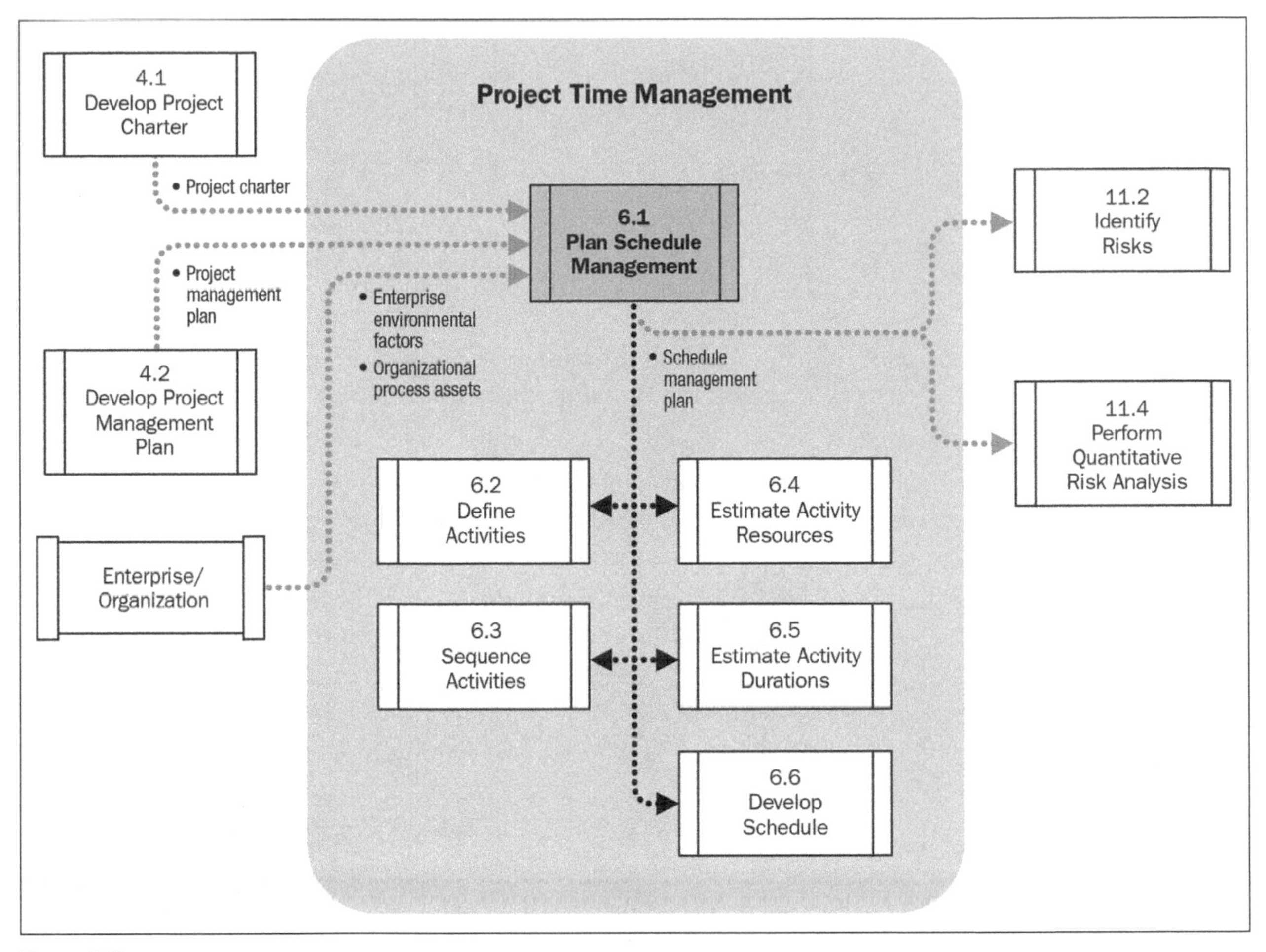

Figure 6-2
Plan Schedule Management Data Flow Diagram
Source: PMBOK® Guide—Fifth Edition

provides the approval requirements that need to be planned into the schedule.

The enterprise environmental factors that influence this process include the company culture and infrastructure as well as resource availability, published information on productivity data, and any work authorization systems the organization has.

Organizational process assets used in this process include information from previous, similar projects, templates, policies, and procedures. Policies and procedures that address scheduling, change management, and risk management are particularly helpful. Scheduling software is an organizational process asset that can help with schedule development, monitoring and control, and progress reporting.

TOOLS AND TECHNIQUES

Meetings with various stakeholders can provide expert judgment with regard to the most appropriate scheduling tools and methodologies to use. During these meetings the group can apply analytical techniques to discuss and select the best methods to sequence activities, which estimating methods to use, techniques to optimize resources, and options for compressing the schedule.

OUTPUTS

The schedule management plan is a component of the project management plan. It describes the process for managing the scheduling processes including:

- The scheduling methodology and tools that will be used to develop the schedule model
- Level of accuracy (this will evolve as the project progresses)
- Units of measure (hours, days, weeks)
- Control thresholds, such as when to implement preventive or corrective action based on the amount of float or buffer that is available
- Measurement methods such as schedule variance (SV) and schedule performance index (SPI)
- Guidelines for developing the network diagram
- Estimating methods for resource utilization and duration estimating
- Frequency and formats for schedule status reports

Schedule Management Plan. A component of the project or program management plan that establishes criteria and the activities for developing, monitoring, and controlling the schedule.

The schedule management plan should also define what a schedule change is. For example, if an activity starts three days late, but has three weeks of float, would you consider this a change? If an activity comes in early, would you document that as a change? What if the resources on an activity change? Identifying these activities during planning will assist in monitoring and controlling the schedule in the future.

Define Activities

Define Activities is the process of identifying the specific actions to be performed to produce the project deliverables. Notice that the definition says the "specific actions to be performed." This tells us that activities are action or verb oriented. When we were developing the scope, we were looking at deliverables (nouns) in the work breakdown structure (WBS). This is where we take those deliverables and define the actions necessary to create them. Figure 6-3 shows the inputs, tools and techniques and outputs for the Define Activities process. Figure 6-4 shows a data flow diagram for the Define Activities Process.

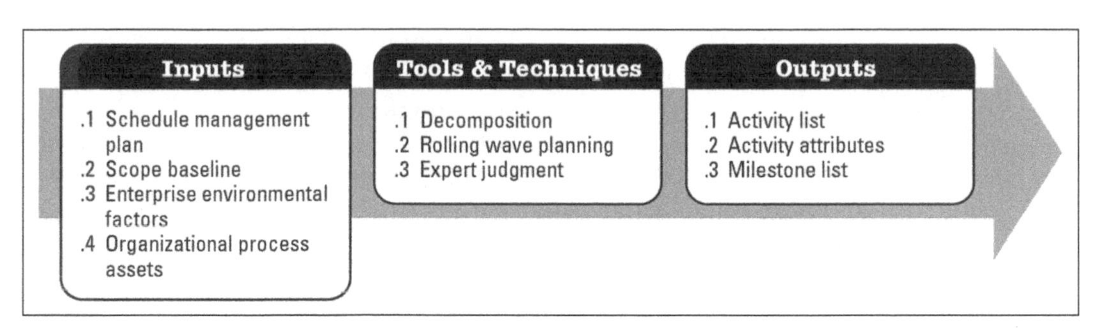

Figure 6-3
Define Activities: Inputs, Tools and Techniques, Outputs
Source: PMBOK® Guide—Fifth Edition

INPUTS

The schedule management plan contains information on the level of detail that should be carried on the schedule. It may also determine how rolling wave planning will be utilized. This will impact the degree of detail that future work is decomposed to. For example, the schedule

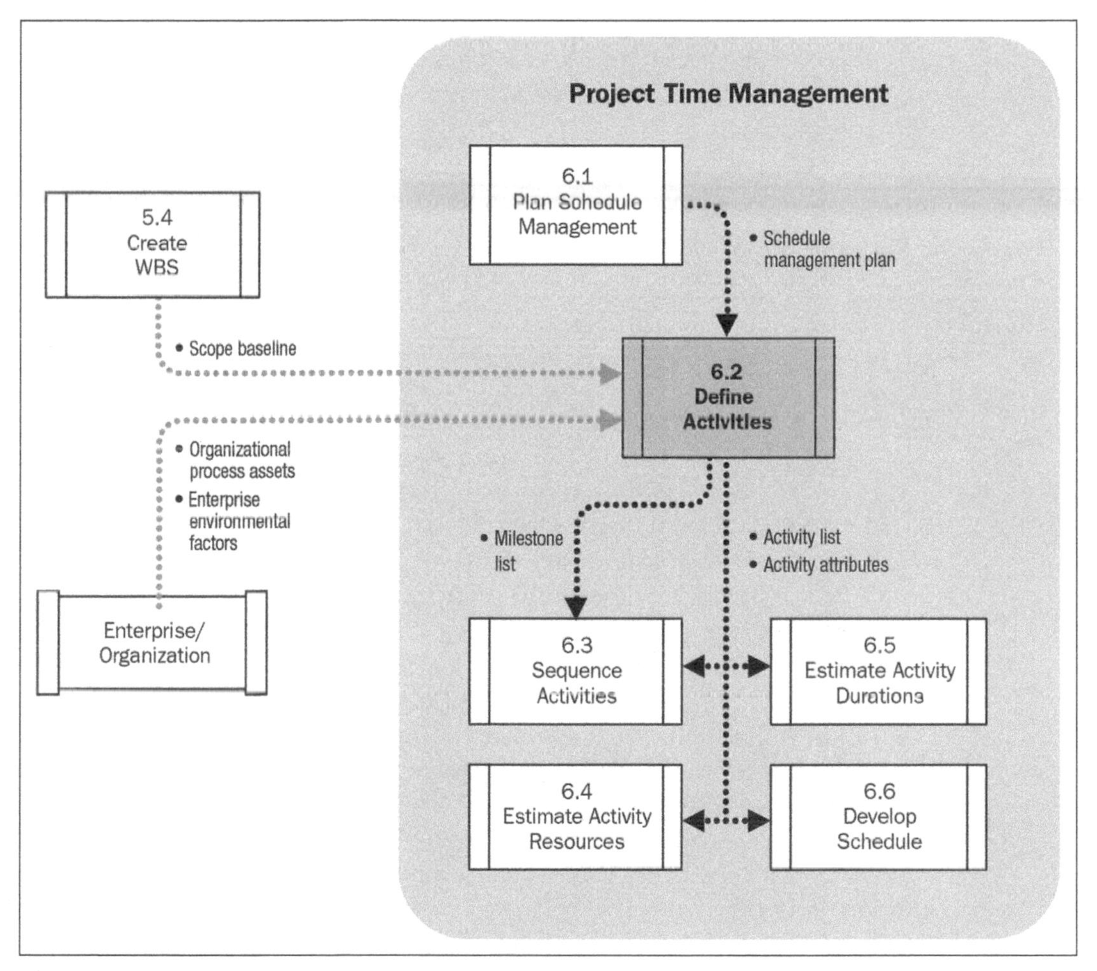

Figure 6-4
Define Activities Data Flow Diagram [6-4_FPO.pdf]
Source: PMBOK® Guide—Fifth Edition

management plan may indicate that work in the 90-day time horizon should be decomposed into activities and work that is 90 to 180 days in the future should be decomposed into work packages.

The scope baseline contains the project scope statement, the WBS, and the WBS dictionary. The scope statement has information on project assumptions and constraints that you should consult when defining activities. For example, an assumption might be that certain deliverables will be procured from outside vendors. If this is the case, your activities should reflect the necessary actions needed to identify vendors, create bid documents, put together a source selection committee, award and manage the contract. Constraints in the scope statement may identify specific development processes, verification procedures, or regulations that need to be accounted for in the project activities.

The WBS contains the work packages. Work packages are the transition from the WBS deliverables to the schedule activities. You will decompose each work package into the activities needed to create the completed work packages.

In the beginning of the project there won't be much information in the WBS dictionary. But as you go along, the information in there can assist in progressively elaborating the project activities.

The scheduling tool, the scheduling methodology, and the degree of maturity and knowledge about scheduling are enterprise environmental factors that influence the way activities are defined and recorded.

Organizational process assets include prior activity lists and schedules, policies, procedures, and templates for developing schedules and lessons learned that can assist you in identifying project activities.

TOOLS AND TECHNIQUES

Decomposition in this process is similar to the decomposition described in the Create WBS process. The difference is that the output is a list of activities, as opposed to work packages. However, the concept of breaking down work packages into activities is basically the same as breaking down deliverables into work packages.

Rolling wave planning is specifically applied as a technique here because on large projects it is impossible to identify all the activities needed several years in the future. This is an example of how planning goes on throughout the project, not just at the beginning. When you are employing rolling wave planning in defining activities, a rule of thumb is to define the activities at least 60 to 90 days in the future.

If you work on projects that are similar in nature, using a template can be a great way to save time and to ensure you don't omit any activities.

Expert judgment is a necessity in defining the activities for most projects. Your team members, subject matter experts, the people who will be doing the work, and people who have done similar work in the past are invaluable in making sure you identify all the work needed to create the deliverables.

OUTPUTS

In addition to a list of all the activities needed to complete the project, an *activity list* will generally include a numeric identifier that is a continuation of the coding structure developed when creating the WBS. Most software programs do this automatically for you.

Activity attributes are detailed information about the activities. Some of this information can be recorded when identifying the activities while other information is defined in future scheduling processes. You should tailor the attributes you need to collect and record so they meet the needs of the project. For some projects, there is no need to develop the activity attribute information past the information needed in the other time management processes.

ACTIVITY ATTRIBUTES

- Activity identifier or code
- Activity name
- Activity description
- Predecessor and successor activities
- Logical relationships
- Leads and lags
- Imposed dates
- Constraints
- Assumptions
- Required resources and skill levels
- Geography or location of performance
- Type of effort

Milestone. A significant point or event in a project, program or portfolio.

A *milestone list* identifies all the key milestones in the project. It can include the start and end of each life-cycle phase, the completion of key deliverables, passing certain benchmarks or tests, or obtaining a signoff. In some cases milestones are required, either contractually or by regulations. For example, a customer may require signoff on all major deliverables.

Most scheduling software will automatically turn an event into a milestone if you enter 0 in the duration column.

Achieving that signoff is a key milestone. Another example is getting an occupancy permit for a construction project. Other milestones, such as completion of the requirements definition phase, are more oriented to a specific project and classifying them as a milestone is optional.

As the team defines the activities they often identify new requirements, risks, costs, and other information that requires *project management plan updates*.

See Appendix A for a sample Activity Attribute form.

Sequence Activities

Sequence Activities. The process of identifying and documenting relationships among the project activities. Once you have identified all the activities needed to deliver the work packages, you need to put them in order. Many people will use a manual method to work out the high-level relationships and dependencies among activities before loading the information into software. You can use a manual method for sections of the project or at a milestone level before moving to the detailed sequencing. Figure 6-5 shows the inputs, tools and techniques and outputs for the Sequence Activities process. Figure 6-6 shows a data flow diagram for the Sequence Activities Process.

INPUTS

The scheduling methodology that was documented in the schedule management plan will influence how activities are sequences and therefore how the network diagram is developed. The activity list

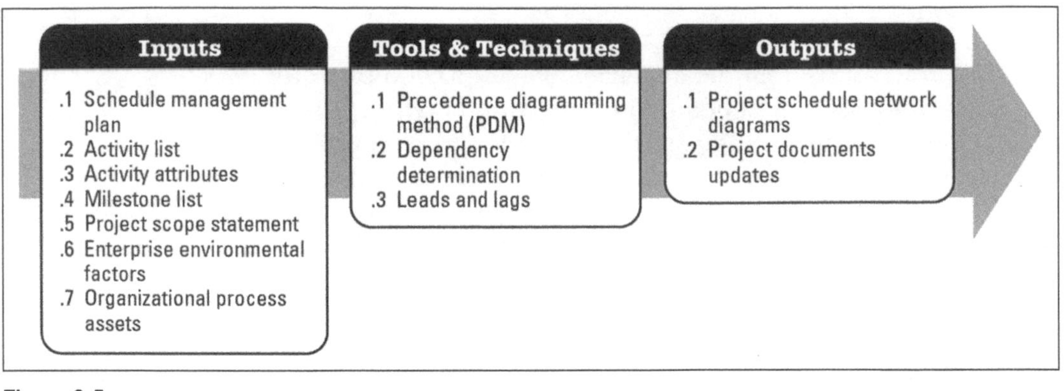

Figure 6-5
Sequence Activities: Inputs, Tools and Techniques, Outputs
Source: *PMBOK® Guide*—Fifth Edition

provides the activities you are going to sequence. If you developed information about the activities and recorded them in an activity attribute form, you may have entered information on predecessor and successor activities. That information is used here.

When you are sequencing your work, you may do it at the activity level, the milestone level, or both. Many project managers will

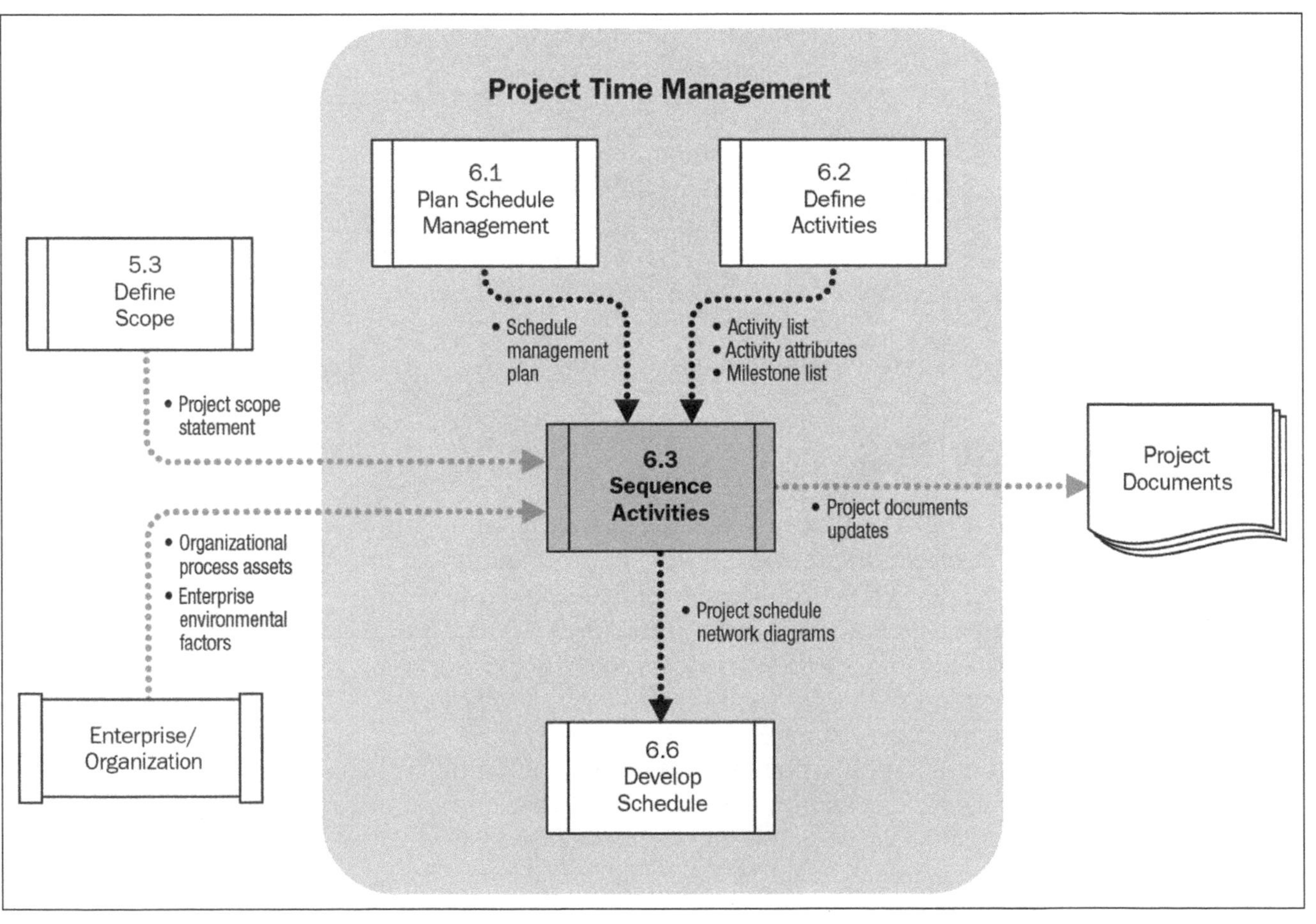

Figure 6-6
Sequence Activities Data Flow Diagram
Source: *PMBOK® Guide*—Fifth Edition

sequence the milestones from the milestone list to provide a high-level view of the project and use that to report actual status against the schedule. They will then use the detailed activity list to help manage the day-to-day work.

The project scope statement may have information about the project deliverables that determines the order in which certain activities need to be performed. This can also be documented in the assumption or constraint section.

The project management information system, scheduling tool and industry standards are enterprise environmental factors that influence how you sequence activities. Information from prior projects, network diagram templates, and lessons learned are all *organizational process assets* that can be used to reduce the effort involved in sequencing activities.

TOOLS AND TECHNIQUES

The precedence diagramming method (PDM) is a technique used to link activities together. Activities are represented by boxes (called nodes) that are connected by arrows. This is also called Activity on Node (AON) diagramming. There are different types of relationships among activities which PDM can demonstrate:

Predecessor Activity. An activity that logically comes before a dependent activity in a schedule.

Successor Activity. A dependent activity that logically comes after another activity in a schedule.

Finish-to-start (FS). This is the most common type of relationship. A finish-to-start relationship indicates that the predecessor activity must finish before the successor activity can start. It is shown like this.

Finish-to-finish (FF). This relationship states that the predecessor activity must finish before the successor activity can finish. It is shown like this:

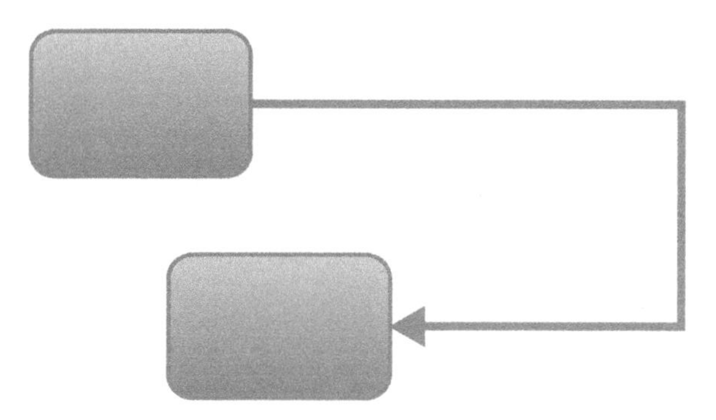

Start-to-start (SS). A start-to-start relationship indicates that the predecessor activity must start before the successor activity can start. It is shown like this:

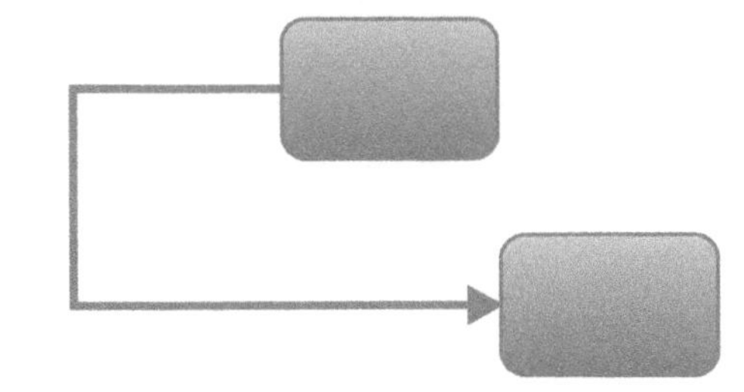

Start-to-finish (SF). Start-to-finish is the least used relationship. It indicates that the successor must start before the predecessor can finish. It is shown like this:

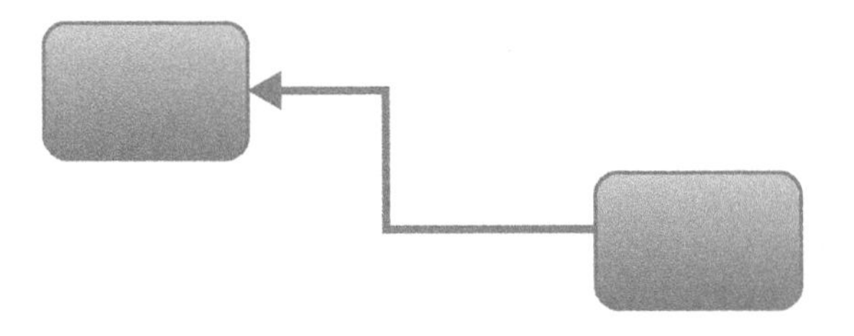

The type of relationships described here can be based on required sequencing, preferential sequencing, or external factors. This is called *dependency determination*. The *PMBOK® Guide* describes four types of dependencies:

Mandatory dependencies. Mandatory dependencies are based on the nature of the work being performed. It is also referred to as hard logic. As an example, you can't edit content that has not yet been written. So there is a mandatory dependency between writing and editing.

Discretionary dependencies. Discretionary dependencies are based on a best practice or a preference in sequencing. For example, it is best to fully complete gathering requirements before beginning design work. However, it is possible to start some of the design work before all the requirements are complete. Therefore, it is a discretionary dependency. When compressing the schedule you should look at the discretionary dependencies for opportunities to fast track by modifying the sequencing.

External dependencies. External dependencies involve factors that are external to the project. They include hand-offs between projects, delivery from an external source, signoff or approval from an external source, and so forth. For example, you need an occupancy permit before you move in. The work may be all done, but until that permit is signed off, you may not move into a new space.

Internal dependencies. Internal dependencies involve an internal relationship that the team does not have control over. For example, another project team is scheduled to use a piece of equipment. Therefore, your use of the equipment is dependent on the other team completing their work. Another internal dependency is resource constraints. For example, if Anna is working on two activities at once and she can complete either one first, it is discretionary as to which activity is the predecessor and which is the successor.

Applying leads and lags can modify the relationship between activities by imposing a delay or an acceleration between activities. For example, you may want to start editing ten days after you start writing. This lag would be indicated as SS+10D. In other words, the start of editing is dependent on the start of writing, and it needs to wait to begin until ten days after writing begins.

Another example is starting to develop test plans ten days before all the programming is complete. This lead would be indicated as FS–10D.

Lag. The amount of time whereby successor activity is required to activity must be delayed with respect to a predecessor activity.
Lead. The amount of time whereby a successor activity can be advanced with respect to a predecessor activity.

If you do projects that have multiple components of similar work you might want to create a schedule network template to accelerate the effort of sequencing work. For example, if you are creating a website with 20 pages, you would probably have the same type of work for each page. The content development could create a "subnetwork" or "fragment" network diagram that was repeatable for each page. If you do projects that are similar to one another you may start with a generic template for the entire network and modify it to meet the needs of your particular project.

LEADS AND LAGS

A lag is always shown with a plus (+) sign. A lead is always shown with a minus (–) sign.

OUTPUTS

The project schedule network diagram (also known as simply "network diagram") is a visual display of the entire project's activities and their relationship to one another. It can be done at a summary milestone level, or for the project as a whole. It shows all the relationships and the type of relationship among activities. An example of a network diagram is shown in Figure 6-7.

Project document updates can include the activity lists, milestones, activity attributes, and the risk register.

Estimate Activity Resources

The process of estimating the type and quantities of material, human resources, equipment, or supplies required to perform each activity.

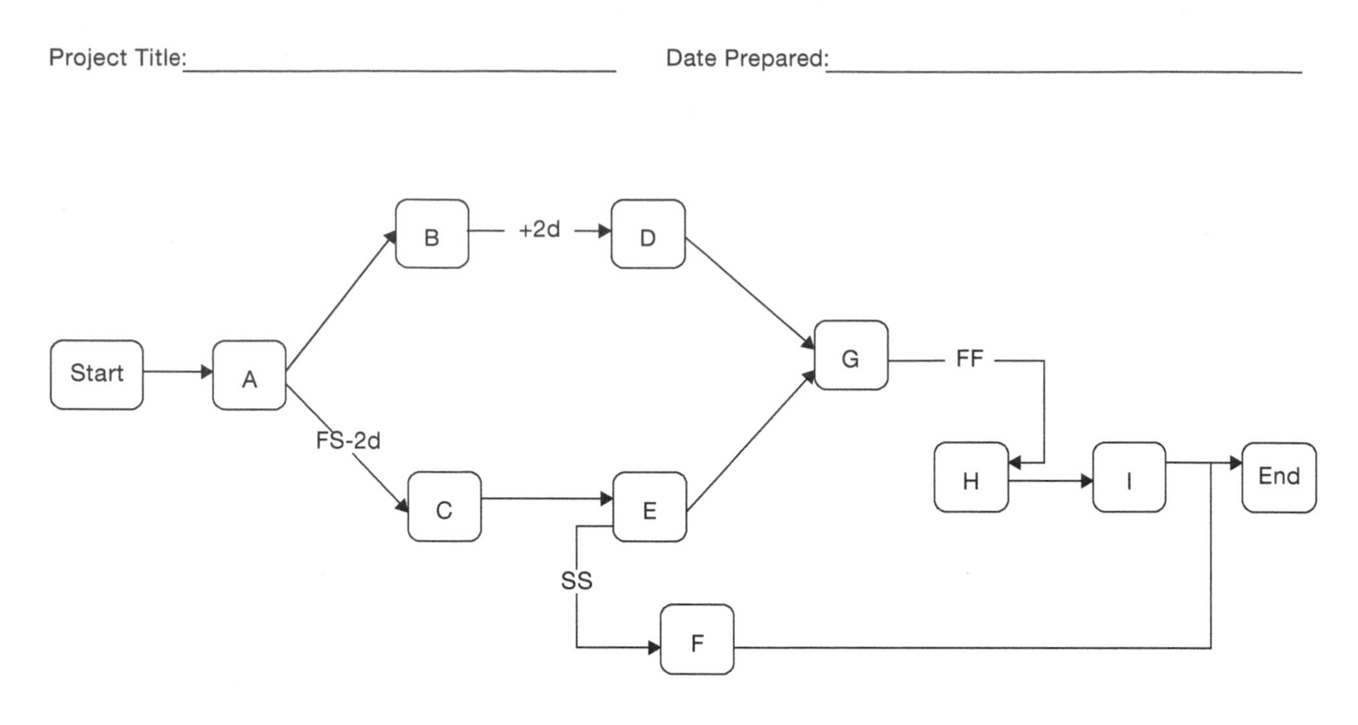

In this Network Diagram:
There is a 2-day lead between the completion of A and beginning of C.
There is a 2-day lag between the completion of B and beginning of D.
There is a start-to-start relationship between E and F.
There is a finish-to-finish relationship between G and H.

All other relationships are finish-to-start.

Figure 6-7
Network Diagram
Source: Snyder, *A Project Manager's Book of Forms,* John Wiley & Sons 2013

Information about resources will help you develop the duration estimates, the project schedule, cost estimates, procurement plans, and staffing plans. Figure 6-8 shows the inputs, tools and techniques and outputs for the Estimate Activity Resources process. Figure 6-9 shows a data flow diagram for the Estimate Activity Resources Process.

INPUTS

The schedule management plan provides guidance on resource estimating methods, the level of accuracy needed at specified phases of the lifecycle, and units of measure, such as hours, days, weeks, or some other measure.

The activity list identifies all the activities that will require resources. Information on activity attributes provides details such as the required resources and skill levels, the location of performance, and assumptions and constraints.

Resource calendars can provide information such as when specific pieces of equipment are available or when key employees

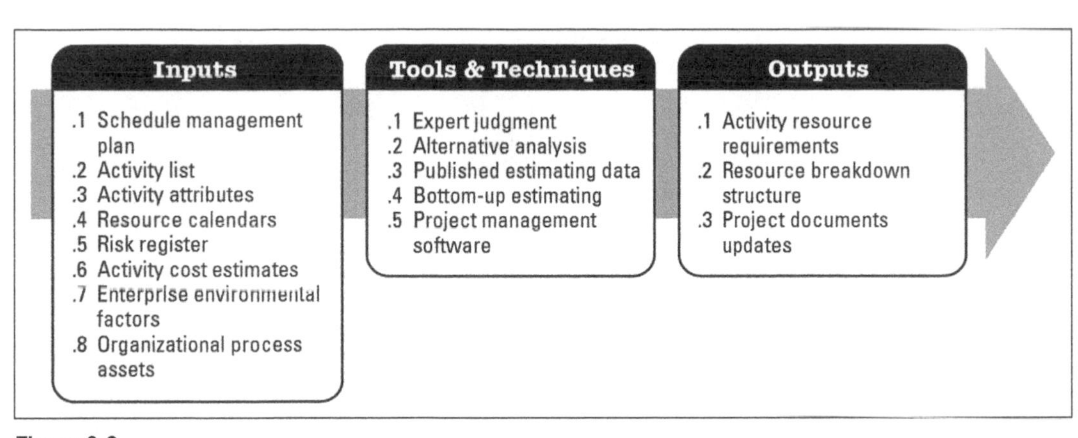

Figure 6-8
Estimate Activity Resources: Inputs, Tools and Techniques, Outputs
Source: *PMBOK® Guide*—Fifth Edition

are available. Identifying this information up front can assist in re-planning later if a key resource is obligated on other projects. The relationship between resource calendars, resource estimates, and the human resource plan is iterative, as shown in Figure 6-10. The resource calendars are an input to the process of estimating the

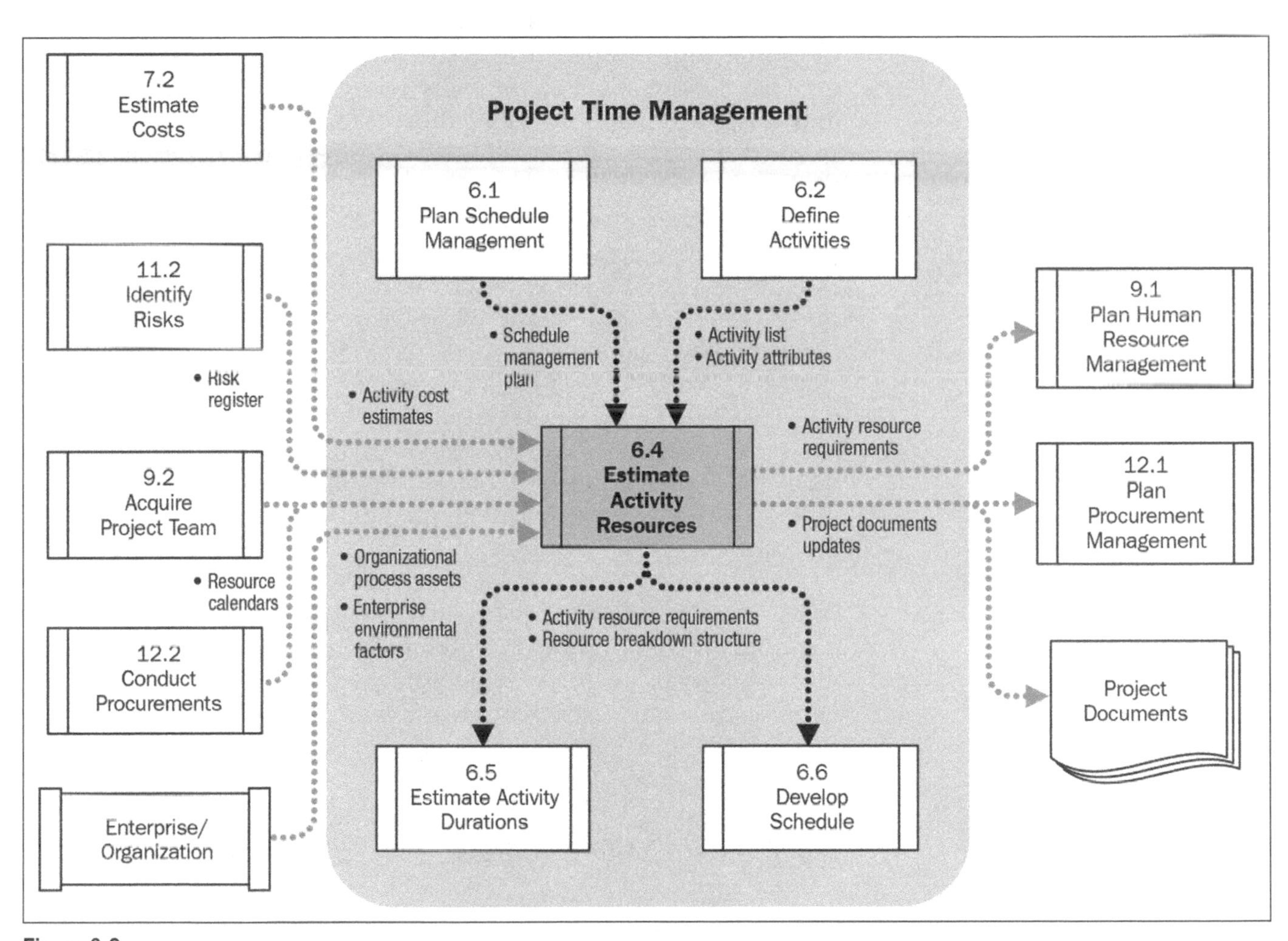

Figure 6-9
Estimate Activity Resources Data Flow Diagram
Source: *PMBOK® Guide*—Fifth Edition

Figure 6-10
Resource Estimating Iterations

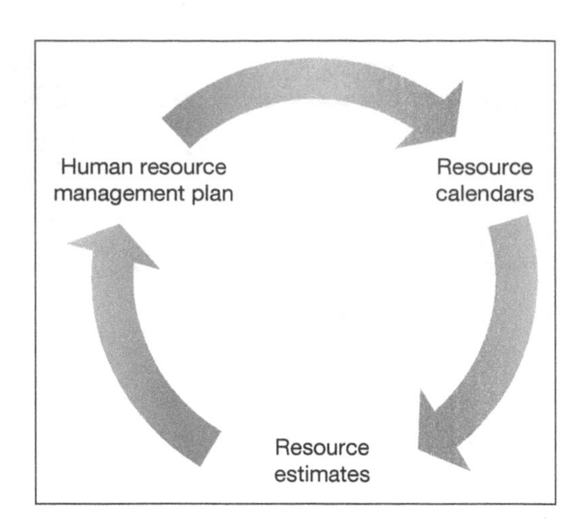

resource requirements, which in turn serve to update the human resource management plan. The human resource management plan may cause adjustments in the resource calendars, which cause a refinement in the resource estimates, and so on until an optimum solution is agreed upon.

Activity cost estimates contain the costs for resources. The cost of internal versus external resources, or lease versus buy decisions, or which skill level to use all impact cost. The project manager may need to make some tradeoffs between resource requirements and cost alternatives to fulfill those requirements. You may find yourself going through several cost and resource iterations to find the right balance between resource requirements and activity cost estimates.

The risk register contains information on risks related to resource skills, availability, rate of work, and so forth. Any risks associated with resources should be considered when developing resource requirements and estimates.

Resource availability is an enterprise environmental factor that needs to be considered. For example, if your organization only has two classrooms and you need to train 200 people on a new product, then you may need to look outside the organization, or spread the training over a longer period. The same is true for skill sets. In most organizations there are areas that tend to create a bottleneck for projects because there is a limited availability of a key resource or a specific skill set. This is a factor that constrains your project.

Organizational process assets include contracts with suppliers that allow project managers to create purchase orders for materials. Often there are policies that provide guidance on outsourcing labor that will come in handy when you are working on obtaining all the materials, equipment, and skills needed to complete the project.

TOOLS AND TECHNIQUES

Team members, subject matter experts, and consultants can provide expert judgment to help determine the type and amount of materials, the types of equipment, the necessary skill sets, and any other types of resources needed to complete the project work. These same people can help you analyze alternative approaches for completing the work. For example, you can determine if it is better to apply expert-level resources for a higher hourly rate, and get the work done quicker, or whether entry-level people should do the work for less cost, and over a longer period of time.

With a little research you may be able to find published estimating data that will help you identify quantities and costs of materials for certain types of work. This is most common in the construction fields.

Where there is not a lot of information on prior projects, you will need to do a bottom-up estimate at the work package and activity level to determine the detailed amounts and types of resources required to complete the work. This type of estimate is more difficult and time-consuming, but ultimately, more accurate than other types of estimates. However, the work needs to be fairly well defined to do a bottom-up estimate. This type of estimate usually requires that you document your basis of estimates along with any assumptions used in building the estimate.

Project management software, such as scheduling software or spreadsheets can help you organize the resources you will need. You can include information on the amount, cost, availability, and any other information needed to keep your resource requirements organized.

OUTPUTS

As described in the section on inputs, activity resource requirements will get more detailed throughout the planning process. When you first start estimating the resources you need you may only have an idea of the type of resource you will need, but you may not know the number, the skill sets needed, the level of expertise, or any special certifications you will require. By the time you are ready to baseline your schedule and cost estimates you should be able to identify the type and quantity of resources by work package and summarize it by type of resource. In addition, you should have full documentation of the basis of estimates and assumptions used in determining the resource requirements.

It may be helpful to create a resource breakdown structure for the category of resources, and then the type and grade. For each entry you would indicate the quantity needed. See Appendix A for a sample Resource Breakdown Structure.

You will at least want to update project documents such as your activity attributes and WBS dictionary. You may need to revise information in your activity list, resource calendars, and risk register as well.

Estimates

Most sponsors and customers tend to want aggressive estimates. They want the project done as soon as possible. If you are too aggressive with your estimates you will start out behind schedule and never be able to catch up. The more aggressive the estimate, the higher the risk you have of running over the schedule, and probably tho budget as well.

Estimate Activity Durations

The process of estimating the number of work periods needed to complete individual activities with estimated resources. Estimating activity durations is concerned with determining the actual work hours needed to complete the work (effort) and the number of work days it will take from start to finish (duration). Estimates are progressively elaborated throughout the planning cycle and the project. In the beginning of the project, when the details are still fairly vague, there is a wide range in the estimate, and the confidence level is fairly low. However, as the planning progresses, requirements are well understood, and detailed design information is available, the estimates will get more accurate, with a tighter range and a higher confidence level.

Refining Estimates

For projects that have stable requirements, common technology and a team that has experience, you will probably be able to get good estimates early in the process, and they will most likely be relatively accurate and won't need a lot of refinement. Conversely, projects that are using new technology and have evolving scope and requirements will need many iterations of estimating. Estimates for this type of project will have a wide range and they are likely to change over the life of the project.

Figure 6-11 shows the inputs, tools and techniques and outputs for the Estimate Activity Durations process. Figure 6-12 shows a data flow diagram for the Estimate Activity Durations Process.

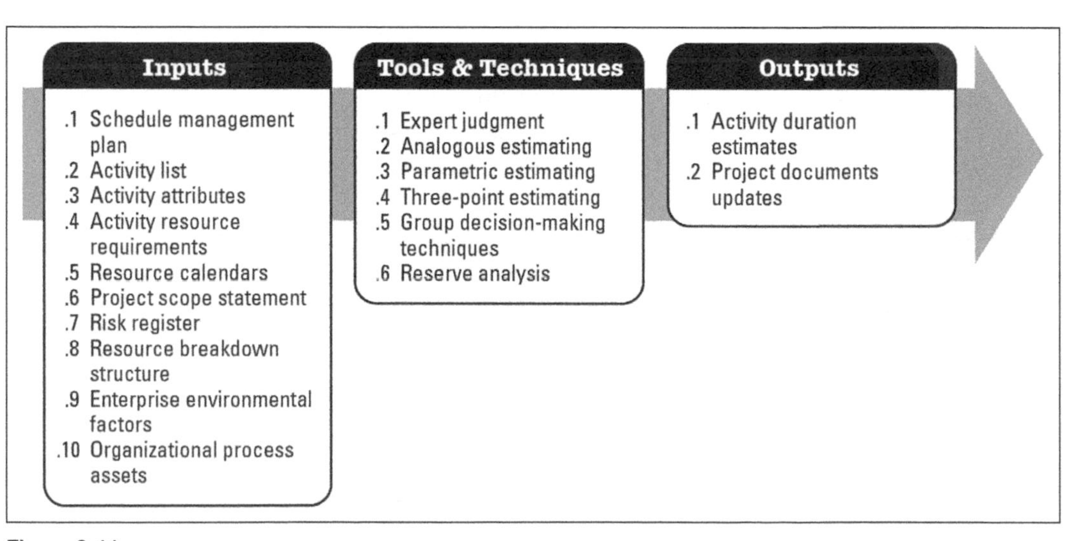

Figure 6-11
Estimate Activity Durations: Inputs, Tools and Techniques, Outputs
Source: *PMBOK® Guide*—Fifth Edition

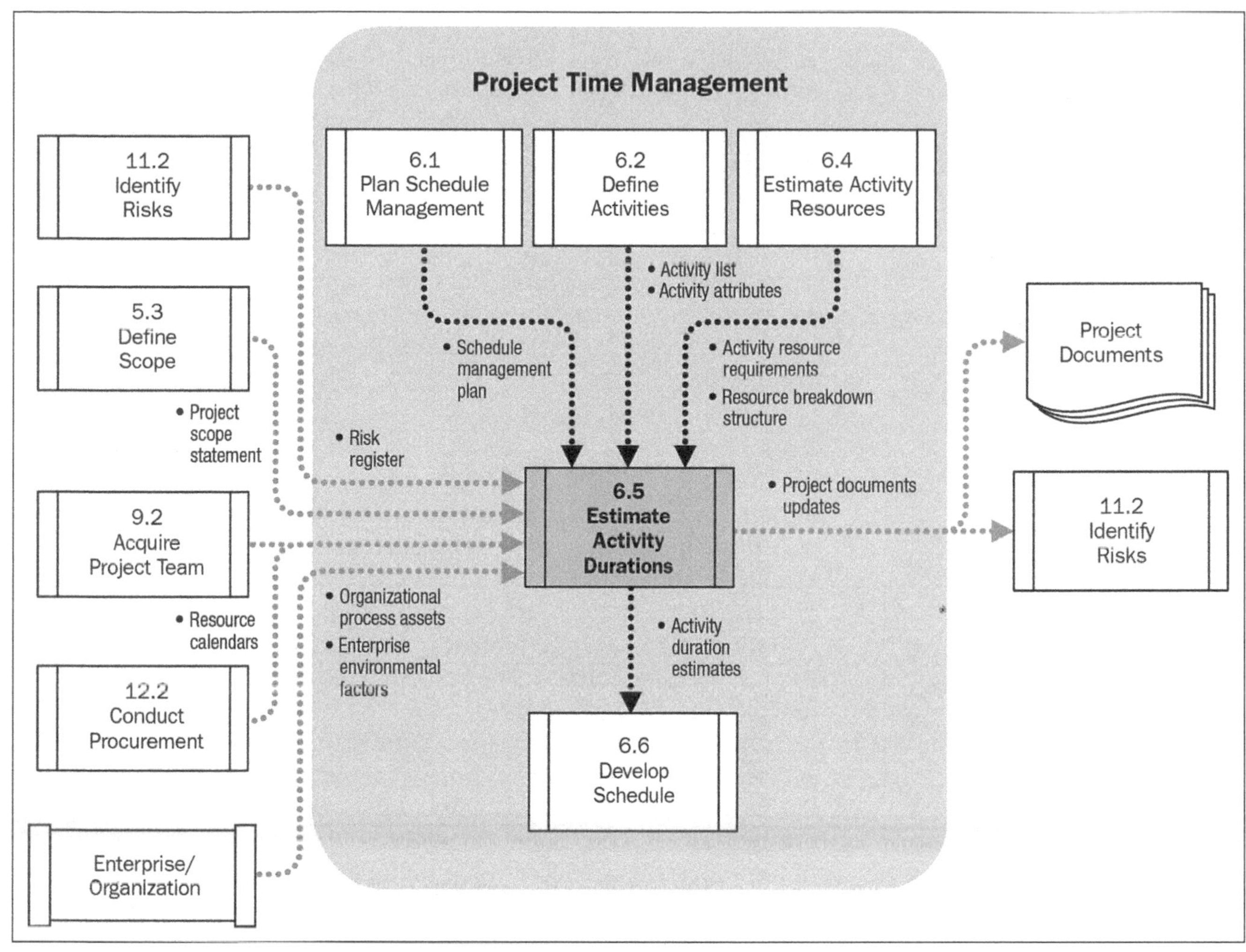

Figure 6-12

Estimate Activity Durations Data Flow Diagram

Source: PMBOK® Guide—Fifth Edition

INPUTS

The schedule management plan provides direction on the method of estimating and the level or accuracy required for the duration estimates. The activity list and activity attributes describe the work necessary to complete the activity. The resource requirements can provide information on the types of resources, the skill level, equipment type, and as the planning process progresses, the basis of those estimates.

Resource calendars provide information on when each type of resource is available. The calendars may indicate that certain people are only available ten hours per week, or that specific equipment can only be used 50 hours before it requires service. The combination of the resource requirements and the resource availability is an important determining factor in the duration estimates.

The project scope statement has assumptions and constraints that should be referenced. There may be environmental conditions or contractual requirements that need to be considered when developing a duration estimate.

The risk register may have risks associated with resource availability, skill sets, unclear requirements, or other information that can impact the effort or duration of activities. The resource breakdown structure provides information on the number and type of resources that are planned for the project.

Enterprise environmental factors include databases for estimating, independent cost estimating departments, and published estimating data. The organizational process assets contain historical information from past projects as well as lessons learned. It is helpful to compare the original estimates on past similar projects with the actual durations. This will assist in reducing the likelihood of repeating mistakes.

TOOLS AND TECHNIQUES

Expert judgment is a critical component in developing sound duration estimates. It is best to get estimates from the people who will actually be doing the work. If they have done a lot of similar work in the past their estimates are likely to be fairly accurate. For work that is cutting-edge, you will have to rely on subject matter experts, consultants, and research. In these situations, your estimates are more likely to have a wider range.

Analogous estimating uses information from prior similar projects to produce an estimate. Analogous estimates are usually done at the beginning of the project when the level of detail is minimal. They can be done at the overall project level, or for project deliverables. To develop an analogous estimate you identify the parameters that drive the project duration, and compare the current project with past projects using those same parameters. You would then create an estimate based on similarities and differences with past projects.

Analogous estimates can be misleading if the two work products only appear similar, but in fact are really quite different. For example, you can't assume that all software development is alike. You have to make sure the projects are similar in fact, not just appearance.

Analogous Time Estimate

Assume you need to create a technical documentation manual. The page count will be about 100 pages. You did a similar project last year and the page count was 150 pages. Last year's project took you 240 hours. You can take the 150 pages and divide it by 1.5 to get a duration estimate for this project. (150 is 1.5 times as large as 100.) By doing this, you estimate the hours for this project to be 160 hours.

Parametric estimating uses a mathematical relationship to determine the duration. This is useful if the work is repetitive in nature. You identify the effort it takes to do one of something, then multiply that effort times the number of units you are developing.

Parametric Time Estimate

If you have 50,000 square feet of flooring to install and your historical information shows that it takes six hours for each 1,000 feet, you can determine that 6 x 50 = 300 hours of effort.

Parametric estimates can have multiple parameters and so it is useful to have software when you create this type of estimate. A parametric estimate is most useful for a production environment, or when there is quantifiable repetitive work. Additionally, the work must be scalable.

Three-point estimating is used to account for uncertainty and risk in a project. In many situations the duration of an activity depends on many distinct pieces coming together, and as we all know, this doesn't always happen. To account for the uncertainty inherent in project work, you develop estimates based on the best case, the most likely, and the worst case scenarios. Then you take an average of these numbers to come up with an estimate that accounts for the range of estimates.

One popular way of modifying a three-point estimate is to weight the most likely estimate heavier than the best case and worst case scenarios. Many people use the following formula:

This is sometimes called a PERT estimate. The result is the "expected duration."

$$\frac{\textit{Best case} + 4\ \textit{most likely} + \textit{worst case}}{6}$$

Three-Point Time Estimate

Assume you are talking to a teammate about how long it will take to com plete a verification test. He says, "If everything works right the first time, it should take 20 hours. But nothing ever works right the first time. In fact, I have seen this type of test take up to 50 hours. But, I think it should take about 26 hours." You can take those numbers and, using the PERT equation, you come up with:

$$\frac{20 + 4(26) + 50}{6}$$

Therefore, your expected duration is 29 hours.

Group decision making techniques include the Delphi technique, or a modification of the Delphi technique. The Delphi technique can produce estimates that are more accurate, particularly when the type of work is leading-edge without a lot of historical basis for estimates, or when the type of work is new to the organization doing the work. For work without a lot of historical data the project manager would identify a group of people to help develop an estimate. The group of people can include experts in the field, team members, consultants, or whoever is the best resource given the situation. The project manager then sends the group the information on the project, such as a scope description, risks, resource availability, and so forth.

Individuals in the group submit their estimates indicating which basis of assumptions they used to develop them. The project manager collates and organizes the various estimates and sends out the entire span of estimates, assumptions and basis of estimates to the group to review and revise their estimates based on the collected information. Generally, the second round of estimates has a smaller spread than the first round. The project manager may conduct three or more rounds until the group of experts has arrived at consensus, or until the estimates have stabilized. At that point the project manager may take an average, or select an estimate based on the information provided.

For a true Delphi technique the people providing the estimate would remain anonymous. In other words, no one would know who else was providing estimates. This minimizes the influence of "name recognition" or "position power" when individuals are developing and discussing their estimates.

To modify the technique the project manager can conduct the first round of estimates anonymously and then get the group together to discuss the estimates and the assumptions and work to come to consensus together.

When developing activity durations you may look at the schedule reserve you need to assure an on-time delivery. This is reserve analysis. Many project managers carry reserve that is added to the date they are driving their team toward. This is the date they promise the deliverables to their customers. One way of determining reserve is to look at the difference between the most likely duration and the expected duration and set that amount of time aside to respond to unforeseen circumstances or risks. This is different than padding. It is responsibly calculating the duration based on what you think you can deliver, and accounting for uncertainty.

OUTPUTS

Activity duration estimates can be expressed in effort and duration. To convert effort hours to duration, take the effort and divide it by the number of resources and their availability. Duration estimates may be accompanied by an indication of the range of the estimate, the confidence level in the estimate, and the basis of the estimate.

Effort. The number of labor units required to complete a schedule activity or work breakdown structure component, often expressed in hours, days or weeks.

Duration. The total number of work periods (not including holidays or other nonworking periods) required to complete a schedule activity or work breakdown structure component. Usually expressed as work days or work weeks.

Converting Effort to Duration

If a project team member tells you it will take 100 hours of effort to complete an activity, and she has two people who are available 30 hours a week, how many days in duration will it take?

$$\frac{100 \text{ hours}}{(2 \times 30)} = 1.67 \text{ weeks or } 8.35 \text{ days}$$

Project document updates can include activity attributes and assumptions used in developing duration estimates.

See Appendix A for a sample Activity Duration Estimates Form.

Develop Schedule

The process of analyzing activity sequences, durations, resource requirements, and schedule constraints to create the project schedule model. This process is the culmination of the five previous processes. All the schedule data is combined into a schedule tool and you get your first look at the project schedule. Often the first schedule is not acceptable, and thus begins an iterative series of revisions and negotiations to balance the needed delivery date with available resources, estimated durations, schedule risks, and other project data to arrive at an agreed upon schedule. Once the schedule is agreed to, it is baselined. The process of re planning and managing the schedule will continue throughout the project. Figure 6-13 shows the

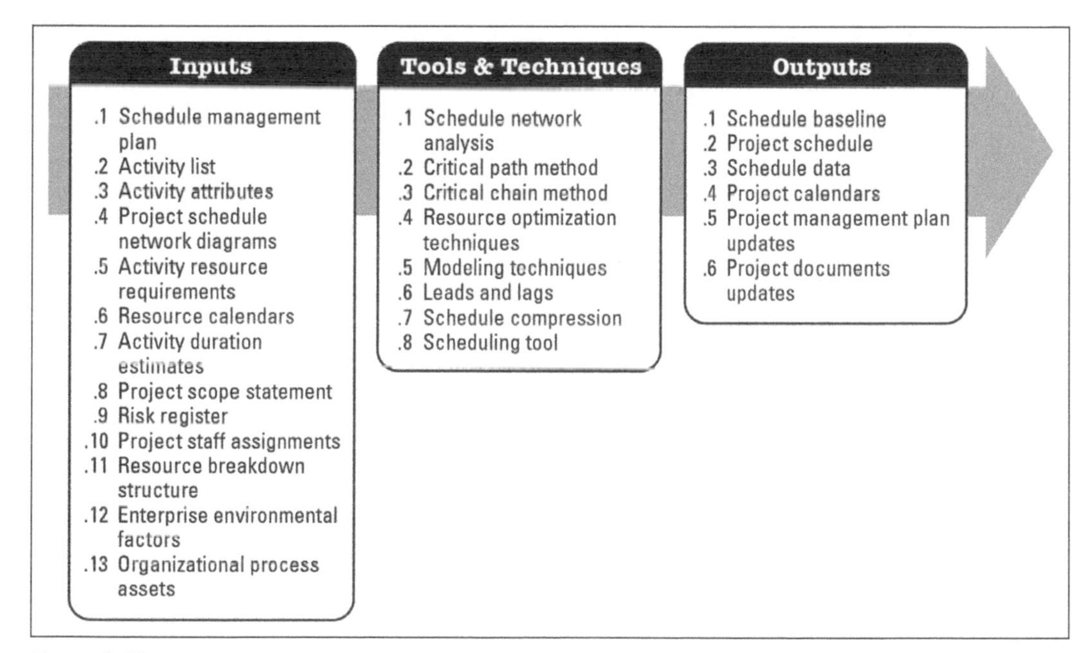

Figure 6-13
Develop Schedule: Inputs, Tools and Techniques, Outputs
Source: PMBOK® Guide—Fifth Edition

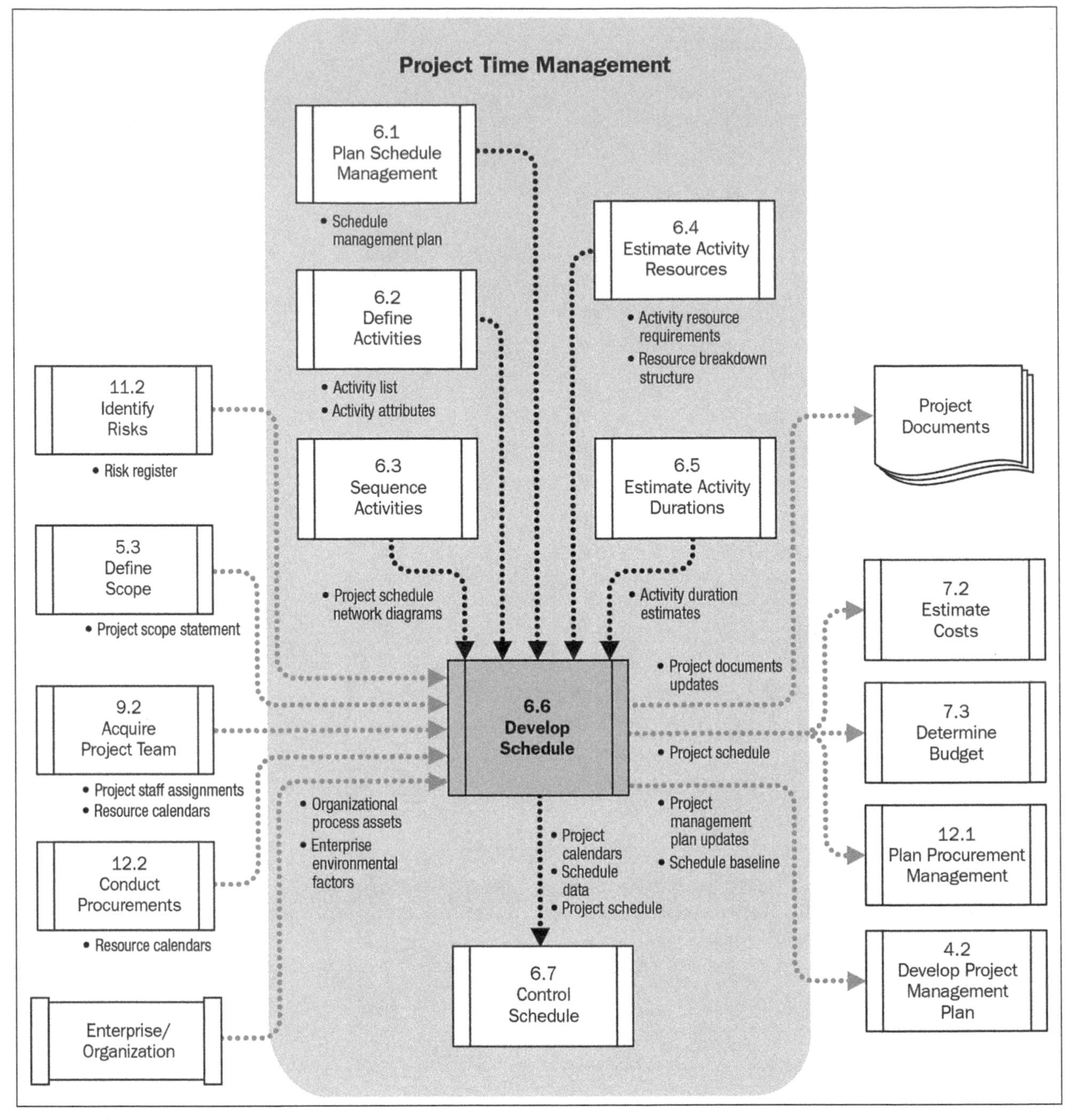

Figure 6-14
Develop Schedule Data Flow Diagram
Source: PMBOK® Guide—Fifth Edition

inputs, tools and techniques and outputs for the Develop Schedule process. Figure 6-14 shows a data flow diagram for the Develop Schedule Process.

INPUTS

The key outputs from the previous processes are used to develop the schedule. These include: the schedule management plan, activity list, activity attributes, project schedule network diagrams, activity resource requirements, and activity duration estimates. Project staff assignments and resource calendars describe who is assigned to each activity and their availability. The resource breakdown structure provides summary information on the types and numbers of resources that are available to assist in what-if scenario analysis activities to help optimize resource utilization.

The project scope statement has assumptions, such as whether work will be done in-house or outsourced, and constraints, such as interim and final delivery dates. The risk register provides information on schedule risks regarding resources, milestone deliverable dates, and other risk factors that might impact the schedule development.

The scheduling tool is an enterprise environmental factor that determines how the schedule will be entered. Organizational process assets such as scheduling policies and procedures, as well as project schedules from previous projects and lessons learned are useful when developing the schedule.

TOOLS AND TECHNIQUES

Schedule network analysis is really an umbrella term for reviewing the schedule, figuring out how to apply the various other techniques, and determining the best approach for arriving at an acceptable schedule. It is usually done with the scheduling tool used by the organization.

There are various scheduling methodologies in use today to calculate a workable schedule. By far, the most widely used methodology is called the *critical path method*. This method entails determining the range of times within which the activities can occur. This is done by determining the early start and finish dates for each activity and the late start and finish dates for each activity. Those represent the range of dates that each activity can start.

The mathematical difference between the early and late dates is called float. It represents the amount of time an activity can slip from its early start or finish date, and not impact the due date of the project, or a schedule constraint (such as a mandatory milestone date). An activity with float has some scheduling flexibility. However, keep in mind that the future activities may be affected by shifting the start or finish date within the float. You will have to understand the implications to other activities that share the path to ensure you are not creating resource issues. Those activities with no float are on the critical path. The critical path usually has zero float, but it can also be interpreted as that path with the least amount of float. That means that there can be negative float. In this situation, you must find a way to compress the schedule, or the project will be late.

Another kind of float is called "free float." Free float is the mathematical difference between the early finish of a predecessor task, and the early start of the successor task. This happens when there are multiple paths converging into one. The last activity on the path(s) that are not critical have free float. Free float is the amount of time an activity can slip and not impact the very next activity. Activities with free float have the most flexibility because if they start within the free float time, there should be no schedule implications for any other activities in the schedule.

Early start date (ES). In the critical path method, the earliest possible point in time when the uncompleted portions of a schedule activity can start based on the schedule network logic, the data date, and any schedule constraints.

Early finish date (EF). In the critical path method, the earliest possible point in time when the uncompleted portions of a schedule activity (or the project) can finish based on the schedule network logic, the data date, and any schedule constraints.

Late start date (LS). In the critical path method, the latest possible point in time when the uncompleted portions of a schedule activity can start based upon the schedule network logic, the project completion date, and any schedule constraints.

Late finish date (LF). In the critical path method, the latest possible point in time when the uncompleted portions of a schedule activity can finish based on the schedule network logic, the project completion date, and any schedule constraints.

Total float. The total amount of time that a schedule activity may be delayed or extended from its early start date without delaying the project finish date, or violating a schedule constraint.

Free float. The amount of time that a schedule activity can be delayed without delaying the early start date of any successor or violating a schedule constraint.

Critical Path Calculation

Figures 6-15 through 6-18 show a network diagram with durations, then a forward pass, then a backward pass, then the identification of the critical path, total float, and free float.

Critical chain methodology uses more aggressive estimates, but protects the network by using buffers. The resource-loaded schedule shows the longest path through the project as the critical chain (as opposed to the critical path). The critical chain is protected by placing project buffers at the end of the chain. For paths that feed into the critical chain, a feeding buffer is used so no feeding chains negatively impact the critical chain. The buffers act as a protection for the project because more aggressive duration estimates are used.

There are other assumptions that go along with the critical chain methodology as well. All resources are expected to work full-time during their tasks. They are not expected to be on more than one project at a time and they should only focus on the project work. Resources are considered the "bottleneck" in projects, so all effort is expended on exploiting those resources. "Exploiting" in this sense

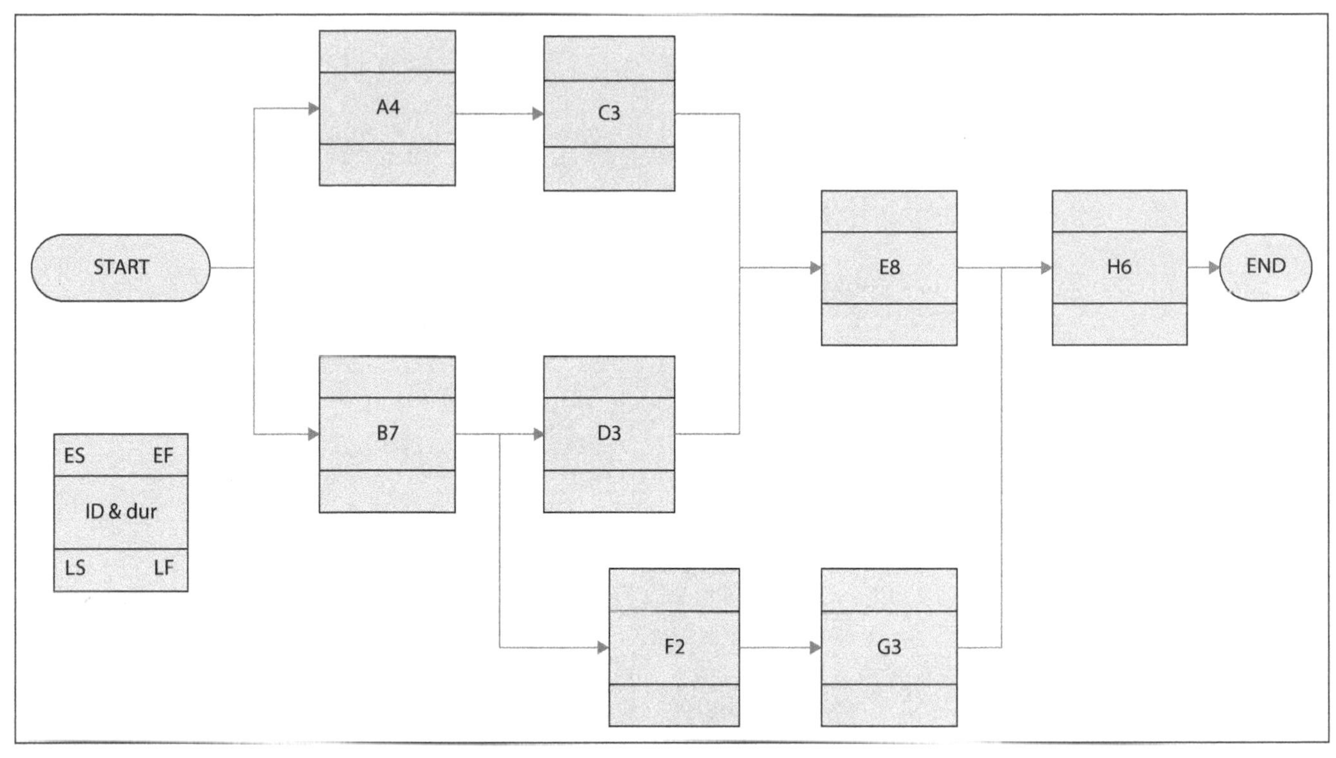

Figure 6-15
Network Diagram with Durations

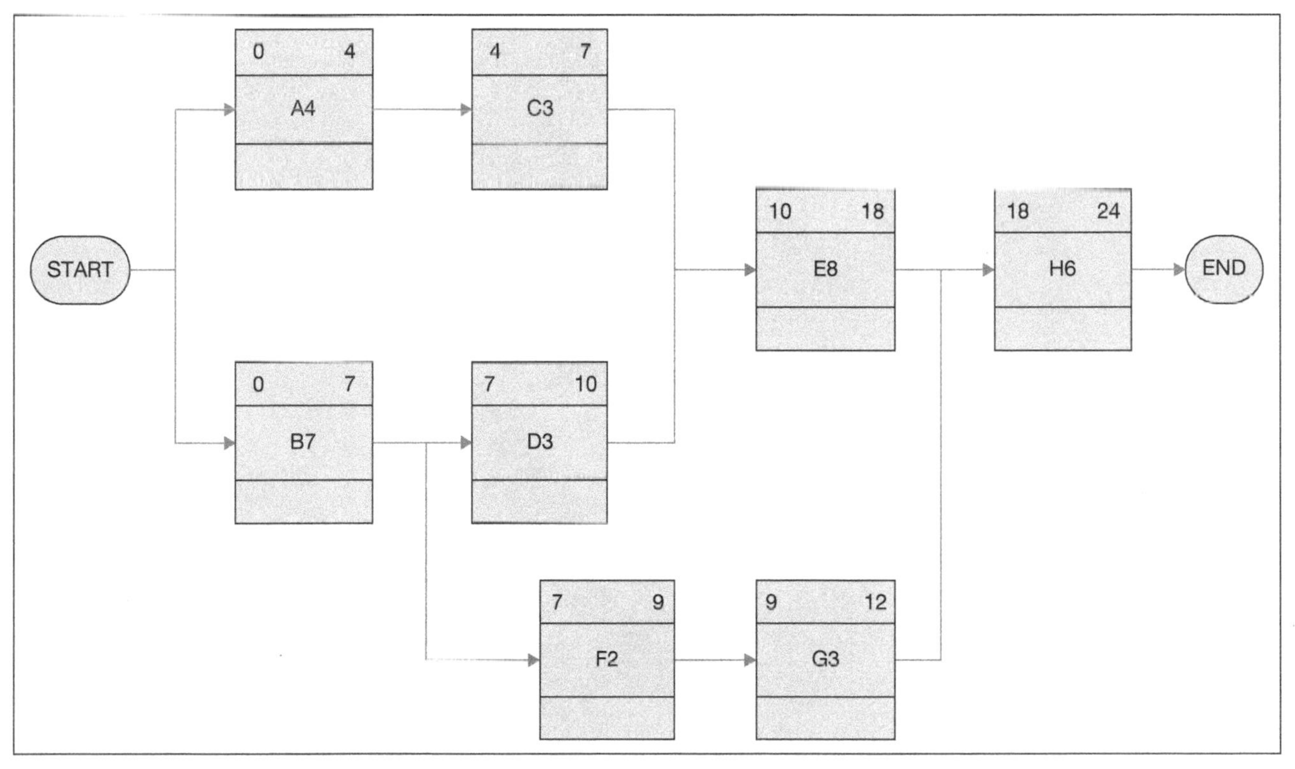

Figure 6-16
Forward Pass

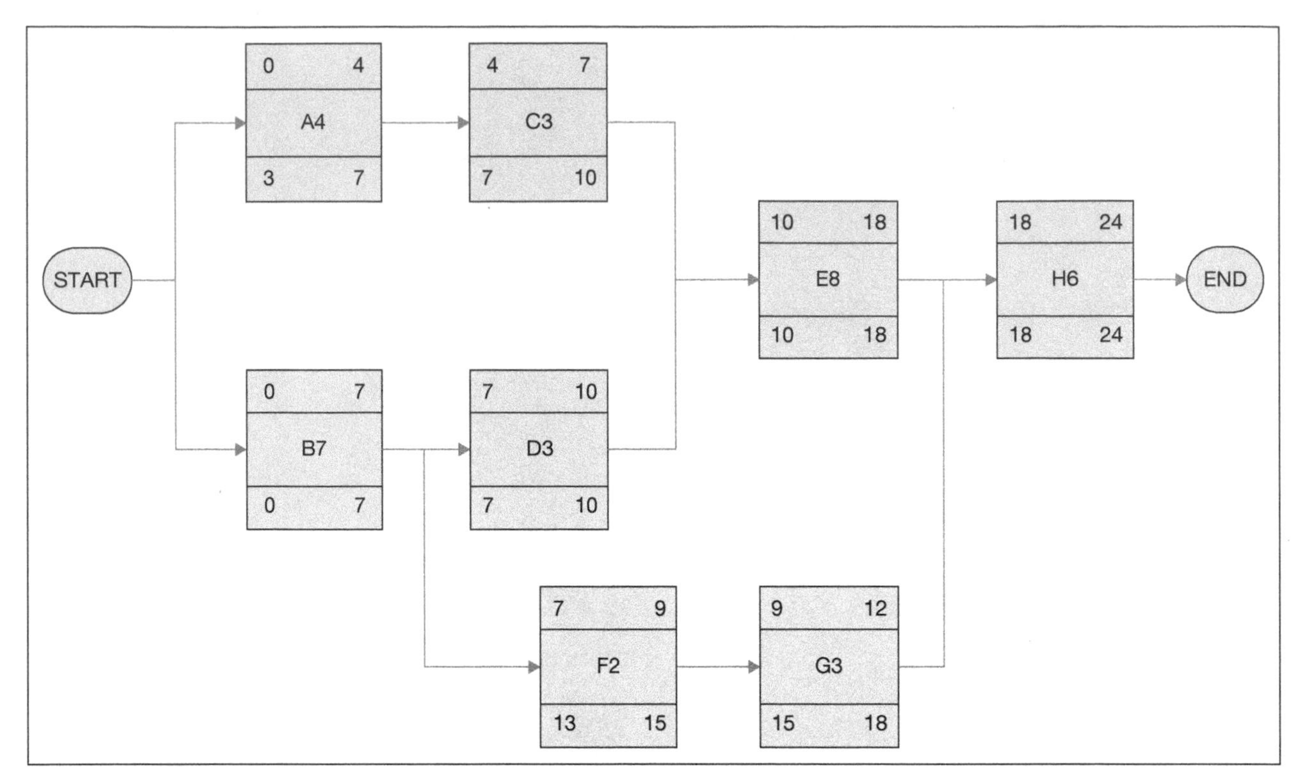

Figure 6-17
Backward Pass

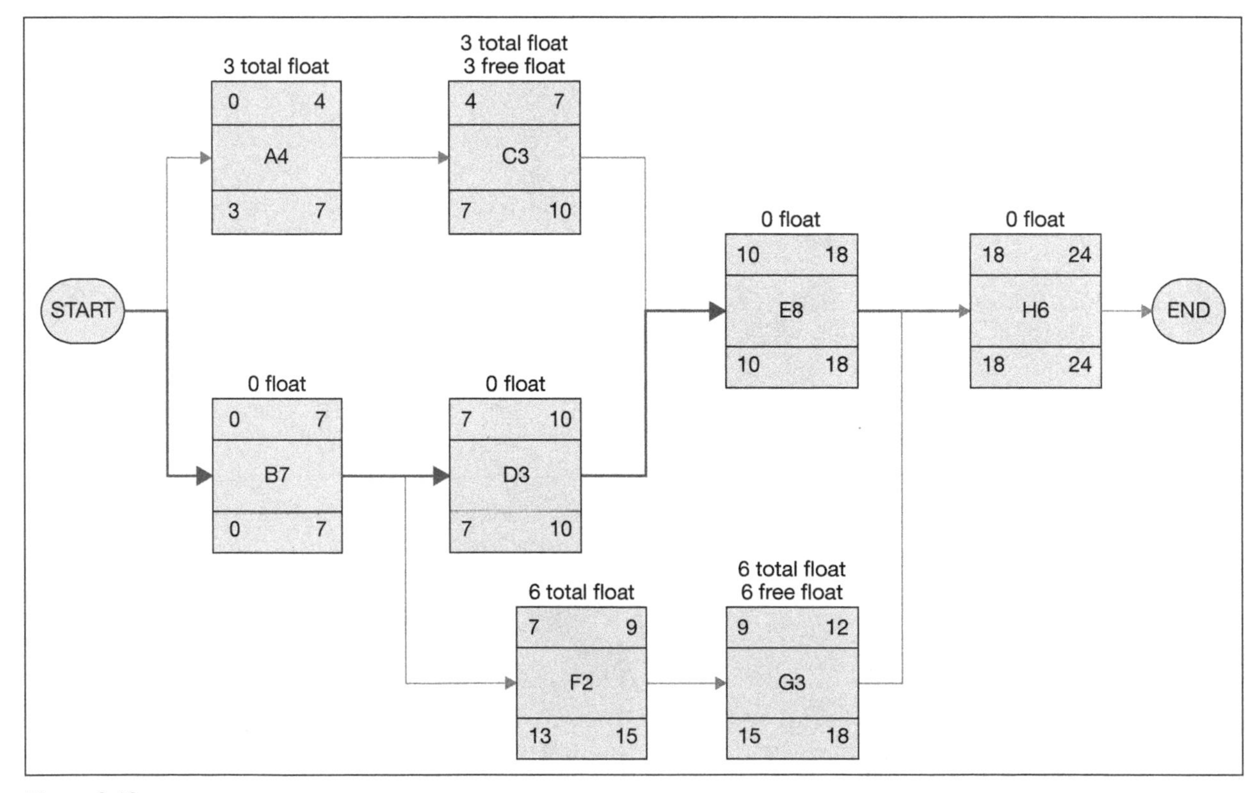

Figure 6-18
Critical Path with Float Identified

means using them fully for the project and making sure no other work distracts them.

In critical chain methodology, the duration estimates are never padded, or given a conservative number. The duration estimate should be at the 50 percent confidence rate. This means there is a 50 percent chance the activity will be done on time and a 50 percent chance it won't. The buffer is there to absorb the risk of the more aggressive estimates. The whole focus of critical chain is to optimize resources to get projects through the pipeline faster.

Figures 6-19 and 6-20 show a comparison of the critical path method and the critical chain method. Notice with the critical

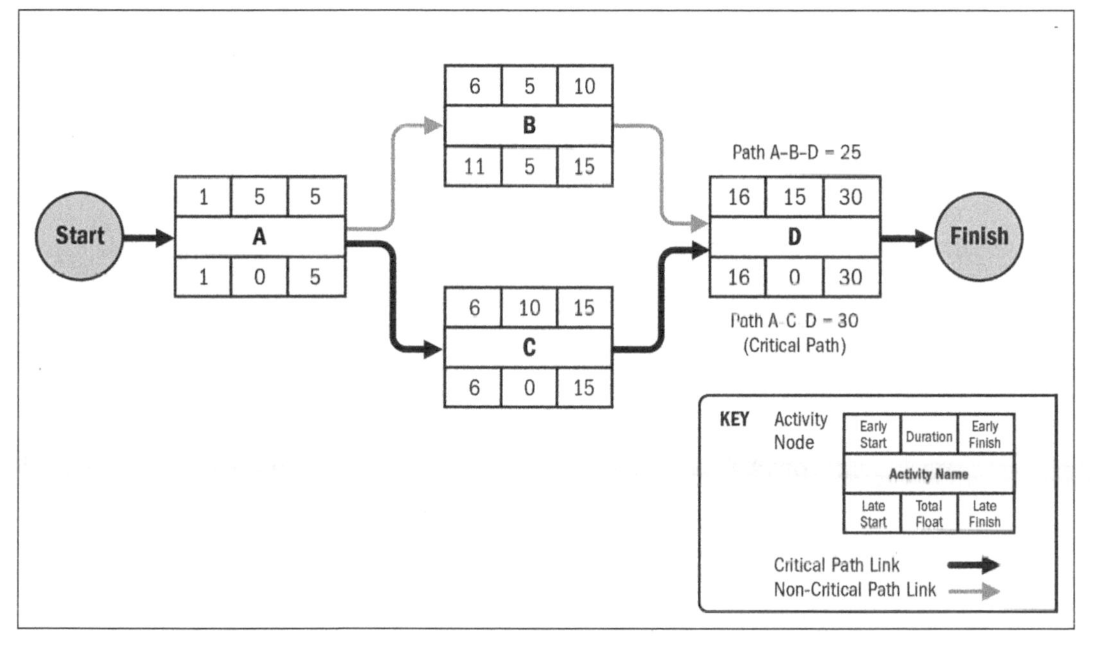

Figure 6-19
Example of Critical Path Method

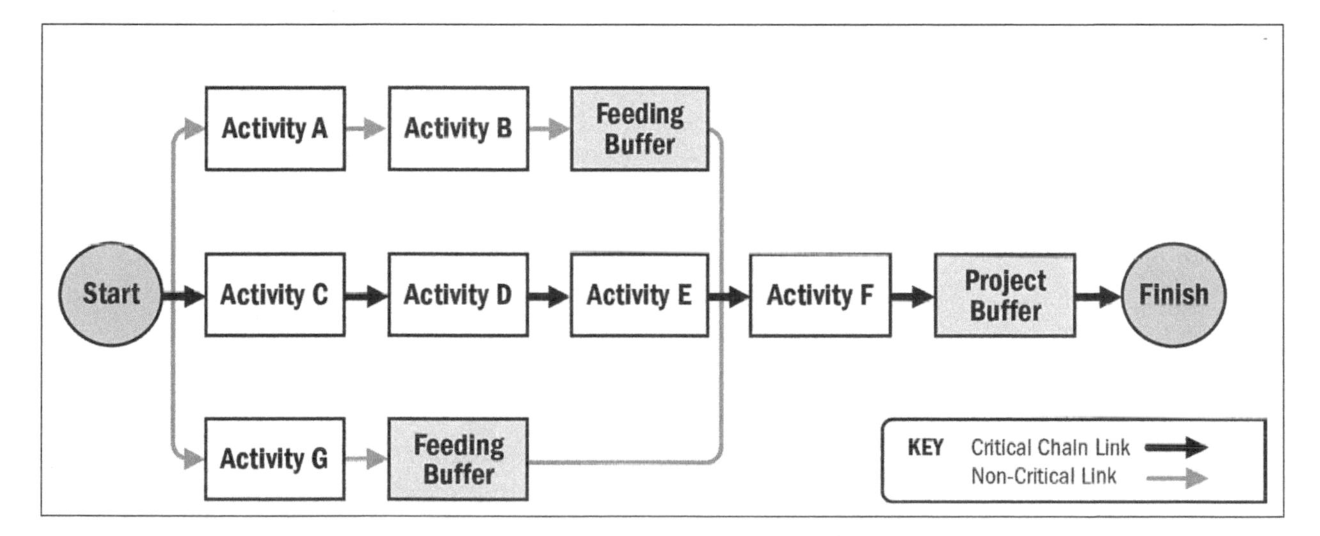

Figure 6-20
Example of Critical Chain Method

path method that activities take longer and there is no buffer. With the critical chain method, each activity's duration is more aggressive, but there is buffer at the end of the chain.

Generally the first iteration of the critical path can be done without regard to resource constraints. Once resources are loaded, along with their availability, the critical path usually changes. At this point you can apply resource optimization techniques. For example, you should take into consideration the amount of time a resource is actually available to work. If resources are working on several projects at once their availability will be limited. In some cases you may find several activities occurring simultaneously by the same person. If the work requires that person be working on each task full-time to meet the duration, then the schedule has to be modified by bringing in additional resources, or pushing the schedule out to allow all the activities to be completed.

Adjusting the start and finish dates of activities to accommodate the existing resources is called resource leveling. However, if the durations are not determined based on the amount of effort, you will need to consult with the resource to determine their ability to get all the work done. If the resource is overutilized then you may need to bring in additional resources, prioritize activities or extend the schedule.

Another resource optimization technique is resource smoothing. Resource smoothing is adjusting activities within their early and late start and finish dates to allow the most work to get done without changing the dates. Resource smoothing avoids peaks and valleys of 60 hours of work scheduled one week and 20 hours the next.

Modeling techniques include what-if scenarios and Monte Carlo simulations. A what-if scenario analysis creates versions of the schedule based on either uncertain events, or based on various durations of activities. For various versions of events, you might look at your risk register and assume one of the events occurs that delays the schedule. You can analyze the impact that the event and the response would have on the overall schedule.

A Monte Carlo analysis is used to look at the impact of the uncertainty associated with activity durations. Remember in the Estimate Activity Durations process we looked at three-point estimates. Those estimates are loaded into simulation software and thousands of iterations of the project are run assuming random durations within the range of best case and worst case scenarios.

The resource balancing scenario shows how you can use what-if scenario analysis to optimize resource utilization.

Resource Balancing Scenario

Based on the initial schedule you have a resource problem for a deliverable that is scheduled to take three weeks. There are three resources scheduled to work on the deliverable full-time: Jake, Martin, and Tina. The initial resource utilization diagram is shown in Figure 6-21.

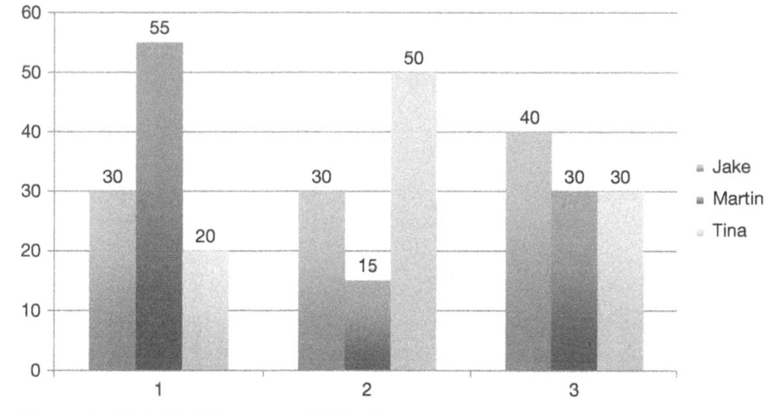

Figure 6-21 Initial Resource Utilization

Notice that Martin is overutilized on week 1 and Tina is overutilized on week 2. If there are no delivery constraints other than the deliverable being due in three weeks, the resources can be optimized by leveling the utilization across the three weeks. The resulting utilization is shown in Figure 6-22.

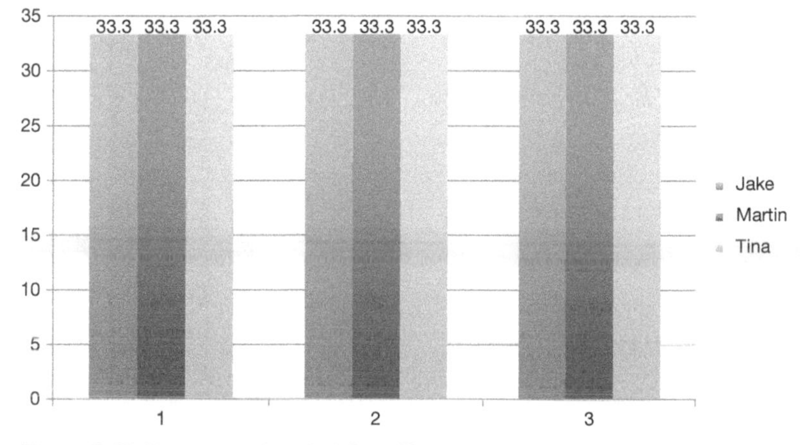

Figure 6-22 Resources Leveled Over Time

Another option is to level the work across resources. However, to do that you need to ensure that the resources have comparable skill sets and that the ramp-up time, or work transition time, is not excessive. Figure 6-23 shows how the work for week 1 was spread evenly across the three resources.

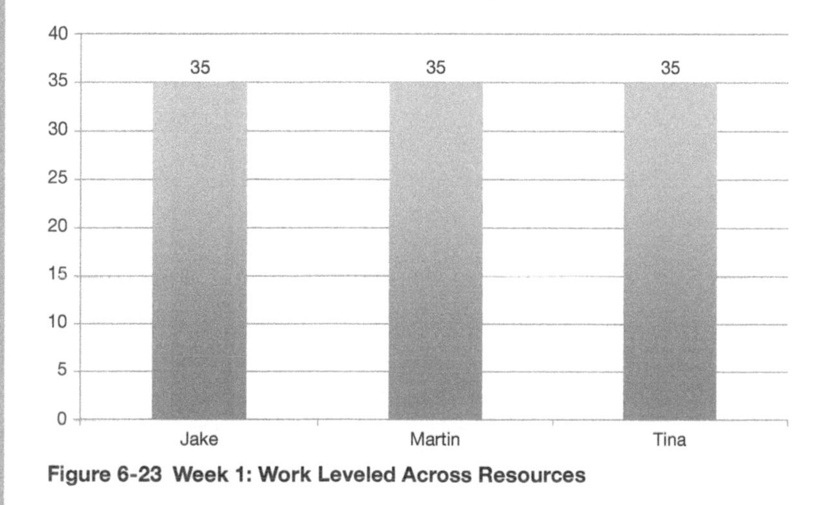

Figure 6-23 Week 1: Work Leveled Across Resources

In the Sequence Activities process we introduced leads and lags. When we are trying to create an acceptable schedule we may change the relationship between activities or apply leads and lags to adjust the schedule.

All the above data is entered into a scheduling tool to develop a schedule model. Most times when we come up with a schedule model it is longer than the customer would like. Two ways to compress the schedule without reducing quality or scope include crashing and fast tracking. Crashing involves adding resources, overtime, extra shifts, or expediting shipping to accelerate the schedule. It usually incurs cost so assess the impact to the budget before finalizing the schedule.

Fast tracking is overlapping activities, or conducting activities in parallel that would normally be done sequentially. This could include changing a finish-to-start relationship to a finish-to-start with a lead, or a finish-to-finish with a lag, or a start-to-start with a lag. You have to be careful that you don't overload your resources when fast tracking. This method can increase your risk on the project so assess the revised network for increased risk before finalizing the schedule.

Crashing

When crashing, look to those activities on the critical path. It does no good to crash activities with float! Then determine which activities will get you the most compression for the least amount of cost.

Fixed Duration

Not all activities can be crashed. Some activities have a fixed duration no matter how many resources work on them or how hard they work. An example of an activity with a fixed duration would be a test to determine how long it will take a component to fail, or a training class.

OUTPUTS

The schedule baseline is the schedule the team agrees to and commits to. This is usually developed toward the end of the planning process when the requirements are fully defined, the resources are loaded, and project risk responses are built into the schedule. Any change in project or product scope needs to be represented in a new schedule baseline.

The project schedule shows the planned start and end dates for activities. Notice this is different from the critical path analysis that shows the early start and finish dates and late start and finish dates providing a range of dates. It is also different from critical chain that schedules using the latest possible start and finish dates. The project schedule takes that information into consideration and develops a schedule with the planned start and finish dates. For complex or large projects, it is not uncommon to have different types of schedules, for example:

- A target schedule that shows the target start and finish dates. These may be more aggressive than the planned start and finish dates.
- A best case or worst case schedule that shows how activities would flow if everything worked well, or if nothing worked well.
- A milestone schedule that shows the start and finish dates of project phases and/or key deliverables and significant events.
- A network diagram with durations to highlight the project logic; this may be developed specifically for the critical path.
- A modified Gantt chart that shows bars that indicate duration and the network logic connecting them.

Schedule data includes all the data that is loaded in the schedule tool to produce the project schedule and the baseline schedule. It includes much of the information from the activity attributes, and may also include reserves, alternative dates, such as target dates, best case and worst case dates, resource histograms, and so forth.

Once the schedule is baselined, you can create the project calendars that indicate working days (as opposed to holidays and weekends), and shifts. For example, if you are working on a project with multiple vendors, particularly if they are in different countries, your calendar would show if a particular vendor and their resources has a two-week nonworking period over the winter holidays.

The project management plan updates can include the schedule management plan or other plan documents depending on the project. There are a number of project documents that may need to be updated, including, but definitely not limited to:

- Activity attributes
- Risk register
- Resource requirements
- Duration estimates
- Cost estimates
- Project calendar

Creating a Baseline

Consider baselining at a level above the detailed schedule, perhaps even at the milestone level. The baseline is your promise. You don't want to have to explain every time an activity doesn't start or end on the baseline date. But you do want to make sure you can commit to key milestones.

Chapter 7

Planning Cost

TOPICS COVERED

Project Cost Management

Plan Cost Management

Estimate Costs

Determine Budget

Project Cost Management

Project Cost Management includes the processes involved in planning, estimating, budgeting, financing, funding, managing, and controlling costs so that the project can be completed within the approved budget.

Along with the project schedule, which was addressed in the previous chapter, the project budget is one of the most important documents in the project. Cost estimates are developed for cost categories and work breakdown structure (WBS) elements and are then applied across the schedule to develop a funding curve and a cost baseline.

The cost management processes can also analyze costing options by weighing the lifecycle cost of investing more in the project development in order to have a lower cost in maintaining the product, or investing less up front and having to spend more on maintenance and upkeep.

Plan Cost Management

Plan Cost Management is the process that establishes the policies, procedures, and documentation for planning, managing, expending, and controlling project costs. When planning how to manage the project costs the team will take into consideration how costs will be estimated, the accuracy level needed in the budget, and the measurements that will be used to monitor the budget status. Figure 7-1 shows the inputs, tools and techniques and outputs for the Plan Cost

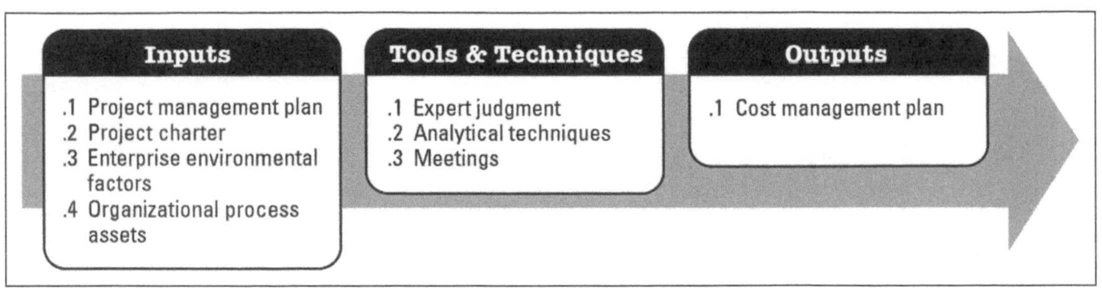

Figure 7-1
Plan Cost Management: Inputs, Tools and Techniques, Outputs
Source: PMBOK® Guide—Fifth Edition

Management process. Figure 7-2 shows a data flow diagram for the Plan Cost Management process.

At the start of the project the project management plan will be more of a framework and contain high-level information. However, having it as an input provides a consistent method of developing subsidiary management plans. As the project progresses, the project management plan will contain the scope baseline, the schedule baseline, and other more robust and detailed information that can be used to iteratively refine the cost management plan.

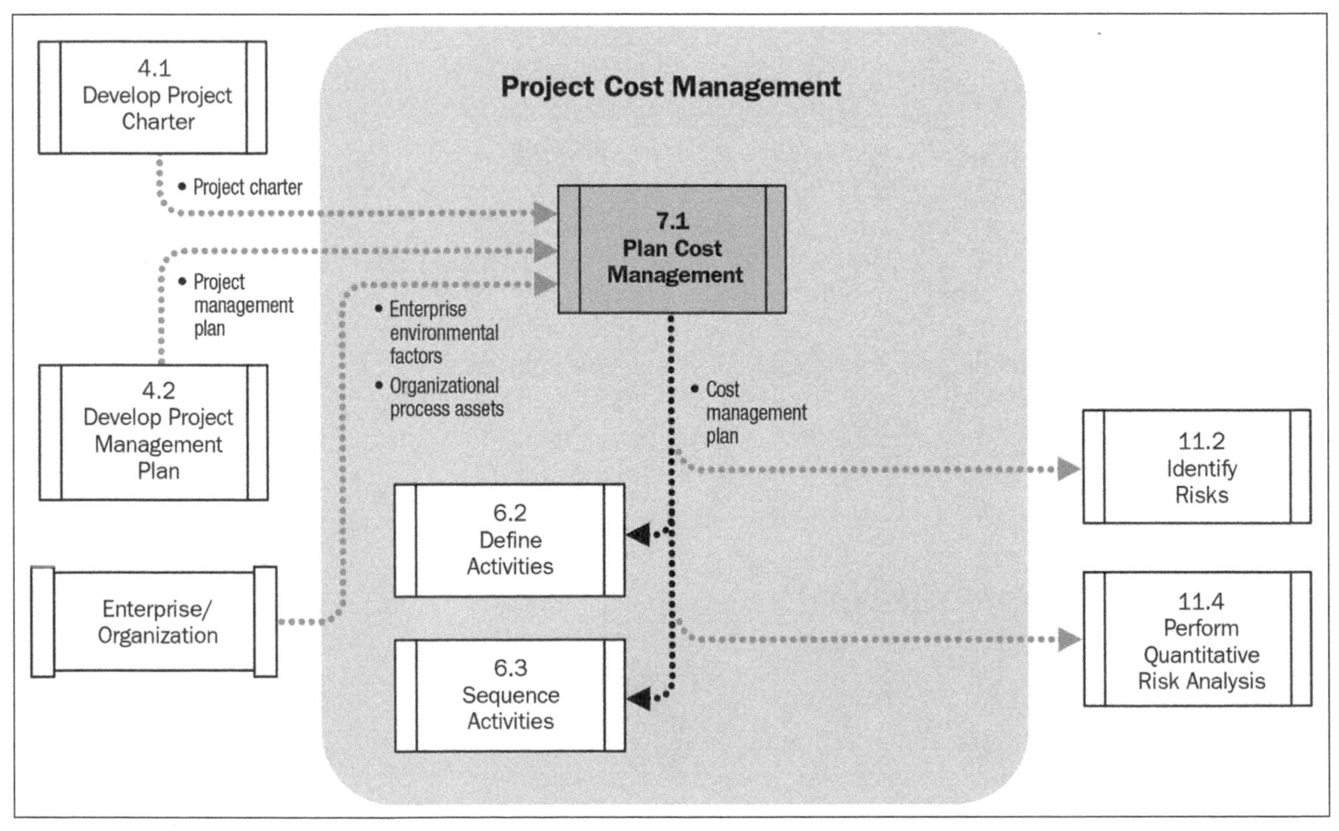

Figure 7-2
Plan Cost Management Data Flow Diagram
Source: PMBOK® Guide—Fifth Edition

The *project charter* usually contains a high-level budget that is used prior to developing a detailed bottom-up series of cost estimates and project budget.

Enterprise environmental factors that influence the cost management plan include the availability of resources in the marketplace, published resource rate information, and information systems that are used in tracking and reporting costs.

Organizational process assets used in this process include information from previous similar projects, templates, policies, and procedures.

TOOLS AND TECHNIQUES

Expert judgment during the cost management planning process typically includes expertise that helps determine the appropriate estimating methods to use and the viable cost management and control techniques. Experts can provide analytical techniques to help determine funding implications for options such as make, lease, or buy for equipment. Analytical techniques are often used when looking at the financial viability of a project or an approach to a project. Common financial analytical techniques include return on investment measurements, internal rate of return, net present value, and discounted cash flow calculations.

Meetings are often used to discuss the various estimating, funding, and management methods that will be used on the project.

OUTPUTS

The *cost management plan* is a subsidiary plan of the project management plan that describes how the costs will be planned, monitored, managed, and controlled on the project. Contents may include:

Level of accuracy. Will costs be estimated and collected in hundreds of dollars, thousands of dollars, or on some other basis?

Units of measure. Will costs be planned and tracked using hours, days, dollars, euros, or some other unit?

Linkages. How will project control accounts interface with accounting codes?

Control thresholds. How much variance is allowed before corrective or preventive action needs to be taken? If you are using a dashboard or stoplight rating system, what is considered green, yellow, and red?

Rules of performance measurement. How will you track performance, and at what level? If you are using earned value management techniques, what are the rules for employing various methods of measurement, such as 50/50, level of effort, weighted milestones, and so forth?

Reporting. How often will you report progress, in what format, and to whom? This information should correspond to information in the communications management plan.

Cost Management Plan. A component of a project or program management plan that describes how the project costs will be planned, structured, monitored, and controlled.

Estimate Costs

The process of developing an approximation of the monetary resources needed to complete project activities. Estimating implies a prediction; estimates are not facts. As the scope of the project is defined in more detail, the cost estimates are more accurate and the range of estimates narrows. In projects with a lot of research and development and a high degree of complexity, it is not uncommon for the initial estimates to have a range of ±50 percent. By the time the details are defined, the schedule is baselined, and project risks are accounted for, the range may be closer to ±10 percent.

Cost Estimating Is Sensitive!

Cost estimating is one of the most sensitive topics in project management. Customers and sponsors want to know how much it will cost very early in the project so they can make good decisions. However, getting reliable estimates early in the project is often not possible. This often leads to putting out estimates that are not accurate (often less than is really needed), and then being held accountable for meeting the estimate.

Additionally, many customers and sponsors will look at an estimate and ask you to arbitrarily cut it by some percent. The fastest way to go over budget on a project is to underestimate or underfund the project from the beginning.

In your cost management plan include a section on estimating guidelines. Define the ranges of accuracy that are expected for each phase in the lifecycle. This will help communicate the level of accuracy that stakeholders can expect as you move through the planning phase and subsequent phases.

Figure 7-3 shows the inputs, tools and techniques and outputs for the Estimate Costs process. Figure 7-4 shows a data flow diagram for the Estimate Costs process.

INPUTS

The cost management plan provides guidance on the types of cost estimating techniques you can use to develop your cost estimates for the project. The plan also defines the expected level of estimate accuracy depending on where you are in the lifecycle. Earlier in the lifecycle you would expect to have a wider range of estimates. Once

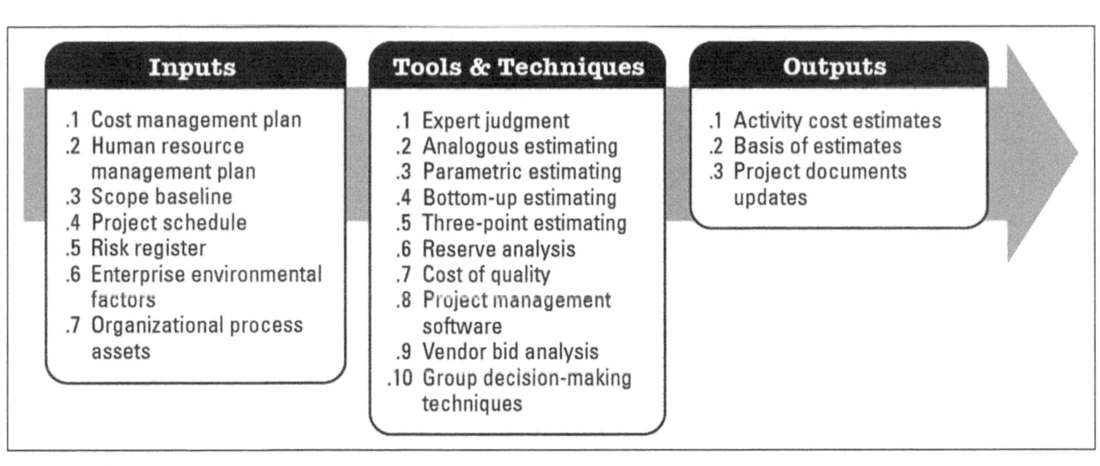

Figure 7-3
Estimate Costs: Inputs, Tools and Techniques, Outputs
Source: *PMBOK® Guide*—Fifth Edition

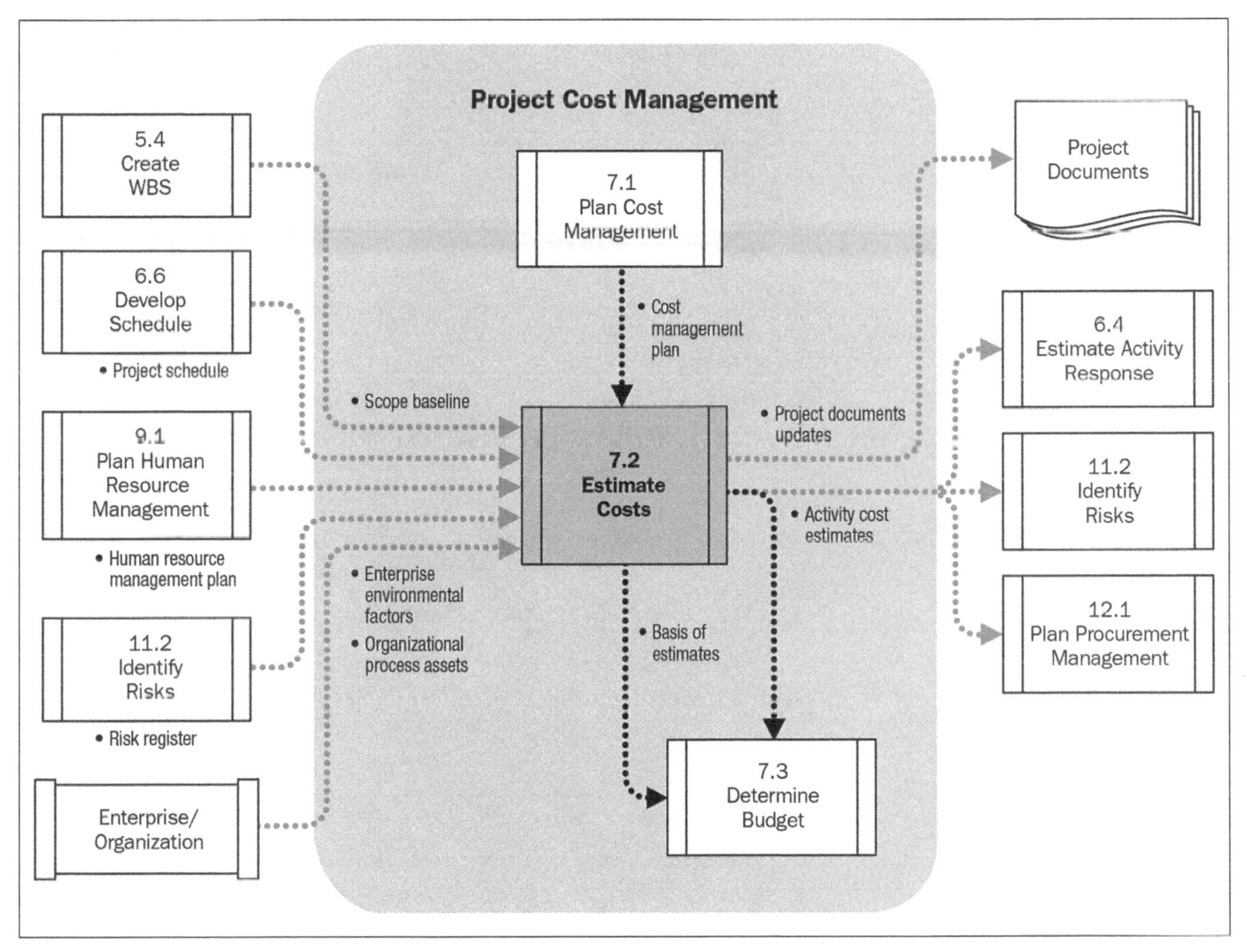

Figure 7-4
Estimate Costs Data Flow Diagram
Source: *PMBOK® Guide*—Fifth Edition

you have detailed requirements and have completed a risk management plan, you should have a much tighter range of estimates.

The scope baseline, comprising the project scope statement, WBS, and WBS dictionary, contains the full scope of the product and project work. You should review at least the following information from the scope baseline documents:

- Product description
- Product deliverables
- Work packages
- Technical description of work
- Acceptance criteria
- Assumptions
- Constraints

The human resource management plan identifies the resources, their charge rate, and sometimes information such as training or certificates needed for the project. Information from the project schedule (whether finalized, or still in the development stage) will have the resource requirements and the duration the resources will be required. This information helps in accumulating costs when rates are given on an hourly, daily, or other unit basis.

The risk register lists the identified risks, planned responses, and reserve amounts. The cost of implementing responses and the monetary reserve need to be incorporated into the cost estimates.

Enterprise environmental factors that influence cost estimates include the availability of resources and skill sets in the market. The availability of material can have a significant impact on cost. Published cost databases, developed either in-house, or via subscription, are also useful in getting rate and cost information. If the organization has a particular cost estimating or cost tracking system, this will constrain how cost estimates are developed and recorded.

What Goes in a Cost Estimate?

Different organizations have different expectations about what should be included in a cost estimate. Some organizations include overhead, some include indirect costs; others only consider direct labor. Make sure you understand what is expected for your project.

Cost estimating policies, procedures, and guidelines are organizational process assets you can use to help develop the estimates. Additionally, information from prior projects and lessons learned are very helpful. Make sure the project information you are looking at from prior projects is timely. Estimates from a project completed eight years ago are rarely relevant today.

TOOLS AND TECHNIQUES

Expert judgment in cost estimating can help you get the most accurate estimates because subject matter experts will have insight into labor rates, material costs, the types and amount of resources you will need to complete the work, risk factors, and other variables that can influence the cost of resources. Subject matter experts can also help determine the best methods to develop cost estimates, and provide assistance in applying the following techniques.

Analogous estimating uses information from prior similar projects to produce an estimate. Analogous estimates are usually done at the beginning of the project when the level of detail is minimal. They can be done at the overall project level, or for project deliverables. To develop an analogous estimate you identify the parameters that drive the project cost. Then you compare the current project with past projects using those same parameters. You would then create an estimate based on similarities and differences with past projects. Some of the common cost drivers include: project size, complexity, duration, product weight, speed, and functionality.

Analogous Cost Estimate Example

Assume you are putting in a new parking lot. It will have 500 parking spaces. You have done several parking lots in the past so you have some history to compare this one to. The most similar one you have done recently had 300 spaces and cost $15,000. Since the previous one is only 70 percent the size of this one, you take the $15,000 and divide it by .7 to develop an estimate of $21,429.

Like analogous duration estimates, analogous cost estimates can be misleading if the two work products only appear similar, but in fact are really quite different.

Parametric estimating uses a mathematical relationship to determine the cost. This is useful if the work is repetitive in nature. You identify the cost it takes to develop one unit then multiply that effort times the number of units you are developing. This only works if the cost can be estimated using a consistent unit of measure.

Parametric Cost Estimate Example

If you need 200 laptops and each laptop is $500, the total cost would be $100,000.

Don't Forget This!

Some areas that can be overlooked in developing cost estimates include:

- Travel
- Certification
- Shipping and handling
- Licenses
- Regulatory requirements
- Legal requirements
- Permits
- Security costs
- Inflation estimates
- Cost of money

Parametric estimates can have multiple parameters so it is useful to have software when you create this type of estimate.

Bottom-up estimating is used when you have a good understanding of each element of work. Bottom-up estimating is done at the work package level of the WBS. Often the estimating information you will need is recorded in the WBS dictionary. In bottom-up estimating you would determine the labor effort and multiply that times the labor rate. Then determine the equipment you need and the cost for the equipment. Determine the material needs and multiply that by the material rate and other direct costs and the amount of each. Sum the totals for the labor, equipment, direct costs, and add any additional cost elements, such as indirect costs, travel, overhead, and the like. The resulting total is the cost for the work package.

After developing bottom-up estimates for all work packages you sum the estimates and determine any needed risk reserves or other costs and develop a project cost estimate. Obviously you must have very detailed information to develop this type of estimate. Bottom-up estimating is a very time-intensive process, but it is also the most accurate.

Three-point estimating is used to account for uncertainty and risk in a project. Like three-point duration estimates you develop estimates based on the best case, the most likely, and the worst case scenarios. Then take the average of those numbers to come up with an estimate that accounts for the range of estimates.

One popular way of modifying a three-point estimate is to weight the most likely estimate heavier than the best case and worst case scenarios. Many people use the following formula:

$$\frac{\textit{Best case} + 4\ \textit{most likely} + \textit{worst case}}{6}$$

A three-point estimate is sometimes called a PERT estimate. The result is the "expected cost."

Three-Point Cost Estimate Example

In reviewing cost information from past projects you determine the best case scenario for airfare is $350 for a round-trip ticket. The highest it has been is $1,000. The most common cost per ticket is $600. You can take those numbers and, using the PERT equation, you come up with:

$$\frac{360 + 4(600) + 1000}{6}$$

Therefore, your expected cost is $625.

Reserve analysis takes into consideration the complexity and risk in the project as well as where the project is in the lifecycle. The amount of reserve can be increased or reduced based on the project progress. Some organizations have policies that indicate the amount of reserve that should be available at each phase in the lifecycle.

Cost of quality analysis is done as part of the Plan Quality process and the relevant costs are entered here. It basically determines the cost of quality investment in defect prevention, quality appraisal, and internal and external failure. Costs can include training, quality assurance and control, scrap, rework, and replacement parts.

In order to develop a detailed cost estimate you will need some type of project management estimating software. This can be as simple as a spreadsheet, or it can include statistical tools, simulation tools, and estimating tools. Such tools help develop more robust estimates and also speed up the process. They can be used later to monitor and control costs during the project.

Vendor bid analysis is used when the project, or sections of the project, will be obtained from external sources. A fixed price bid is easy to incorporate into the cost estimate. However, cost reimbursable contracts, or contracts with incentive and award fees can be more challenging to estimate.

Group decision making techniques, such as the Delphi technique described in the previous chapter, can be applied to cost estimates as well as duration estimates.

OUTPUTS

Activity cost estimates can be categorized by cost type, such as labor, equipment, material, facilities, contingency reserve, indirect costs, and the like. They can be presented at the detail level by line item or summarized by cost category.

The cost estimates should be backed up with documentation that provides the assumptions used and the *basis of the estimates.*

Documentation should also include the range of estimates and the confidence level of the estimate.

Basis of Estimates and Assumptions Example

Basis of estimates can include 40 hours of drafting at $75 per hour.

Assumptions for cost estimates can include the assumption that you will be able to reuse existing code, or that work will be done during regular business hours without overtime.

Project document updates can include activity attributes and the risk register.

See Appendix A for a sample Cost Estimating Worksheet.

Determine Budget

The process of aggregating the estimated costs of individual activities or work packages to establish an authorized cost baseline. The cost baseline is what the project cost performance will be measured against when monitoring and controlling the project. The cost baseline includes the cost estimates and contingency reserves for identified risks. The budget is different from the cost baseline in that the budget includes the cost baseline plus management reserve, which is used for unplanned changes to the project scope or cost. The budget is time-phased to show the cost of the project scope over time. Figure 7-5 shows the inputs, tools and techniques and outputs for the Determine Budget process. Figure 7-6 shows a data flow diagram for the Determine Budget process.

INPUTS

The cost management plan provides information on how project costs will be managed and controlled. The *scope baseline* provides information on the technical details of the scope, as well as any identified funding constraints or limitations. For example, projects that

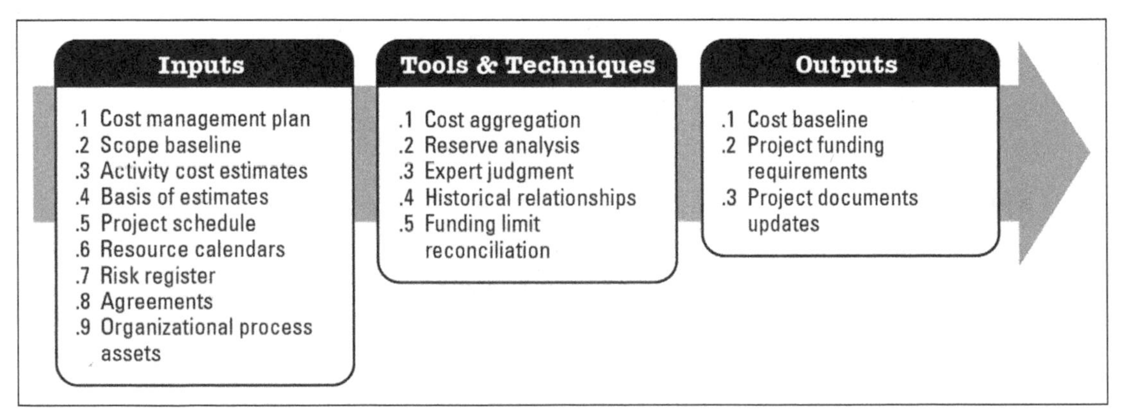

Figure 7-5

Determine Budget: Inputs, Tools and Techniques, Outputs

Source: PMBOK® Guide—Fifth Edition

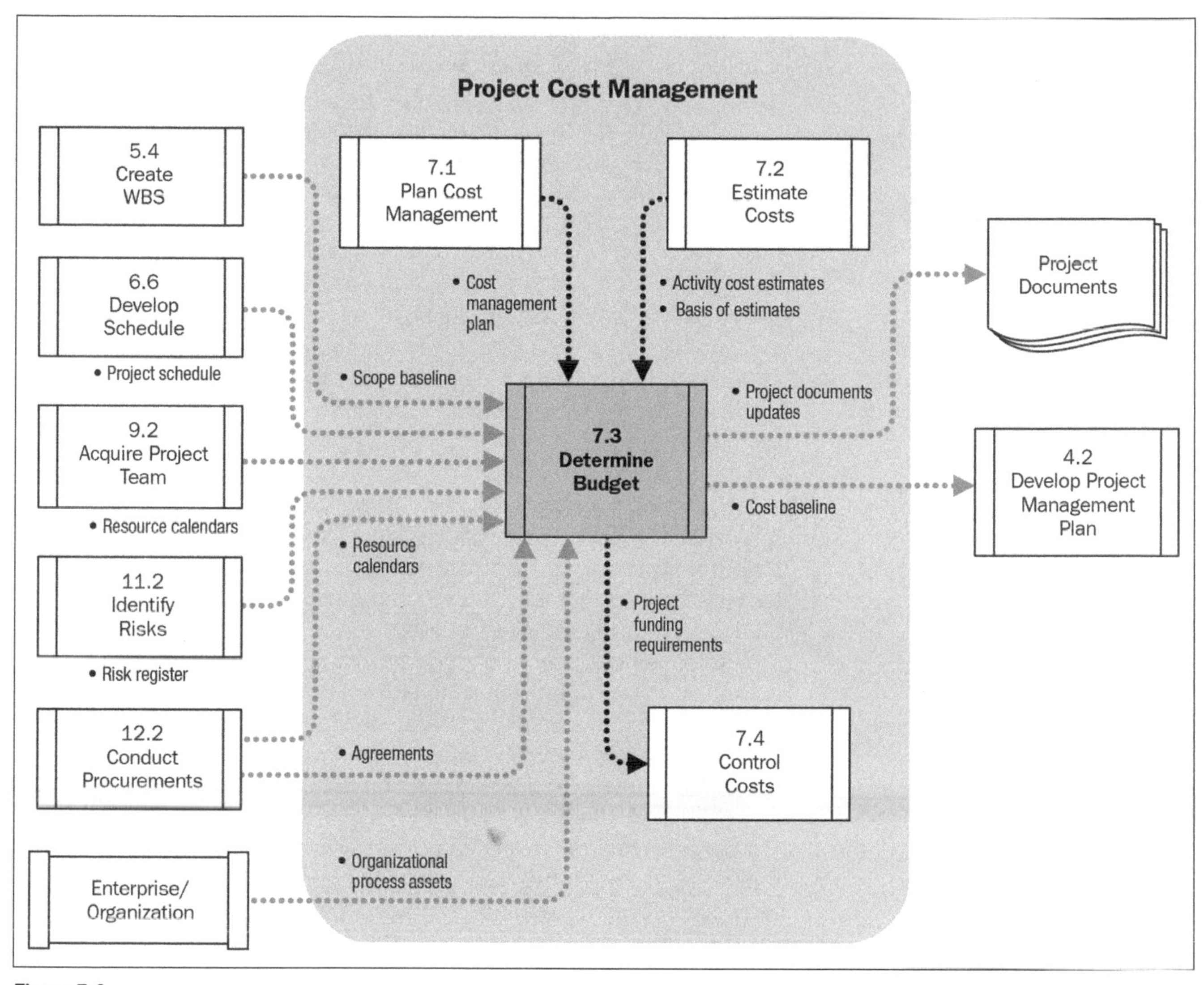

Figure 7-6
Determine Budget Data Flow Diagram
Source: PMBOK® Guide—Fifth Edition

span multiple years may have a set budget for a year, and may have to make the work fit into the funding for the year. The activity cost estimates and the basis of estimates from the Estimate Costs process provide the quantitative cost estimates by category.

The project schedule and the resource calendars show how activities and resources will be used over time. The schedule and resource calendars provide information that allows costs to be aggregated into specified time periods, such as weekly, monthly or quarterly. The risk register has costs associated with risk responses and the expected profile for expending contingency reserve.

Agreements contain information about when deliverables and payments are due. Organizational process assets include budgeting procedures, worksheets, budgeting software, previous projects, and lessons learned.

TOOLS AND TECHNIQUES

Expert judgment is needed for budgeting skills as well as cost aggregation, historical relationships, and reserve analysis. *Cost aggregation* is summing up the costs of each individual work package to determine project costs as a whole.

When conducting a *reserve analysis* for a project you should look at the risk register and determine the amount of contingency needed to bring the risk of cost overruns to an acceptable level. For example, if you built your cost estimates based on the most likely cost of items, based on historical data, and added in a set percent for cost growth you would have a certain degree of confidence that you could achieve the project scope for that amount. However, in looking at the degree of uncertainty and complexity, and knowing your organization's risk tolerance for budget overruns, you might decide you should have a contingency reserve of 10 percent to reduce the probability of overrunning. The 10 percent contingency reserve is only used to respond to risks that occur on the project—it is not padding. If it is not used, it is returned to the organization. In some cases the project manager identifies contingency funds, but the sponsor maintains control over the funds.

Management reserve is for unplanned in-scope work. This can include an extra round of testing and fixing problems if a component does not meet the quality criteria, or if you did not identify some work necessary to complete a deliverable. The management reserve is not part of the baseline, but the contingency reserve is.

Historical relationships are similar to the analogous estimates and parametric estimates used in cost estimating. They are used at a high level to check the cost estimates. For example, looking at the historical information on the cost per square foot of a ship or a building are examples of historical relationships.

Funding limit reconciliation is used to make sure the scope planned and the funds available are in sync. As mentioned previously, there are times when a certain amount of funds are allocated for the year. You need to make sure the scope of work planned for the year is not greater than the funds available.

OUTPUTS

The cost baseline is the authorized budget allocated over time for all the project work. It includes contingency reserve, but not management reserve. It is usually displayed as an S-curve. The cost baseline will be used to measure the degree of budget variance as the project progresses.

Project funding requirements define the funds needed on a periodic basis. They include funds for the cost baseline, plus management reserve. While project expenditure usually occurs fairly smoothly along the S-curve, the funding requirements are usually incremental at defined intervals, such as quarterly or annually.

Project document updates can include at least cost estimates, the risk register, and the project schedule.

Cost Baseline

The cost baseline is sometimes called the cost performance baseline. When using earned value management it is called the performance measurement baseline.

Chapter 8

Planning Quality

TOPICS COVERED

Project Quality Management

Plan Quality

Project Quality Management

Project quality management includes the processes and activities of the performing organization that determine quality policies, objectives, and responsibilities so that the project will satisfy the needs for which it was undertaken.

Project quality management is applied to the project and the product. Quality processes and tools and techniques for the *product* are industry and product specific. Quality processes and tools and techniques for the *project* can be applied to most projects the same way. Both must be planned into the project to ensure the result meets the customer expectations and fulfills the project objectives.

There are certain underlying assumptions in quality management that the *PMBOK® Guide* has adopted. They include:

- Customer satisfaction is a key component of project success. Customer satisfaction can be described as conformance to requirements and fitness for use.
- It is better to plan quality in and prevent defects and errors from occurring rather than to find errors during the inspection process.
- The plan-do-check-act process as defined by Deming and Shewhart is the basis for quality improvement, but it is not explicitly identified in the quality processes.
- Much of the investment in quality comes from the organization. Quality processes and procedures, certifications, investment in proprietary quality methodologies (such as TQM or Six Sigma) are the responsibility of the organization. It is the responsibility of the project to follow those processes.

Some of the key quality terms you should be familiar with are:

- Quality
- Grade
- Precision
- Accuracy

Let's take a look at those definitions and provide some examples that explain them.

Quality. The degree to which a set of inherent characteristics fulfills requirements.
Grade. A category or rank used to distinguish items that have the same functional use (e.g., "hammer") but do not share the same requirements for quality (e.g., different hammers may need to withstand different amounts of force).
Precision. Within the quality management system, precision is a measure of exactness.
Accuracy. Within the quality management system, accuracy is an assessment of correctness.

Quality and Grade

If you were to replace all of the organization's cell phones you would be concerned about the quality and grade of the product. You might have Product A that has all the latest technology including cloud computing, GPS positioning, a big screen, and very cool bells and whistles. However, you heard that the network they use has a lot of dropped calls, and the touch pad sometimes doesn't respond very well. In this case, you have a high-grade product, but it has low quality.

Product B, on the other hand, is a flip phone that allows texting, pictures, and different ring tones. However, it does not have a separate keypad, it doesn't have a very big screen, or many other features. However, it has been rated No. 1 in customer satisfaction and has almost no instances of needing to be replaced because of faulty parts, and no reports of dropped calls. This is a high-quality product, but it is low grade.

Precision and Accuracy

Assume your organization wants to place 20 digital billboards at various places onsite. You set up and test the billboards and they pass all the data feed tests. The billboards are programmed to tell the time and outside temperature every three minutes. Everything appears to be fine. You check them a week later and, as you are looking at each of the installation sites, you notice that each temperature reading is different. You know the temperature isn't vacillating all that much in the ten minutes it takes to check them all. You check ten of the billboards and get these readings: 68, 72, 76, 67, 69, 79,

75, 71, 65, and 75. In this case, you have a lot of scatter in the results. This indicates that you do not have precise readings. There is a lot of variance in there. One of them may be accurate, but you don't know which one.

A day later the service representative comes out and tweaks all the billboards. You check the next day, and they all say the same thing—the outside temperature is 54 degrees. Now there is a lot of precision, but the problem is, they are not accurate. You just came in from outside and you are wearing short sleeves and still feel quite hot.

Plan Quality Management

The process of identifying quality requirements and/or standards for the project and its deliverables, and documenting how the project will demonstrate compliance with quality requirements. There are many places to identify quality requirements that are either listed explicitly or implied. You should be looking for quality requirements while you are involved in all the other planning processes. The scope and procurement processes will focus more heavily on *product* quality needs, while the time, cost, risk, and integration processes will focus more on the *project* quality needs. Figure 8-1 shows the inputs, tools and techniques and outputs for the Plan Quality Management process. Figure 8-2 shows a data flow diagram for the Plan Quality Management process.

INPUTS

The project management plan contains the scope baseline, cost baseline, and schedule baseline. The *scope baseline* contains the scope statement which explicitly lists the acceptance criteria for deliverables and the project. It also documents assumptions and constraints. The WBS dictionary identifies technical work associated with each deliverable. This information can help identify quality metrics, testing scenarios, and items for the quality management plan.

The cost baseline and the schedule baseline are useful in establishing quality metrics and measurements for project quality. For

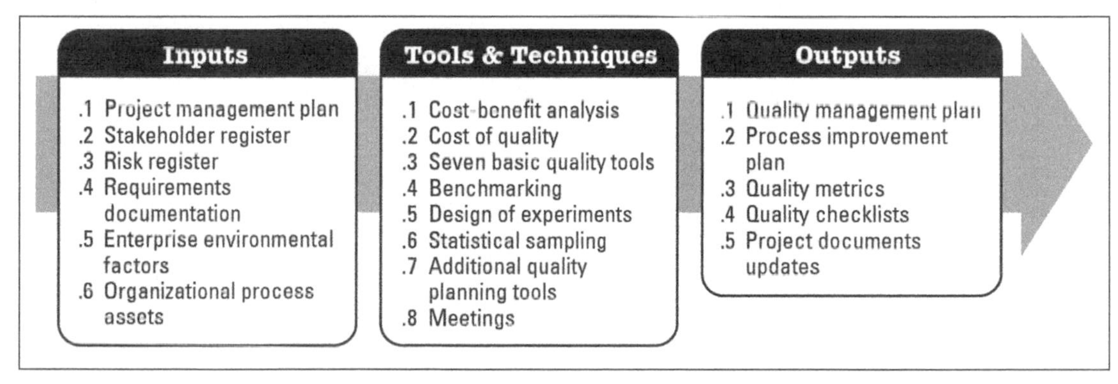

Figure 8-1
Plan Quality: Inputs, Tools and Techniques, Outputs
Source: PMBOK® Guide—Fifth Edition

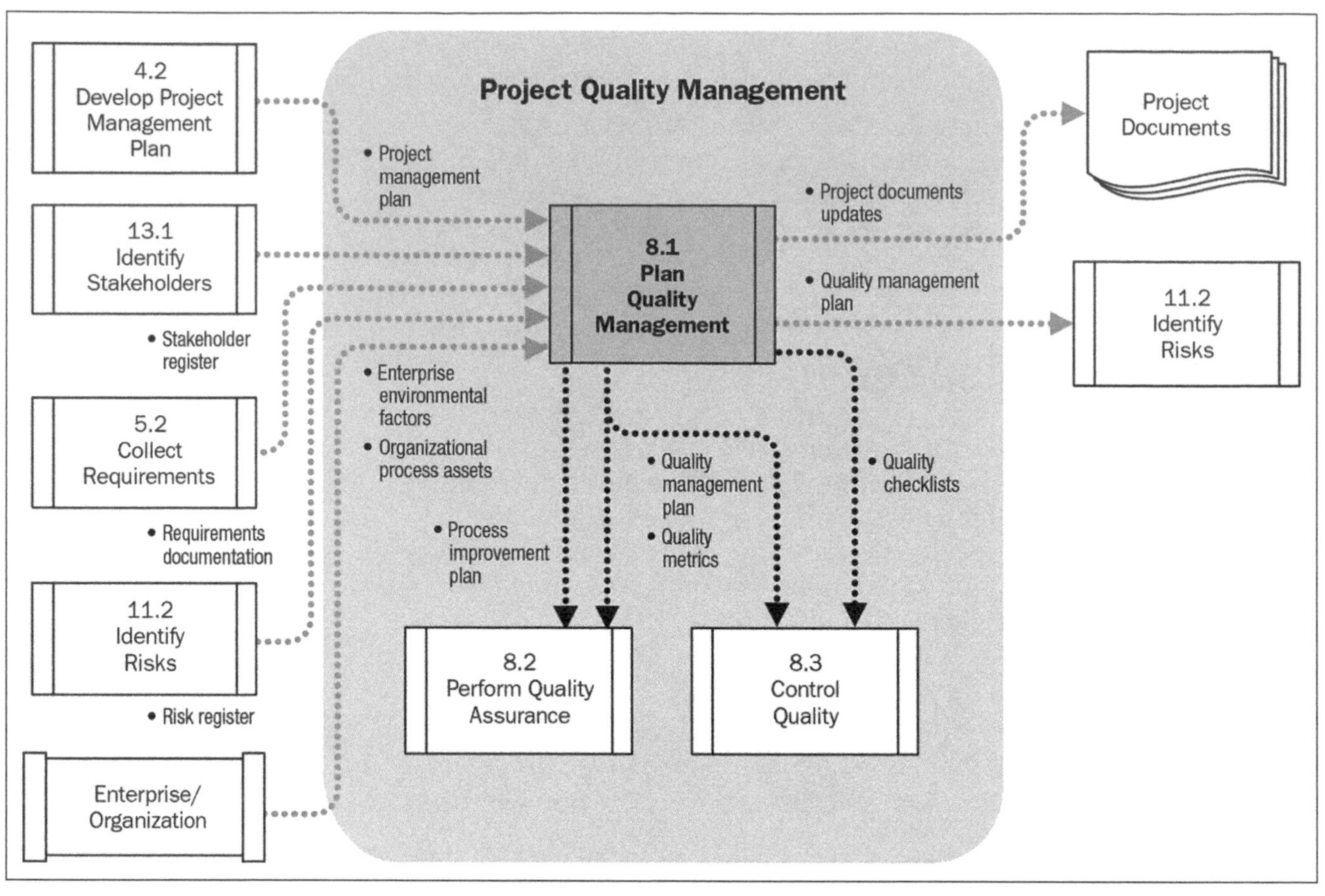

Figure 8-2
Plan Quality Data Flow Diagram
Source: PMBOK® Guide—Fifth Edition

example, you may say that project quality means meeting 90 percent, or 95 or even 100 percent of the schedule milestones on time; or, coming within +/−10 percent of the project budget.

The stakeholder register lists stakeholders who have an interest in quality, such as end users and customers, as well as stakeholders who can influence quality, such as regulatory agencies, and the Quality function in the organization. The *risk register* can be used to identify risks to product and project quality.

Requirements documentation contains detailed performance measurements, product specifications, and product functionality that must be met and tested for the product to be accepted. The requirements documentation should be used to identify the appropriate quality control measures and thresholds that will be used to ensure the requirements are met.

Laws, regulations, and standards are enterprise environmental factors that influence and constrain the quality standards for the project. For example, there may be emission laws, or regulations that determine how you can work with certain types of chemicals or raw materials. Industry standards don't govern how work is done, but they will certainly influence how you manufacture or develop something in order to be marketable. The organizational structure, such as the quality management department, will influence this process as will defect repair areas, rework areas, testing bays, and so forth.

Organizational process assets include the organization's quality policy overall, along with other policies, procedures, templates, checklists, and guidelines for quality. If there is a defect repair tracking system, this is very helpful so the team does not repeat mistakes. Lessons learned and information from prior projects can assist as well.

TOOLS AND TECHNIQUES

The cost of quality and a cost-benefit analysis go hand in hand to determine the optimum investment in quality management activities for a project. A cost-benefit analysis looks to optimize the productivity and stakeholder satisfaction while minimizing costs and rework.

The cost of quality looks at the cost of conforming to requirements and the cost of not conforming to requirements (cost of failure). The cost of conformance includes:

1. Prevention
   - Training
   - Research
   - Documentation
   - Equipment
2. Appraisal
   - Testing
   - Testing equipment and maintenance
   - Inspections

The cost of nonconformance includes:

3. Internal failure
   - Root cause analysis
   - Rework
   - Scrap
4. External failure
   - All internal costs
   - Warranty
   - Shipping
   - Liability and law suits
   - Lost business

If you recall from the introduction, one of the axioms in quality management is that it is better to prevent defects than to inspect and find them. Therefore, investment in prevention and appraisal is generally preferred to costs of nonconformance. However, this must be weighed with the type of product and the cost-benefit analysis.

Cost of Quality

Assume your project is to develop a modification kit for a diesel engine to accept and burn vegetable oil (biodiesel) and maintain a tow capacity of up to two tons. The following are examples of the costs of quality you might encounter.

Prevention Costs

Research: Research on the engine specifications required to achieve the necessary power and whether auxiliary fuel tanks will be necessary. Research on the fuel mixture requirements (amount of conventional diesel versus biodiesel) in order to avoid gelling at low fuel temperatures.

Equipment: Equipment that can provide real-time results on the efficiency and performance of various mixtures at different temperatures.

Training: Training on the equipment and the processes involved in developing the conversion kit, maintaining the equipment, and interpreting the data.

Documentation: Documenting the work flows, processes, procedures, and policies for producing the biofuel mixture.

Appraisal Costs

Testing: Testing includes testing the results of the engine performance to ensure conformance to requirements. This would mean determining if a specific engine and fuel mixture can actually tow two tons.

Testing equipment and maintenance: Testing the engine conversion kit and the fuel mix to make sure it is properly calibrated and the data is precise and accurate, as well as paying to have it properly maintained and serviced over the life of the equipment.

Inspections: Inspections include inspecting the end product as well as auditing the process to ensure compliance and conformance.

Internal Failure Costs

Root cause analysis: If the biodiesel engine fails, the cause of the failure has to be determined. Was it the conversion kit? The engine itself? The process? The person doing the job? The QC process? Developing a response without identifying the source of the failure is useless.

Rework: If all or part of the biodiesel engine can be saved, then it is reworked and tested again.

Scrap: The cost of failed parts and material.

External Failure Costs

Warranty: If a conversion kit ships and fails, and it is covered under warranty, the company will have to either replace or repair the equipment. This may entail travel to the site or having a shop on location to repair equipment.

Shipping: If repairs are not done at the customer site, the manufacturing organization bears the expense of shipping parts and equipment to and from the customer.

Liability and lawsuits: At the very least, the performing organization is responsible to replace malfunctioning parts; however, if those malfunctions cost a loss of customer business, or injury, then the manufacturing organization can be held liable for those costs as well.

Lost business: If the engine is fraught with problems and rework, it is likely that word will spread and other customers will look elsewhere for biofuel conversion kits. This can even spread to other product lines as well.

The seven basic quality tools are used in planning to identify measurements that will later be used in quality assurance, improvement, and control. They are described here and then referenced in the Perform Quality Assurance and Control Quality processes.

A *cause and effect diagram* is used to identify the cause of defects. People often try to treat the symptom of a problem rather than the cause of the problem. This tool can be helpful in identifying the variables and subvariables that can cause problems or defects in a deliverable or a result. Figure 8-3 demonstrates how you can use a cause and effect diagram to identify causes of a budget overrun.

Cause and effect diagrams can be relatively simple like the one shown in Figure 8-3, or they can be more complex with many branches and inter-relationships between the causes. Cause and effect diagrams are also known as Ishikawa or fishbone diagrams.

Flowcharts are used to develop a model of a process by identifying all the steps in the process, putting them in order, and identifying decision points. They are sometimes called process maps. By developing a flowchart of a new process you can identify areas that might cause problems or defects before implementing the process in order to develop resolution strategies. Developing a flowchart of an existing process gives an "as-is" look at the process. The new process is then designed and that is called the "to-be" process. This is a frequently used technique in business process improvement projects.

Checksheets or tally sheets are used to collect frequency data, such as the number of times a defect occurs because of a specified

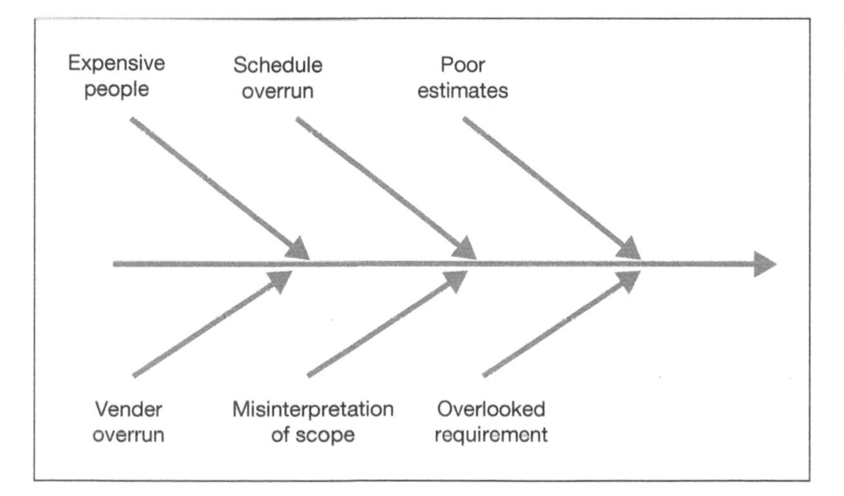

Figure 8-3
Cause and Effect Diagram

variable. The information from checksheets can be used to create a Pareto diagram to identify and prioritize causes of defects.

Pareto charts are bar charts that are used to identify the most frequent causes of an event. The Pareto chart in Figure 8-4 shows the causes of overruns ranked by frequency.

Histograms are bar charts that are usually shaped like a bell curve to show the shape of a statistical distribution. The histogram in Figure 8-5 shows the number of calls into the help desk based on the hour of the day. The data can be used to staff the desk to ensure the service quality levels are met.

Control charts are used to measure the performance of a process to ensure it is in control. When a process is in control the results of the process are consistent with expectations. When it is not in control there are outlier or unexpected results. In quality planning, the processes are designed and the tolerance levels that determine whether the process is in control are set. Control charts are more common in a production or factory environment to measure product results. However, they can be used to measure project results as well.

Figure 8-4
Pareto Chart

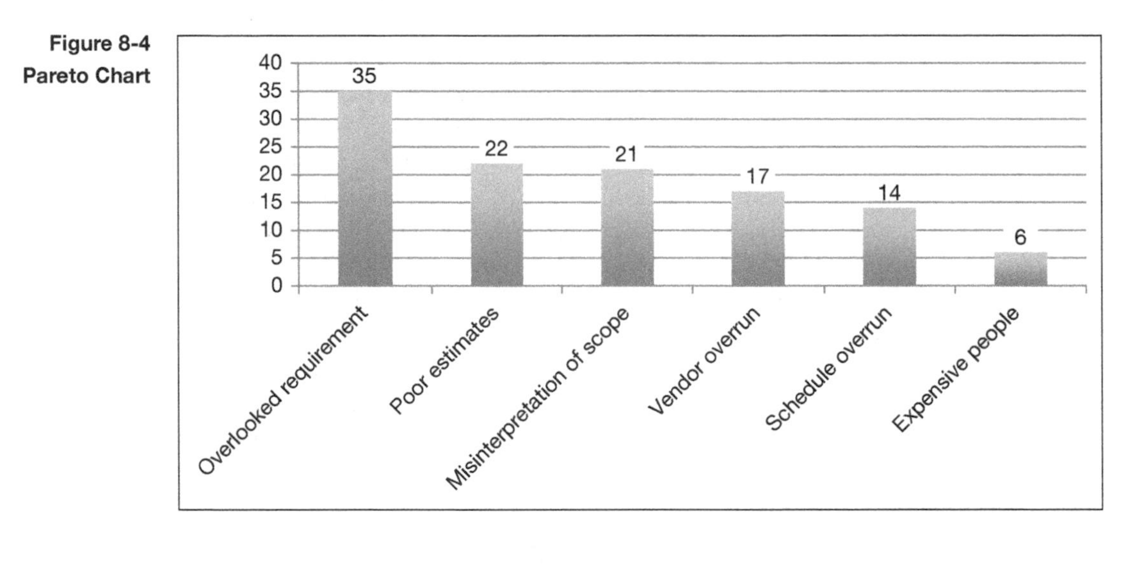

Figure 8-5
Histogram

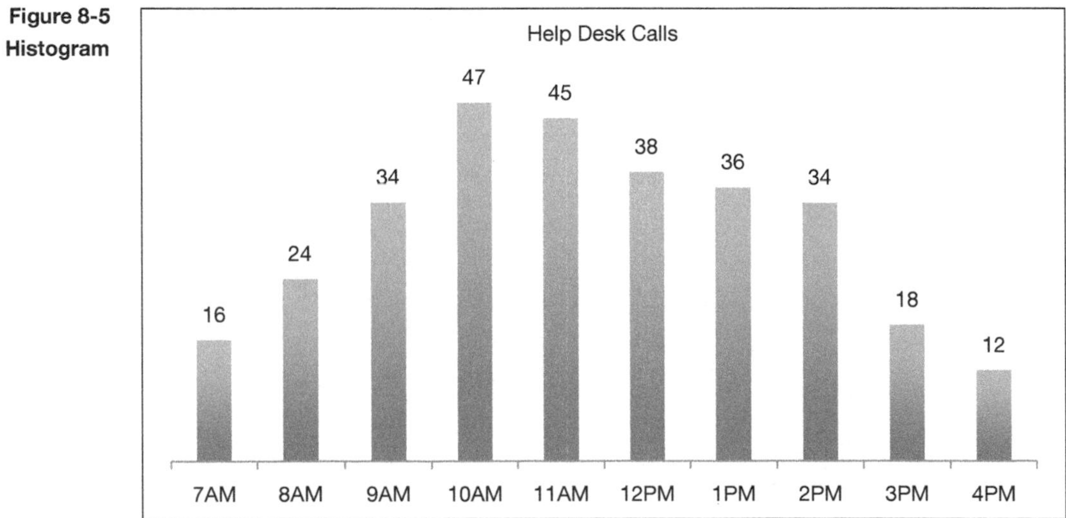

Control Chart

Assume you are managing a project with a critical path and several near-critical paths. Any significant delay in the start of any of the near-critical paths will establish a new critical path and could cause the project to be late. To track the progress you measure the amount of float that is used every week for each of the near-critical paths. For this example, you set thresholds of 0 float days used as the midline expected behavior. You establish control limits of +/−3 days. This indicates that if a path uses less than 3 days of its total float, it is in control, and if it does not accelerate performance by more than 3 days it is in control. (We watch adding float to make sure no one is taking short cuts, or crashing the project which would negatively impact cost.)

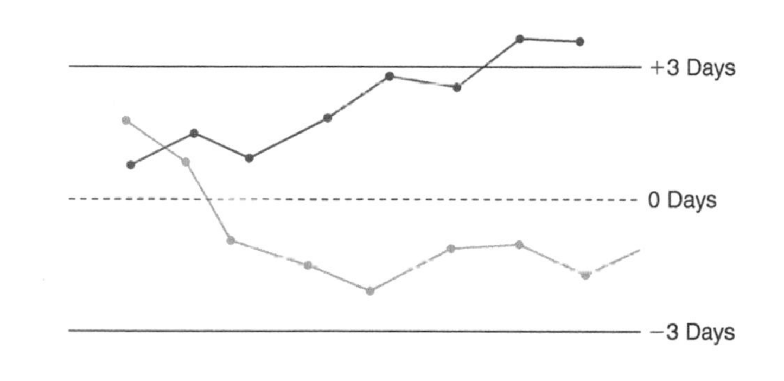

In this example we see that one of the paths has a result that is outside the control limits. This means we need to investigate and determine what happened and how it can be corrected.

A *scatter diagram* plots the relationship of two variables. The variables make up the x and y axis. For example, if you were to plot the amount of budget spent on a task and the remaining duration of the task you would expect to see a negative correlation. The duration should shrink as more budget is spent. You would expect to see a positive correlation if you were tracking resources added to the project and budget expended.

Another tool that can be used is benchmarking. *Benchmarking* can be used to identify a best-in-class process or product. This information is used to build processes to replicate or even exceed the current best-in-class practices. Benchmarks can also be used as performance targets for both product and project quality.

Design of experiments as applied to project quality management is a statistical practice used to determine how multiple variables interact to create the desired effect. This is generally used in engineering design, product development, and production. An example of designing using multiple variables would be to determine the right mixture of ingredients, heat, and time to create a new recipe. Another example is to see how to combine different materials, weight, size, and color to create the most appealing car cockpit.

In both of these situations, you are not looking at the impact of one variable on an outcome, but the combination of multiple variables on the outcome.

Statistical sampling is determining the number of items that will be tested from a group (population) to determine if the group meets requirements. Statistical sampling is a cost-effective way to measure the acceptability of a batch of similar deliverables. The sampling is conducted in the Control Quality process, but the number of items that will be sampled and the timing of the sampling are determined in the Plan Quality Management process.

Proprietary quality management methodologies are often enterprise environmental factors that define how products are designed or how processes are conducted. Examples include Six Sigma, Quality Function Deployment (QFD), and Capability Maturity Models (CMM). If the performing organization has any of these methodologies, they are considered techniques in planning for project quality.

Additional quality planning tools are dependent on the nature of the product, service, or result of the project. You might see brainstorming to identify variables that can impact project quality. You could also use a force field analysis to determine forces that support quality and forces that detract from quality. Nominal Group Techniques can be used to rank and prioritize the information generated from brainstorming.

Meetings are used to develop the quality management plan, discuss which tools and techniques to use for specified deliverables, and brainstorm possible ideas to help improve performance.

OUTPUTS

The quality management plan will be a component of the overall project management plan. It documents the approach for quality assurance, quality control, quality improvement, and process improvement. It can define which quality tools will be used, how they will be used, when they will be used, and the *quality metrics* that will be used with them to report on quality. A metric is a standard of measurement. Quality metrics are specific measurements that describe in detail an attribute that will be measured and the numeric value or tolerance that the result needs to fall within to be acceptable.

Sample Metrics

Defect Frequency Metric. No more than three defects per 100,000 parts.
Customer Satisfaction Metric. An average rating of 7.5 or higher on each of the ten questions, with no single score under 6.0.
Budget Metric. The entire project cost is ±10 percent of the cost performance baseline, with no individual component having a greater variance than 20 percent.

When recording quality metrics, it is helpful to keep a chart of all the project metrics. You can include at least:

- A metric ID
- The item being measured
- The metric or measurement required
- The measurement method

ITEMS IN A QUALITY MANAGEMENT PLAN

Quality roles
Quality responsibilities
Quality assurance approach
Quality control approach
Quality improvement approach

A quality checklist can be developed based on the process flowcharts, cost of quality decisions, acceptance criteria, information from past projects, and so forth. They ensure that a series of steps or a specified set of actions has been performed. A simple checklist can be used for equipment check-in and check-out. More complex checklists can be developed prior to launching a complicated and expensive test. Checklists are developed in this process and applied in the Control Quality process.

Process improvement projects use a *process improvement plan* as part of the overall project management plan. It describes the processes that will be analyzed, the tools that will be used, the metrics that will be applied to measure the steps in the process, and the targets for improvement. Process improvement plans can be developed for a piece of a process, or for an entire end-to-end process. Some projects that are not process improvement oriented may develop one to try and improve the outcome of the project by improving the process to develop the product, service, or result.

Project documents which may be updated as a result of this process include the stakeholder register, as additional stakeholders are identified or as strategies for meeting the needs of existing stakeholders are documented, and the responsibility assignment matrix (RAM). The RAM can be updated to show the type of participation various project stakeholders have in planning, assuring, and controlling quality. The RAM will be discussed in more detail in Chapter 9.

ITEMS IN A PROCESS IMPROVEMENT PLAN

Process description
Process boundaries
Process metrics
Targets for improvement
Process improvement approach
Flowchart of the as-is process
Flowchart of the to-be process

Chapter 9

Planning Human Resources

TOPICS COVERED

Project Human Resource Management

Plan Human Resource Management

Project Human Resource Management

Project Human Resource Management includes the processes that organize, manage, and lead the project team.

Other than the Project Human Resource Management processes and the Project Stakeholder Management processes, all of the other processes in the *PMBOK® Guide* are concerned with documents, artifacts, the environment, and tools and techniques. Project Human Resource Management processes are concerned with the people who actually carry out the work of project management and the work of the project itself. You can do all the technical aspects of project management correctly, but if you can't manage and develop a team, you will have a very hard time delivering a successful project.

In Chapter 3 we talked about two important roles on the project, the project sponsor and the project manager. Two other groups of people are the project management team and the project team. The project management team supports the project manager in managing the project. This group is sometimes known as the core team, the steering committee, or other similar names. They may not be assigned to the project full-time, but they are responsible for helping to set the direction and steer the project. The project management team can have schedulers, business managers, technical subject matter experts, cost accountants, or whatever the appropriate roles are to support the project.

The project team is an extended team and it includes anyone working on the project, including in-house resources as well as vendors, subcontractors, and teaming partners.

Project Manager. The person assigned by the performing organization to lead the team that is responsible for achieving the project objectives.

Project Management Team. The members of the project team who are directly involved in project management activities. On some smaller projects, the project management team may include virtually all of the project team members.

Sponsor. A person or group who provides resources and support for the project, program, or portfolio and is accountable for enabling success.

The information in the planning process for Project Human Resource Management is focused on using tools and techniques to develop parts of the project management plan. You will notice that in the executing phase there will be more emphasis placed on the interpersonal skills needed to effectively manage and guide the team to successful completion.

Plan Human Resource Management

The process of identifying and documenting project roles, responsibilities, required skills, reporting relationships, and creating a staffing management plan. This information, like most project information, is progressively elaborated throughout the planning and executing process groups. At the start of a project you may plan at the job-title level only. As the scope is elaborated, you can start to identify the level of skill or expertise needed. Once the project is in motion you may be able to assign a specific resource. On projects that are shorter in duration, these steps may occur simultaneously, but for long-term projects, this type of elaboration is common. Figure 9-1 shows the inputs, tools and techniques and outputs for the Plan Human Resource Management process. Figure 9-2 shows a data flow diagram for the Plan Human Resource Management process.

INPUTS

The project management plan provides the project life cycle and the approach to accomplishing the project work. The life cycle

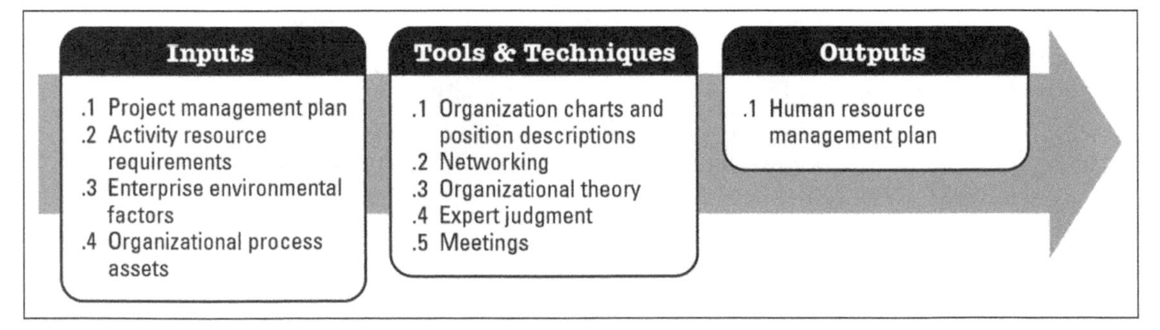

Figure 9-1
Plan Human Resource Management: Inputs, Tools and Techniques, Outputs
Source: *PMBOK® Guide*—Fifth Edition.

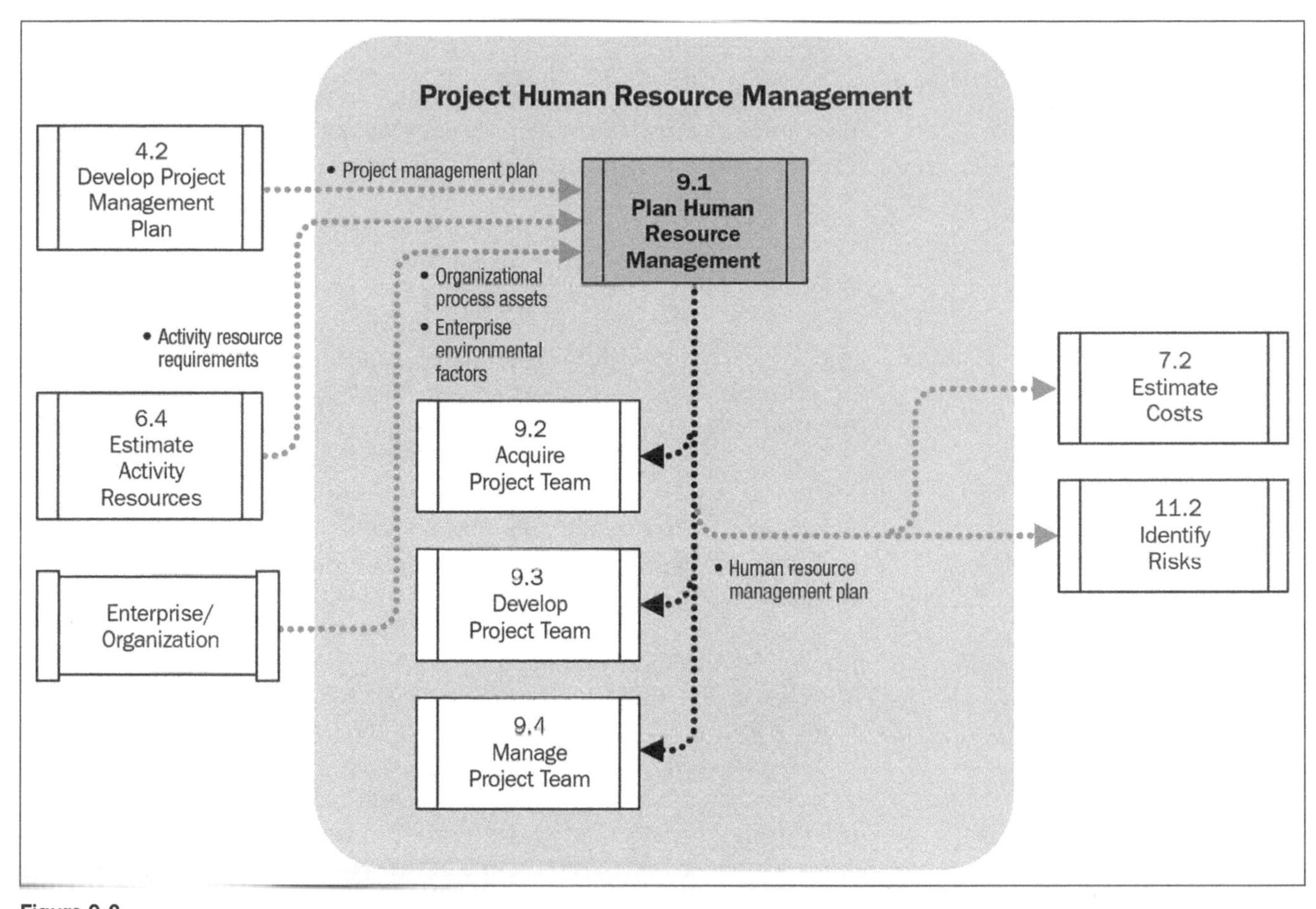

Figure 9-2
Plan Human Resource Management Data Flow Diagram
Source: PMBOK® Guide—Fifth Edition.

and approach can provide insight into when a particular type of resource will be required, and whether certain skills or abilities are needed during the project. For example, if the project is going to use an iterative life cycle, and include agile techniques, the human resource plan would include roles such as scrum master and product owner. The organizational chart would reflect the roles and team structure that is consistent with agile development.

The activity resource requirements that were developed as part of the Estimate Activity Resources process are used to identify the skills and roles needed to develop a plan for acquiring human resources for the project. As mentioned above, this will get more detailed as the project progresses.

Enterprise environmental factors can restrict options available for staffing the project. For example, if the organization is functionally oriented, and resources are full-time on operations work, the amount of time available to work on projects is minimal. Or, if there is a lack of depth in a particular skill set, that could require hiring subcontractors to fill roles. Sometimes the Human Resource department has policies for bringing in vendors, or in many organizations subcontractors need to go through a background check due to the proprietary or competitive information associated with the project.

All of these situations restrict and limit the project manager's options when staffing the project.

Organizational process assets that can help facilitate developing the human resource plan include standardized roles and responsibilities descriptions, organizational charts, and information from prior projects.

TOOLS AND TECHNIQUES

Position descriptions describe the roles and responsibilities associated with each position on the team. They should list a description of the type of work that will be done, the areas of responsibility, the authority vested in the position, and any required qualifications or competencies. Presenting this information to the team member helps clarify what is expected of them on the project. Position descriptions should note specific areas of responsibility that relate to the project management knowledge areas, such as quality management and risk management. For example, a system engineer is not only responsible for the job of developing the system requirements, design, and architecture, but they will have certain responsibilities in the quality control and risk management as it pertains to the system.

Organizational charts are used to depict the organization structure for the project team. An organizational chart is typically laid out in a hierarchical structure to demonstrate reporting relationships. Some projects will use the resource breakdown structure that shows the various types of skills needed across the project. This was described in the section on Estimate Activity Resources.

CONTENT IN A ROLES AND RESPONSIBILITY DESCRIPTION

- Role Description
- Authority
- Responsibility
- Qualifications
- Competencies

Another practice is to develop an organizational breakdown structure (OBS) that shows the department, roles, and individuals, and matrix it with the WBS that shows the deliverables, control accounts, and work packages. The intersection is where you identify the accountable party and other resources for each WBS control account or work package (depending on how much detail you have). Those points of accountability are then documented in a responsibility assignment matrix (RAM).

Organizational Breakdown Structure. A hierarchical representation of the project organization that illustrates the relationship between project activities and the organizational units that will perform those activities.

Resource Breakdown Structure. A hierarchical structure of resources by category and type.

Responsibility Assignment Matrix (RAM). A grid that shows the project resources assigned to each work package.

The RAM defines the type of participation each role will have on the project. For example, who is accountable for a particular work package, who will be performing the work, who can sign off on the completed work, who can provide subject matter expertise, and so forth. This ensures that no roles are overlooked to complete the deliverables. The *PMBOK® Guide* uses a version called a RACI chart. This outlines who is Responsible, Accountable, Consulted, and Informed. You can tailor the RAM to meet your needs. You may want to have a primary and backup resource. You may want to include a signoff function. The concept is to have all roles identified and assigned for each element of work. How that can best be applied is based on the individual project. See Appendix A for a sample Responsibility Assignment Matrix.

Networking with team members, other project managers, and people throughout the organization and profession can help identify the best way to set up the project structure. For example, by talking with people who have worked on prior similar projects you may find that people in a certain department work best as a self-directed team. They don't really need heavy direction and management, whereas another area is more comfortable having an identified leader who schedules and assigns the work. This type of analysis is based in organizational theory. Understanding how people are motivated, how they work best together, and how to structure the project to leverage that information will make team management a whole lot easier!

Expert judgment that goes into developing the human resource plan includes information on how to structure the organizational chart, risks associated with acquiring resources, role descriptions, and required skills sets. Frequently there are planning meetings to review the resource requirements and discuss the best ways to fulfill the team needs.

OUTPUTS

The *human resource plan* is a component part of the overall project management plan. It contains at least three sections:

- Roles and Responsibilities
- Project Organization Charts
- Staffing Management Plan

The contents of a role and responsibility document were listed in the Tools and Techniques section. In this section we will define some of the terms in a bit more detail.

Role. A defined function to be performed by a project team member, such as testing, filing, inspecting, or coding.
Authority. The right to apply project resources, expend funds, make decisions, or give approvals.
Responsibility. An assignment that can be delegated within a project management plan such that the assigned resource incurs a duty to perform the requirements of the assignment.

The project organizational chart was described in the Tools and Techniques section.

LIST OF CONTENTS IN A HUMAN RESOURCE PLAN

- Role
- Authority
- Responsibility
- Organizational structure
- Staffing management plan
- Staff acquisition
- Staff release
- Resource calendars
- Training needs
- Rewards and recognition
- Regulations, standards, and policy compliance
- Safety

The staffing management plan contains the details associated with project staff. It includes information on bringing staff on board, whether they are employees, or outside contractors. It also describes how staff will be released after their part of the project is complete. One of the challenges in project work is capturing the knowledge of departing team members. A process for this should be outlined in the staffing management plan.

If there are project-specific needs for training, licensing, compliance, safety, security, intellectual property release, and the like, these are documented. Any criteria for bonuses or rewards should be documented.

Most of the information above is relatively static. In other words, once it is determined and documented, it doesn't usually change. However, many staffing management plans either contain or refer to resource histograms. These tend to be more fluid and dynamic. A staffing histogram can show either the availability of resources for a period of time, or the need for resources by time period. If there is a discrepancy in the need and availability, that will have to be addressed by either leveling resources (and probably extending the schedule) or adding more resources (and probably adding to the cost).

Chapter 10

Planning Communications

TOPICS COVERED

Project Communications Management

Plan Communications

Project Communications Management

Project Communications Management includes the processes required to ensure timely and appropriate planning, collection, creation distribution, storage, retrieval, management, control, monitoring, and ultimate disposition of project information.

The majority of a project manager's time is spent communicating, either orally or in writing. Whether you are giving status reports, negotiating for project staff, conducting a team meeting, managing stakeholder expectations, or listening to a team member, you are communicating. Being an effective communicator is essential to being an effective project manager.

Plan Communications Management

The process of developing an appropriate approach and plan for project communications based on stakeholder's information needs and requirements and available organizational assets. The most effective way to manage stakeholders and stakeholder expectations is via a well-thought-out communication plan. This will ensure you get the right information to the right people in a timely manner.

An unspoken axiom in Project Communications Management is to communicate all the relevant information, and only the relevant information. In other words, "reply to all" is not a communication tactic you want to employ all the time!

Elements of Communication

While the majority of your time is spent communicating, the majority of your communication is not "what you are saying." Studies show that only 7 percent of our communication is based on our words. About 38 percent is our tone of voice. That leaves 55 percent that is nonverbal communication (body language).

Suppose you have a conversation with a team member about his deliverable that is due next week. You ask him if he is still on time and if there are any problems. He pauses, looks down and mumbles that no, everything is okay. Then he sighs.

Even though he *said* things were okay, the mumbling, not looking you in the eye, and the sigh all tell a different story. In this case, you would want to dig a little deeper to find out what the situation really is.

But be careful, this works both ways. As the project manager, when you are communicating, you need to make sure your tone, your words, and your gestures all line up. You can't tell someone you understand they are having a hard time with their deliverable while rolling your eyes!

Figure 10-1 shows the inputs, tools and techniques and outputs for the Plan Communications Management process. Figure 10-2 shows a data flow diagram for the Plan Communications Management process.

INPUTS

The project management plan describes how the project will be planned, managed, controlled and closed. The *stakeholder register* identifies all the people, groups, or organizations that have an interest in the project or can influence the project and the type of information that should be communicated to stakeholders.

Enterprise environmental factors such as the organization's culture and the established communication channels and infrastructure shape how communication activities are conducted in the

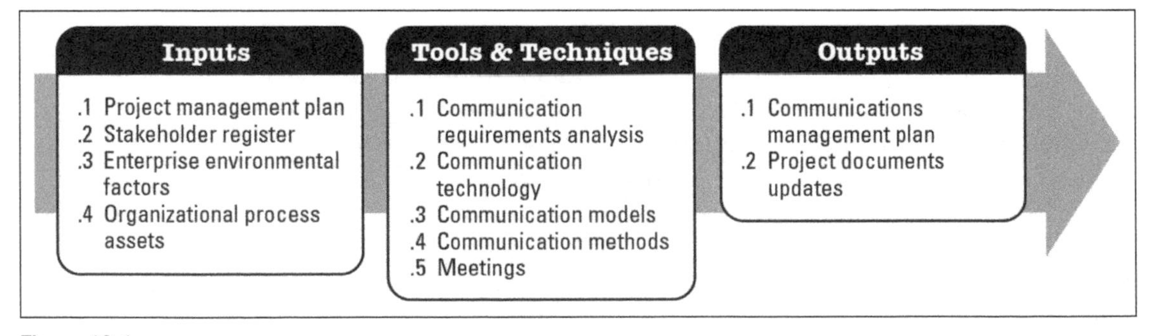

Figure 10-1
Plan Communications Management: Inputs, Tools and Techniques, Outputs
Source: *PMBOK® Guide*—Fifth Edition

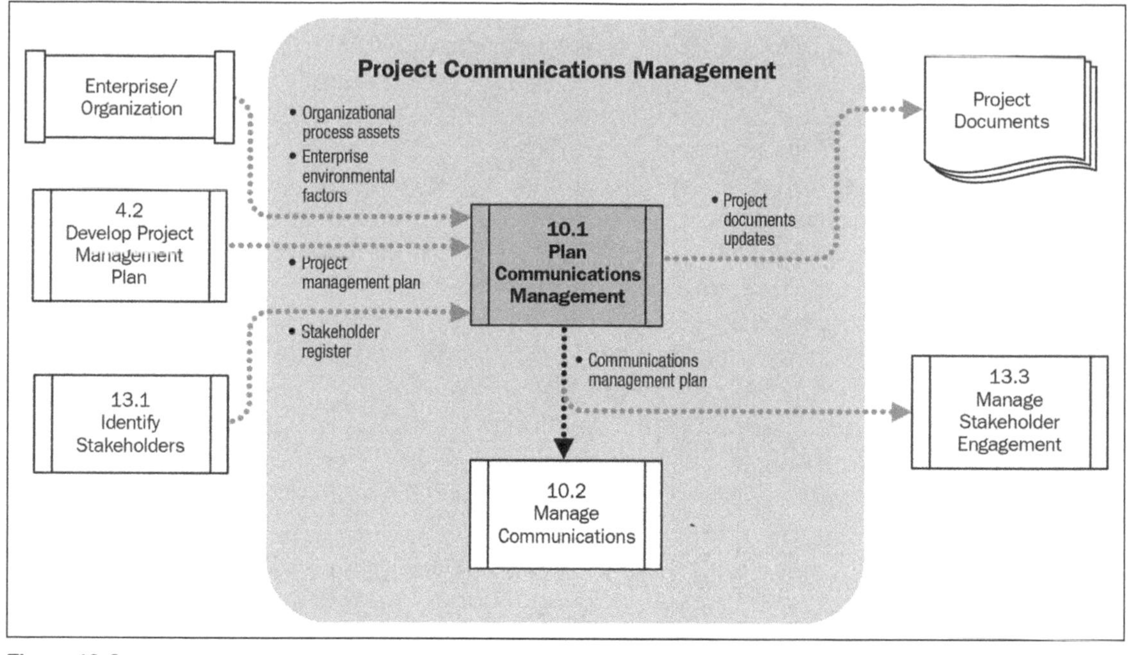

Figure 10-2
Plan Communications Data Flow Diagram
Source: *PMBOK® Guide*—Fifth Edition

organization. Regulatory requirements can also determine communication strategies. For example, if you are a healthcare insurance organization and you are planning on changing benefits, there are regulations that dictate how and when you need to communicate that information to beneficiaries.

Organizational process assets such as communication logs from past projects, communication templates, project files, and communication media are helpful in seeing what can be modified and reused for the current project.

TOOLS AND TECHNIQUES

Analyzing communication requirements often starts with the stakeholder information; however, it can expand to include looking at other project documents such as the OBS, WBS, and procurement management plan. The intent is to identify all the people or organizations that need information, the best way to deliver that information, and the frequency or timing of the need.

Communication technology can be a simple tool to get information distributed, or it can cause unplanned issues. Consider what happens if you want to communicate an "as-is" process to your team members, and you take the time to create a beautiful flowchart, but the problem is, no one else has flowchart software and they can't open it! The same holds true with the schedule. Not everyone has scheduling software. Other issues include people with the same software, but different versions, or Mac versus PC platforms. If there are subcontractors or vendors on your team, this gets even more complex.

Other aspects about communication technology you should consider include the immediacy of the need for information. Does everyone on the team need the ability to have information 24/7? Or are standard working hours sufficient? If the team is geographically disbursed, how will you meet? Via a Web meeting? Teleconferencing? Do you want to see participants? If so, does everyone have technology to allow that? All these considerations need to be thought out and reflected in the communications management plan.

Communication models reflect how you will communicate. A very basic model is presented in the *PMBOK® Guide*. It shows a classic sender-receiver model where the sender is responsible for making the information clear and complete and the receiver is responsible for ensuring the information is received in its entirety, understood, and must acknowledge receipt. The sender then confirms the information was properly understood. This is also known as three-point communication.

The following definitions are not from the *PMBOK® Guide* glossary. They are informal definitions to help you understand the communication model in the *PMBOK® Guide*.

Encode. To translate thought or ideas into a language that is understood by others.

Medium. The method used to convey the message.

Noise. Anything that interferes with the transmission or understanding of the message.

Decode. To translate the message back into meaningful thoughts or ideas.

One of the reasons for three-point communication is to make sure that noise has not gotten in the way of the message. Noise can be background noise, or it can include cultural differences, accents, different use of terminology, and so forth.

Communication methods in this situation refers to how people will get information. In some cases information is shared back and forth, such as a meeting or a conversation. This is called interactive communication. In other cases you will send out status reports, or the updated schedule. This is called push communication. You are pushing the information out to the recipients. Pull communication is when people find the information they want, for example, going to a website, a database, or an intranet.

Communication Equation

The more people involved in communication, the more likely there is to be misunderstanding. The following equation demonstrates how increasing the number of people involved increases the number of communication channels (which increases the likelihood of miscommunication).

$$\text{Number of communication channels} = \frac{N(N-1)}{2}$$

N = the number of people involved in the communication

Let's look at how this works. If you have five people on your team, you have ten communication channels.

$$\frac{5(5-1)}{2} = 10$$

If you double that number, you have 45 channels!

$$\frac{10(10-1)}{2} = 45$$

OUTPUTS

The communications management plan is part of the overall project management plan. A simple version describes who needs what information and when and how it will be given to them. More complex communications plans can include the person responsible for developing the communications, a glossary of terms, flowcharts of communication, constraints on information, and templates for common communications, such as status reports or meeting minutes.

Project document updates can include a circle back to the stakeholder register and stakeholder management strategy. The schedule and budget may also need to be updated to reflect communication activities.

See Appendix A for a sample Communications Management Plan.

Chapter 11

Planning Risk

TOPICS COVERED

Project Risk Management
Plan Risk Management
Identify Risks
Perform Qualitative Risk Analysis
Perform Quantitative Risk Analysis
Plan Risk Responses

Project Risk Management

Project Risk Management includes the processes of conducting risk management planning, identification, analysis, response planning, and controlling risk on a project.

Because projects are unique in nature, there is much more uncertainty in a project than there is in regular operations. Risk is rooted in uncertainty—the more uncertainty, the more risk. Organizations have different perspectives about risk and how much risk they are willing to tolerate. Organizations may be willing to tolerate or accept a greater amount of financial risk if it will reduce the amount of quality or schedule risk. Therefore the risk tolerance for various project objectives is specific to the particular organization and the specific project.

Risk. An uncertain event that, if it occurs, has a positive or negative effect on one or more project objectives.
Opportunity. A risk that would have a positive effect on one or more project objectives.
Risk Tolerance. The degree, amount, or volume of risk that an organization or individual will withstand.

Figure 11-1
Impact of Variable Based on Project Time
Source: *PMBOK® Guide*—Fifth Edition

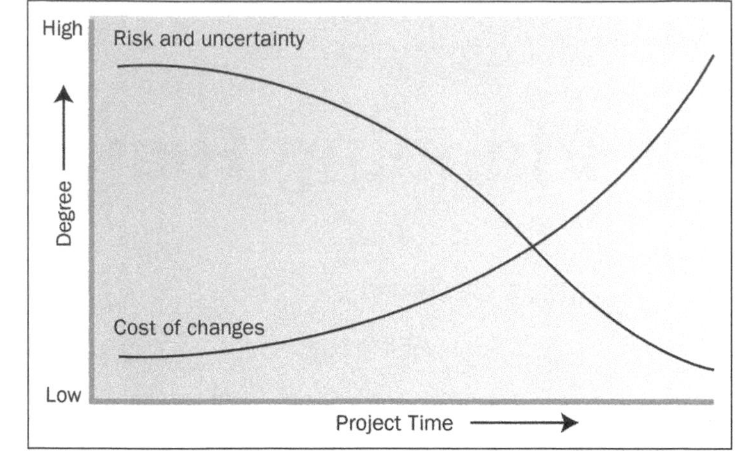

Based on the previous definitions you can understand how risk management is about influencing factors that reduce or eliminate negative risks, and increase or cause opportunities.

Risk management begins when a project is initiated and ends as the project closes out. The nature of the risks will change as the project goes through the life cycle. In the beginning phases there will be more risks associated with understanding requirements and developing accurate estimates. Through the middle phases of the project, risks are likely to be associated with project staffing, development efforts, and product-specific technology and scope challenges. As the project comes to a close, testing, validation, customer acceptance, and project transition present the greatest areas of uncertainty. Regardless of where the project is in the life cycle, the established risk management process should be employed.

Figure 11-1 shows how the number of risks decreases as the project moves through the life cycle, but the cost and impact of risks increase.

For risks that can be identified and documented in a risk register (known risks), the project team will analyze the risk and develop a response plan. For those risks that can't be identified (unknown risks), a contingency fund and contingency in the schedule can be estimated and included in the baseline budget and schedule.

Plan Risk Management

The process of defining how to conduct risk management activities for a project. For smaller and less complex projects the team may use a relatively simple approach to risk management, such as discussing existing risks and identifying new risks in team meetings. However, for long-term, high-cost, and high-complexity projects a thoughtful, well-planned-out approach to risk management is imperative. For this type of project the approach to risk management should be determined in the beginning of the planning process to ensure adequate time and resources are available for effective risk management.

Figure 11-2
Plan Risk Management: Inputs, Tools and Techniques, Outputs
Source: *PMBOK® Guide*—Fifth Edition

Figure 11-2 shows the inputs, tools and techniques and outputs for the Plan Risk Management process. Figure 11-3 shows a data flow diagram for the Plan Risk Management process.

The team should consider and document the approaches for identifying risk, analyzing the probably and impact of risk events, setting thresholds for risk rating, and documenting any tools or techniques that will be used to identify and analyze project risks.

INPUTS

The project management plan contains all the subsidiary management plans such as the scope management plan, schedule management

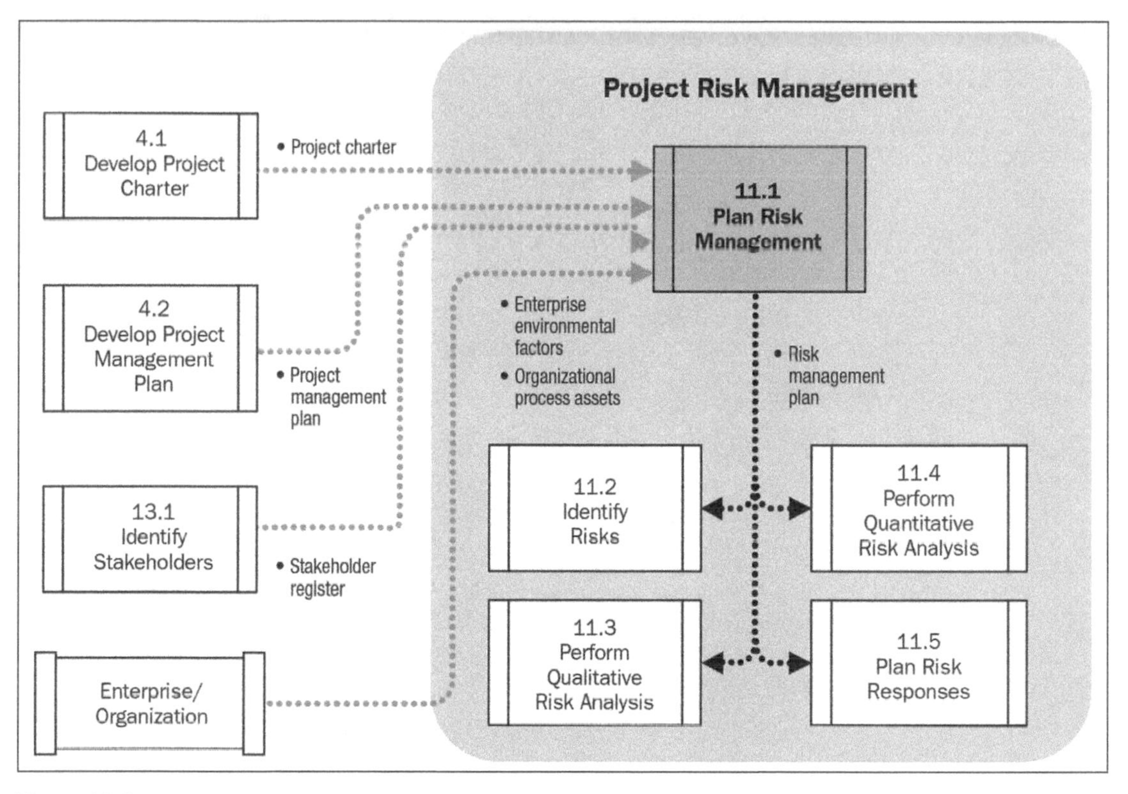

Figure 11-3
Plan Risk Management Data Flow Diagram
Source: *PMBOK® Guide*—Fifth Edition

plan, quality management plan and so forth. Each subsidiary management plan describes how that knowledge area will be managed. The information in the subsidiary management plans can point to shortcomings in areas that the risk management plan may have to account for. The schedule and cost management plans outline how contingency reserve will be allocated. The project management plan also contains the project baselines, such as the schedule baseline which provides information on the life cycle and key reviews. The project charter has information on high-level risks and requirements.

Enterprise environmental factors include the risk tolerance levels for various project objectives. For example, on some projects that have a fixed end date (such as a public event), the tolerance for schedule slip is very low. Risk databases are another example of enterprise environmental factors. In addition, regulatory requirements can introduce constraints that have to be included in risk management planning.

Organizational process assets such as risk register templates, risk management plan templates, and risk matrix templates can be modified to meet the current project's needs. The organization may have policies and procedures that dictate how project risk management should be approached. These assets provide a framework for risk management and help save time in creating something new for each project.

TOOLS AND TECHNIQUES

Planning meetings are attended by project stakeholders who provide expert judgment to analytical techniques such as simulations, definitions or probability and impact and risk scoring. Attendees are those stakeholders who are most affected by the project, such as the customer, the core project team, and if there is someone in the organization with overall risk management accountability, that person often attends or chairs the meeting. If a new or emerging technology is being used, many times someone with an understanding of the technology is in attendance as well. The risk management approach for the project and the contents of the risk management plan are discussed, decided, and documented as a result of these meetings.

OUTPUTS

The *risk management plan* is part of the overall project management plan. For a simple project it may be a few paragraphs that describe roles and responsibilities, timing and frequency of risk management activities, and the approach to cost and schedule reserves. For more complex projects, the risk management plan often starts out by describing the risk management tools, data sources, approaches, and procedures that will be used to manage risk for the project.

The plan generally includes a definition of the roles and responsibilities associated with identifying, analyzing, and responding to risk. This can be documented in any form as described in Chapter 9, Planning Human Resources, for example, a Responsibility Matrix that focuses on risk-related activities, a text format, a hierarchical structure, or whatever makes sense for the project.

The budget for risk management is allocated in terms of funds spent for risk identification, risk responses, contingency reserves, and the processes used to allocate and report on those funds.

The schedule information for risk management includes the frequency of applying the risk management processes, specific points in time during the project life cycle when major risk reviews will occur, and how schedule reserve will be determined, allocated, and reported on.

A thorough job of identifying risks on large projects will lead to several hundred risks. In order to help manage and respond to that many risks, a method for categorizing risks can be documented in the planning process. One method is developing a risk breakdown structure. A risk breakdown structure is similar to a work breakdown structure, but instead of focusing on deliverables it focuses on risk categories. The top level is the project. Beneath that you may have technical risks, external risks, project management risks, and organizational risks. Each of those can be further broken down. The breakdown can continue until the categories are detailed enough to start identifying individual risks instead of categories of risks. Figure 11-4 shows one example of the first three levels of a risk breakdown structure.

Risk Management Plan Contents

Method and Approaches

Tools and Techniques for Each Process

Roles and Responsibilities

Risk Categories

Stakeholder Risk Tolerance

Definitions of Probability

Definitions of Impact by Objective

Probability and Impact Matrix

Risk Management Funding

Contingency Protocols

Frequency and Timing of Risk Management Activities

Risk Audit Approach

Probability

Remember, probability refers to the likelihood of an event occurring. It does not refer to the probability of the event causing a particular impact. You may determine that based on past experience there is a 20 percent likelihood that a specific performance test will fail. Regardless of the impact on cost or schedule or technical success, the likelihood of that event occurring is still 20 percent.

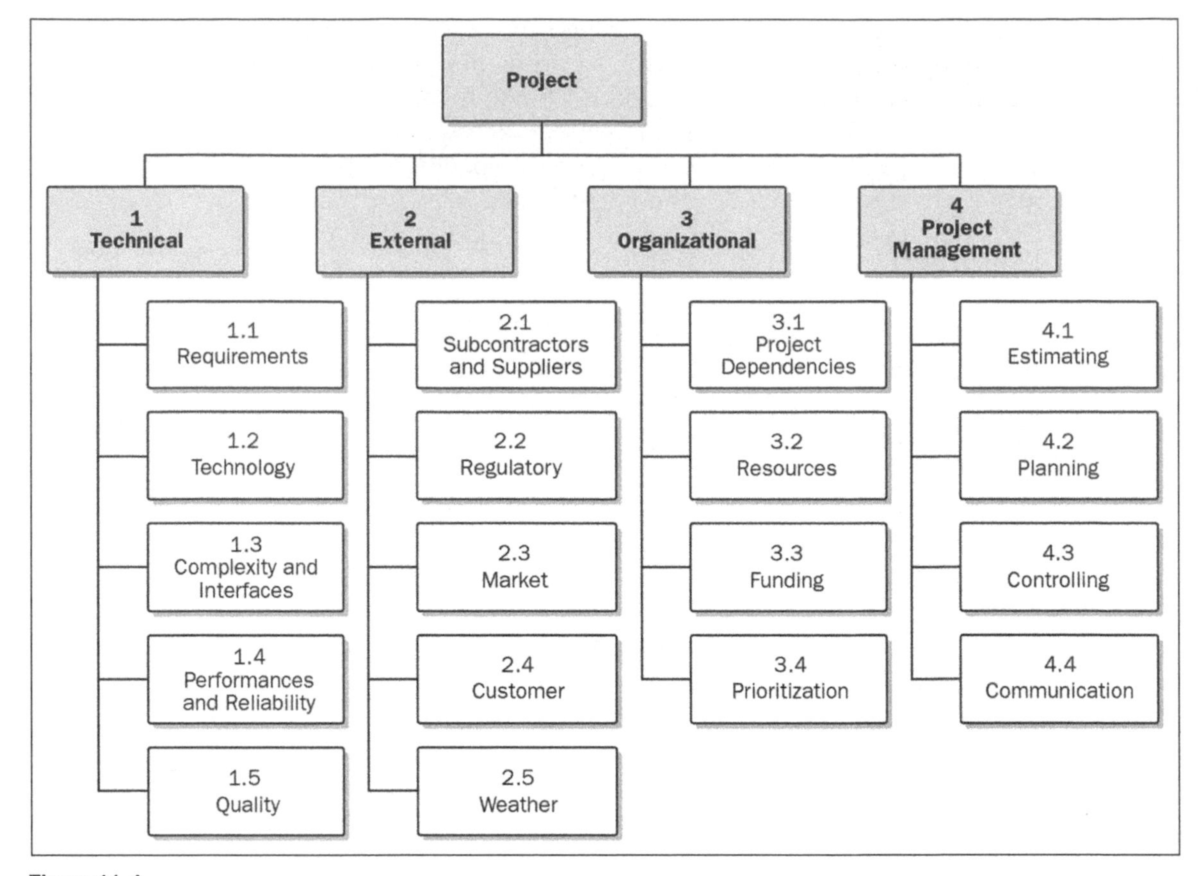

Figure 11-4
Example of a Risk Breakdown Structure
Source: *PMBOK® Guide—Fifth Edition*

When there are multiple stakeholders participating in the risk management process and many risks to analyze, the process can get bogged down in arguing whether a risk has a medium probability of occurring or a medium-high probability, and similar such discussions. For one person "medium" means there is a 50 percent chance something could occur; for another, it might mean a 35 percent chance. Definitions of "impact" can be equally challenging. One of the really helpful things to document in a risk management plan is concise definitions of probability and concise definitions of impact for each objective.

An organization with a balanced risk tolerance may use a rating of a scale of 1 to 5 for probability. It may go up in 20 percent increments, such as 0 to 20 percent, 20 to 40 percent, 40 to 60 percent, 60 to 80 percent, and 80 to 100 percent. Another approach is to show probability as 10, 30, 50, 70, and 90 percent. Yet another approach is to double the outcome each time: 5, 10, 20, 40, 80 percent, and so on. These decisions are made in the analysis meetings and then documented in the risk management plan.

Assessing impact can be even more challenging. For smaller projects, an impact on any objective is rated in the impact matrix. For medium projects, you may decide that a risk is rated low impact if it only impacts one objective, medium if it impacts two objectives, and high if it impacts three or more objectives. For more

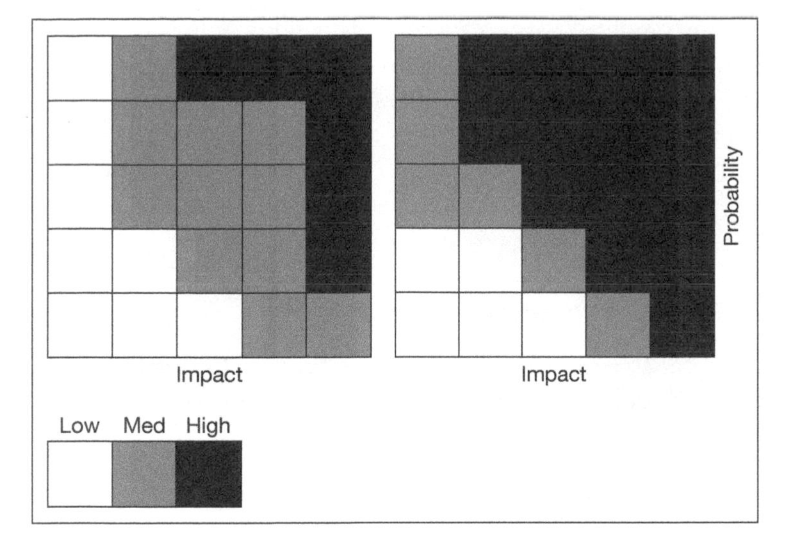

Figure 11-5
Probability and Impact Matrix

complex projects, an impact template is developed for each objective. This allows the team flexibility in prioritizing project objectives. For example, if schedule is a higher priority than cost, then an impact analysis may indicate that a slip of 5 percent on the critical path is a medium impact, whereas a cost overrun of 5 percent is a low impact.

The final aspect of defining probability and impact has to do with how you rank the combination of the two. It is useful to develop a probability and impact matrix to plot the lines where a risk is categorized as high, medium, or low. Figure 11-5 has an example of a probability and impact matrix.

You can see the chart on the right indicates a risk-averse culture (many known risks are rated high). The chart on the left shows a great deal of risk tolerance (only a few known risks are rated high). Using these matrices to define the probability bands, the impact bands, and the level of risk allow the team to tailor the risk process to the needs of the project.

A final component of the risk management plan is risk templates. A description of the risk register fields as well as a template are developed (if none exist) and documented. Some organizations use risk data sheets to collect information on specific risks and then summarize the information in the risk register. The template section can also include a format for risk reporting both for individual risks and for the project as a whole.

Identify Risks

The process of determining which risks may affect the project and documenting their characteristics. As the project progresses new risks are identified, existing risks evolve and change, some risks occur, and some pass without occurring. The process of risk identification goes on throughout the project. All stakeholders should be aware of and communicate risks or opportunities that could impact the project. There will be certain stakeholders who will have a more formal role in risk identification, such as subject matter experts, the project team, risk experts, and customers.

Figure 11-6 shows the inputs, tools and techniques and outputs for the Identify Risks process. Figure 11-7 shows a data flow diagram for the Identify Risks process.

Identifying Opportunities

The same information used to identify threats is also used to identify opportunities. We will not call out opportunities separately, but rather handle them both within the general context of identifying risks.

INPUTS

The *risk management plan* identifies roles and responsibilities for identifying risks, the budget, methodology, and timing for risk identification and risk categories. The stakeholder register is also a good place to identify people who can help identify and understand risks and opportunities for the project.

The *cost management plan* identifies processes for estimating, budgeting, and controlling costs. It documents control thresholds, performance measurement methods, and the approach for managing cost reserves. This information can be reviewed in combination with the *activity cost estimates* to determine if estimates are too optimistic, if the process is appropriate for the project and the life-cycle phase,

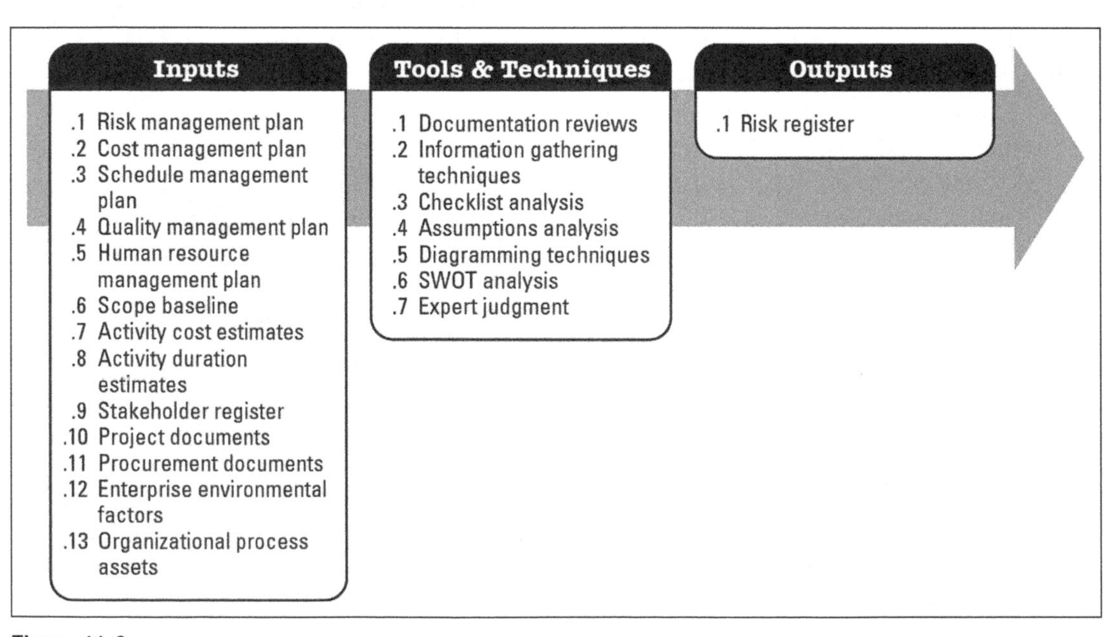

Figure 11-6
Identify Risks Inputs, Tools and Techniques, and Outputs
Source: *PMBOK® Guide—Fifth Edition*

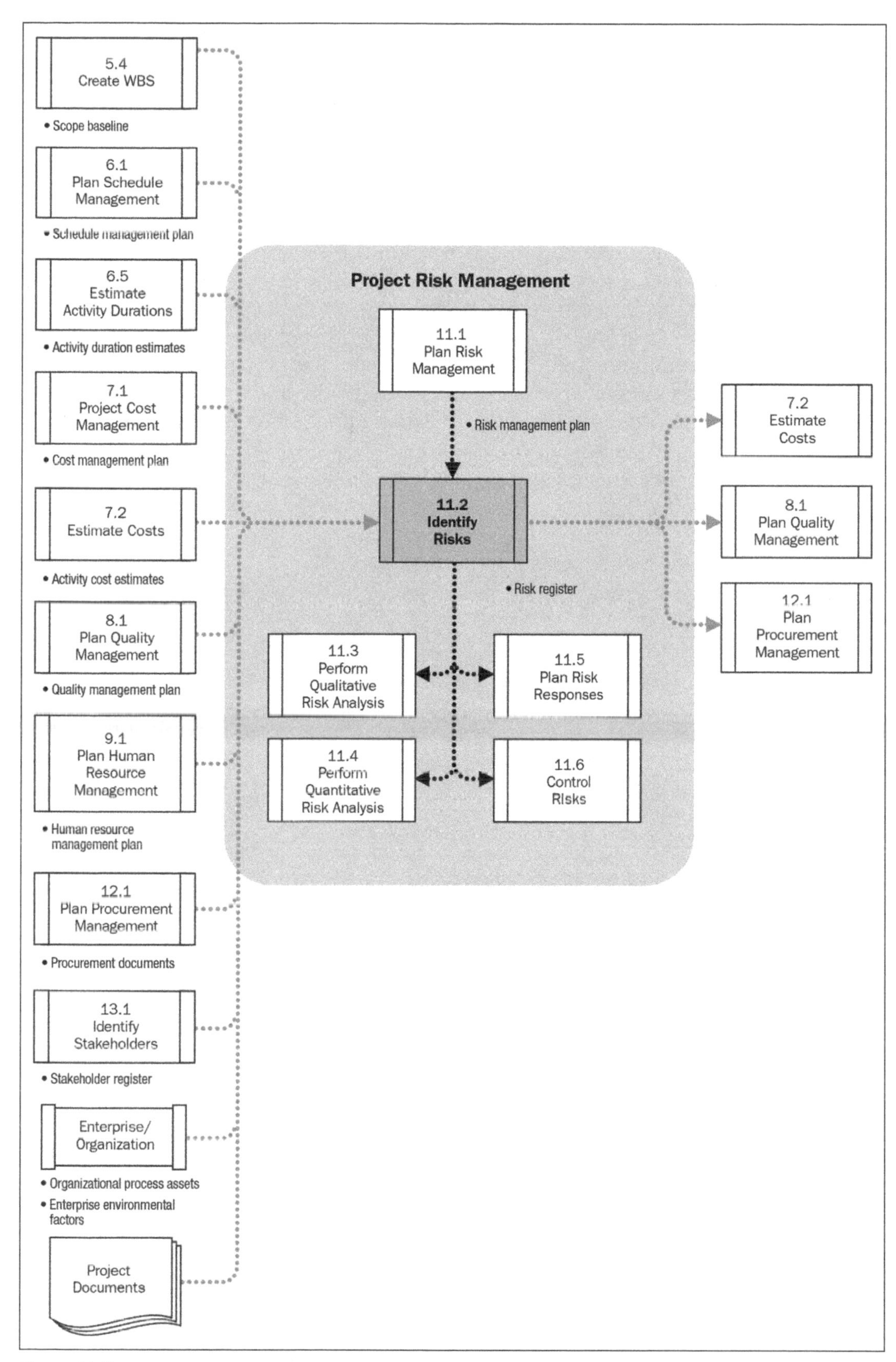

Figure 11-7
Identify Risks Data Flow Diagram
Source: *PMBOK® Guide—Fifth Edition*

and if the techniques for estimating and controlling costs are deemed adequate.

The *schedule management plan* identifies the approach for developing and controlling the schedule. It defines the methodology used to develop the schedule and may document the methods used in developing duration estimates. The *activity duration estimates* are reviewed along with the schedule management plan to determine if the estimates and constraints are too optimistic, if appropriate techniques were used to develop the estimates, and if the schedule reserve is considered adequate.

The *scope baseline* contains all the information about the project scope, including technical details in the WBS dictionary and assumptions in the project scope statement. The *quality management plan* documents the approach for quality assurance, quality control, quality improvement, and process improvement. It can define which quality tools will be used, how they will be used, when they will be used and the quality metrics that will be used with them to report on quality. The scope and quality information is reviewed to determine risks associated with either developing the product, or developing it to the required level of performance.

Information on project staffing, staffing levels, available skills and the staffing approach are contained in the *human resource management plan*

In addition to those specifically named above, any and all *project documents* should be reviewed for risks. Some risks are obvious (such as using unproven technology), while others can only be identified when viewed across the project. Let's look at an example.

Document Reviews

The scope statement may document that a specific deliverable is due on April 1. The assumptions may state that a component for that deliverable is in stock. If you check the schedule, maybe it doesn't align. Perhaps the schedule shows the part is on order and won't be in until April 10, and the deliverable is scheduled for completion on April 20. In addition, since the part has to be ordered, you will need to check the cost estimates to see if that part was included. This example demonstrates that only by reviewing many different documents can you identify some of the project risks.

Project documents that are particularly useful include, performance reports (when available), and requirements documentation. Procurement documents provide information on acquisitions and contract type.

Enterprise environmental factors that help identify risks can include external information and research such as studies and benchmarks. The organizational process assets most often used are lessons learned, prior risk registers, and a risk database if one is available.

TOOLS AND TECHNIQUES

Documentation reviews are used to go over all the documents identified in the inputs. Often reviewing documentation is just the start of the process.

The next step is using more formal information gathering techniques. The easiest way to gather information is to interview stakeholders, with either a structured set of questions, or with open-ended questions. You can also get a group of stakeholders together and brainstorm risks, using the categories to group risks. Brainstorming can be formal, using a trained facilitator, or informal. This is often done to elicit expert judgment and experience from people who have specific knowledge and skills that pertain to the project.

The Delphi technique is designed to have a group of anonymous experts come to consensus on the risks in the project. The steps involved are:

1. A facilitator distributes a questionnaire with a set of questions about the project risk to a preselected group of respondents (usually subject matter experts).
2. Respondents complete the survey and return it to the facilitator.
3. The facilitator compiles and summarizes the responses.
4. The facilitator sends out the summarized responses to the respondents who comment on the information and return it to the facilitator.
5. This cycle continues until either consensus is reached or a decision is made that the compiled information is good enough.

Root cause analysis is an information-gathering technique that seeks to identify the underlying source or cause of a problem or a risk. Diagramming techniques, such as cause and effect diagrams, flowcharts, and influence diagrams are often used to map formal or informal processes to see how events impact one another and lead to risks.

Assumptions analysis not only analyzes the assumption log, but it looks at the assumptions behind the duration and cost estimates, resource estimates, and other project areas. If the assumptions are not documented, or documented poorly, this could indicate that not enough information is known, or that knowledge is retained in a person's head, but not visible to the rest of the team. In some instances, the estimates in the schedule will be different from those in the budget (refer to the example under Documentation Review). Analyzing assumptions can also help identify the impact if certain assumptions are not true.

One way to start identifying risks is to do a *checklist analysis*. This technique uses information compiled from past projects. This should only be used as a starting point. Completing a checklist and thinking that risk identification is complete is a big mistake!

For new product development it might be useful to conduct a SWOT analysis by looking at internal strengths and weaknesses and external opportunities and threats. This can help the organization leverage strengths to take advantage of market opportunities and reduce weaknesses to avoid market threats.

SWOT. Analysis of strengths, weaknesses, opportunities, and threats of an organization, project, or option.

OUTPUTS

The risk register is originated in this process, and it is continually updated during the remaining processes. In this process a description of each identified risk event is entered in the register. It is important to enter enough information about the risk so it is fully understood and can be adequately analyzed and responses can be developed. The call out box below shows examples of vague risk statements and better risk statements.

Risk Register. A document in which the results of risk analysis and risk response planning are recorded.

Risk Statement

Consider the following risk statements:

Schedule is a risk.

Because there is a contractually mandated end date that is aggressive, there is a risk that we will not deliver the product on time.

Or these two:

Budget is a risk.

The labor rates used in the cost estimates are the mid-point labor rates. If we use higher skilled workers the labor rates will increase causing a cost overrun.

In both of the above situations the first statement gives us no real information. The second statement describes the cause of the risk and helps us analyze the probability and impact as well as develop a risk response.

See Appendix A for a sample Risk Register and Risk Data Sheet.

Perform Qualitative Risk Analysis

Perform Qualitative Risk Analysis is the process of prioritizing risks for further analysis or action by assessing and combining their probability of occurrence and impact. For projects with a lot of risks it is important to prioritize the risks to make sure the team is spending time responding to the most critical risks. A great majority of the risks will have a low probability and impact and will not necessitate proactive actions. It is the risks that have a high probability of occurring and/or a high impact if they do occur that the team should focus on. Performing qualitative risk analysis is a rapid way to prioritize risks. In addition to the probability and impact, the urgency of the risk should be incorporated into the risk analysis. Risks that are imminent should be handled before those that are in the distant future.

Figure 11-8 shows the inputs, tools and techniques and outputs for the Perform Qualitative Risk Analysis process. Figure 11-9 shows a data flow diagram for the Perform Qualitative Risk Analysis process.

Analyzing Opportunities

The same information used to analyze threats is also used to analyze opportunities. However, with opportunities, the higher the probability and impact, the better the opportunity. We will not call out opportunities separately, but rather handle them both within the general context of analyzing risks.

INPUTS

At the start of the project when the team is engaged in the Plan Risk Management process they define much of the information that will be used in this process. The risk management plan contains definitions of probability and impact for each project objective and the subsequent rating of high, medium, and low risks are all defined in that up-front planning process. Those definitions and matrices will be applied to the risks identified in the *risk register* to prioritize risks.

The scope baseline with the project scope statement may contain information about the nature of the product that will help

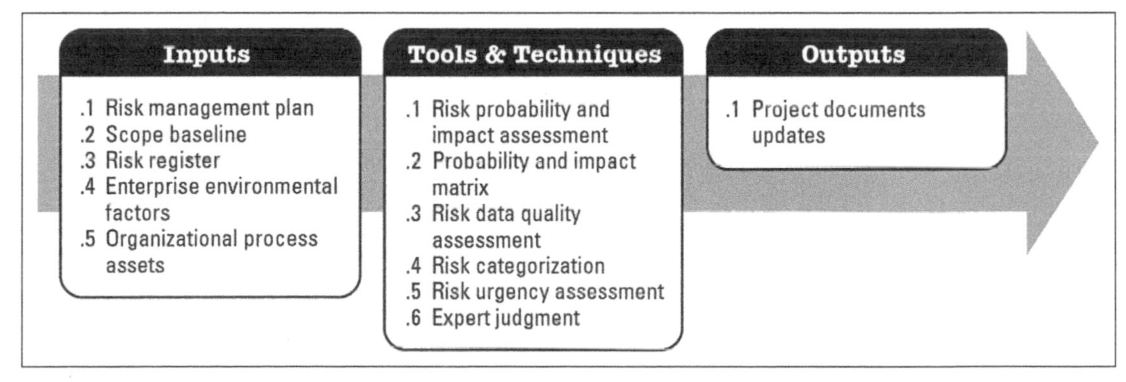

Figure 11-8
Perform Qualitative Risk Analysis Inputs, Tools and Techniques, Outputs
Source: *PMBOK® Guide*—Fifth Edition

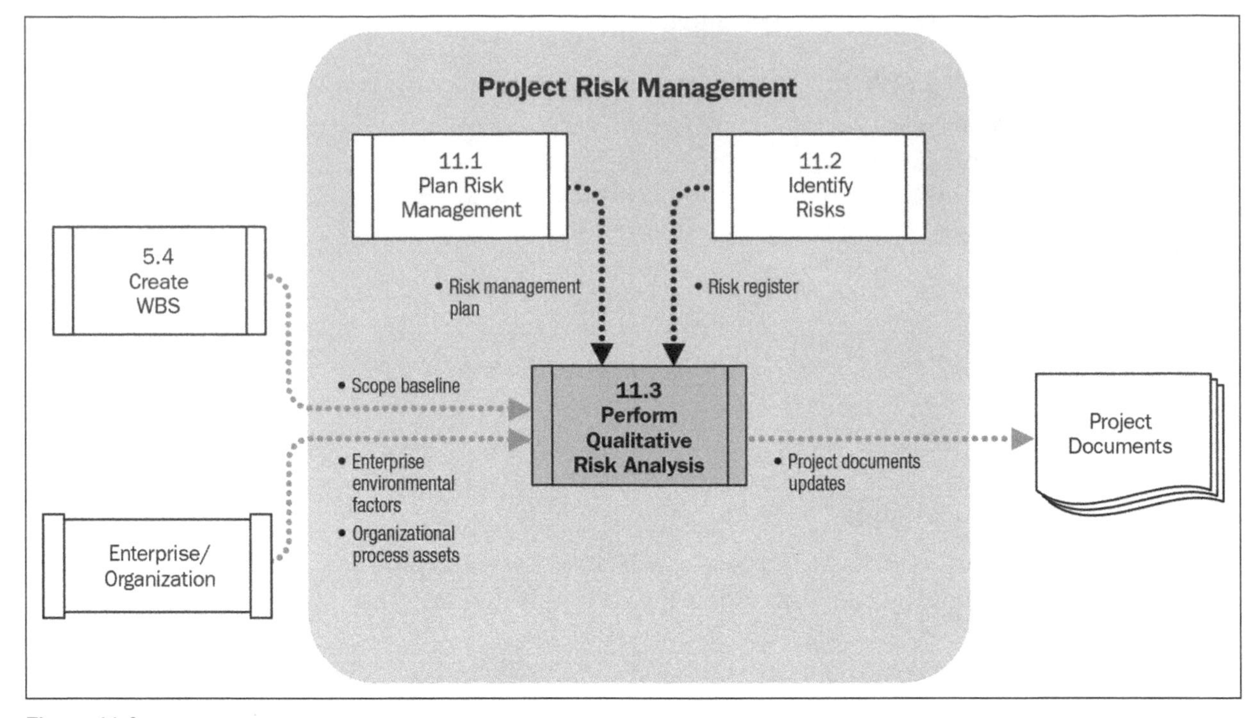

Figure 11-9
Perform Qualitative Risk Analysis Data Flow Diagram
Source: *PMBOK® Guide*—Fifth Edition

determine the probability of risks (new technology means problems in development and testing are more likely), and the impact (hard constraints increase the impact of missed deadlines or budget overruns).

Enterprise environmental factors that influence this process include a risk database. Information from previous projects are organizational process assets that are helpful in this process.

TOOLS AND TECHNIQUES

The people responsible for risk management conduct a risk probability and impact assessment for each risk (or those they are responsible for) using the definitions of probability and impact defined in the risk management plan. This information can then be plotted on a probability and impact matrix. As discussed in the Plan Risk Management process, the matrix is shaded to reflect the relative severity of risks for each objective.

Probability and impact matrixes can use numerical ratings, such as 1 to 3, 1 to 5, or 1 to 10. Or they can use descriptive ratings, such as high-medium-low or very low through very high. The following sample shows a numeric rating for a 5-by-5 matrix that is evenly balanced with regard to risk tolerance. Using this example, any risk that scored 15 or higher is considered a high risk. A risk that scores from 4 to 12 is a medium risk, and risks less than 4 are low risks.

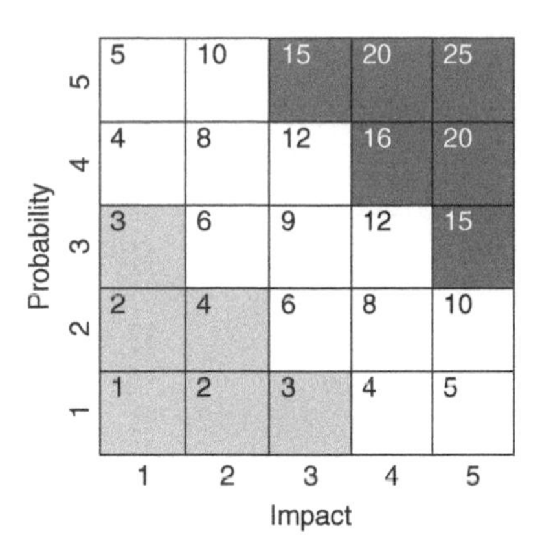

Before assessing the risk the team needs to perform a risk data quality assessment. This is a fancy way of saying you need to make sure the information you have about the risk is accurate and as detailed as possible. Saying a risk has about a 40 percent likelihood of occurring because it sounds about right is a lot less accurate than looking at data for the last 20 projects and noticing that the event occurred on eight of them. Data integrity is important when analyzing risk. If necessary the team should do more research to make sure they fully understand the risk so they can accurately describe and analyze it.

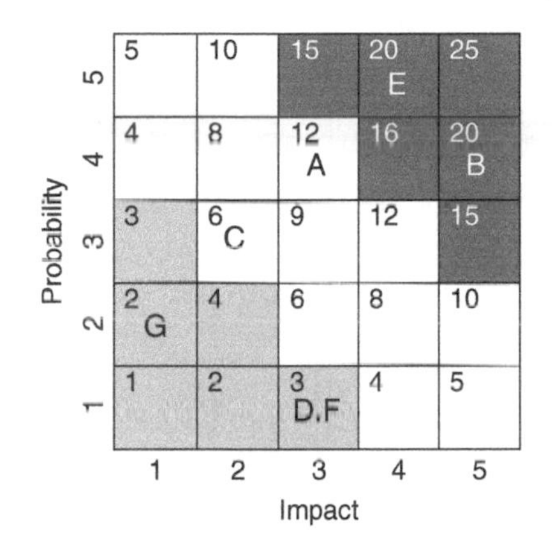

For projects with many risks you may want to *categorize risks* via a risk breakdown structure, by phase, by root cause, or by some other method to get a more meaningful understanding of the risk exposure relative to similar risks.

A risk urgency assessment is used to focus on the near-term risks. For example, let's assume you have risks A through G. Using the same matrix shown earlier, but categorizing it so you were looking at resource risks only, you have the outcome in the grid above.

Given this information it would make sense to address E first, then B, then A, then C, and so forth. However, if A is likely to occur in the next month, and E and B are 9 to 12 months away, you would focus on A first because it is more urgent.

Expert judgment is used to accurately assess the probability and impact as well as the data quality. Another type of expert judgment is someone very skilled in the risk management process.

OUTPUTS

The project documents that are updated include the assumption log and the risk register. The risk register is updated to reflect the probability, impact, and risk score for each risk (the risk score is the probability times the impact). Many times the risk register will be sorted based on score so the most critical risks are at the top of the list. Another way of sorting the list is to sort first by category and then by urgency, then by criticality.

At this point, most of the risks will be rated as low. Low risks go on a watch list. They will be watched in case their probability or impact scores change and require action. Other risks will require a more detailed assessment. Low risks will carry forward into the Perform Quantitative Risk Analysis process. Many risks will go straight to the Develop Risk Response process.

As the project progresses the team may identify risk trends, such as the probability of a certain category of risks is increasing, or more risks in a particular category are moving into the low end of the matrix.

Perform Quantitative Risk Analysis

Perform Quantitative Risk Analysis is the process of numerically analyzing the effect of identified risks on overall project objectives. Not all projects require formal quantitative risk analysis. Many of the modeling and simulation techniques used in this process are appropriate for very large projects that have millions of dollars at risk.

There are two purposes for quantitative risk analysis: to quantify the impacts of individual high-risk events and to make a determination of the overall project schedule and cost risk on the project.

Figure 11-10 shows the inputs, tools and techniques and outputs for the Perform Quantitative Risk Analysis process. Figure 11-11

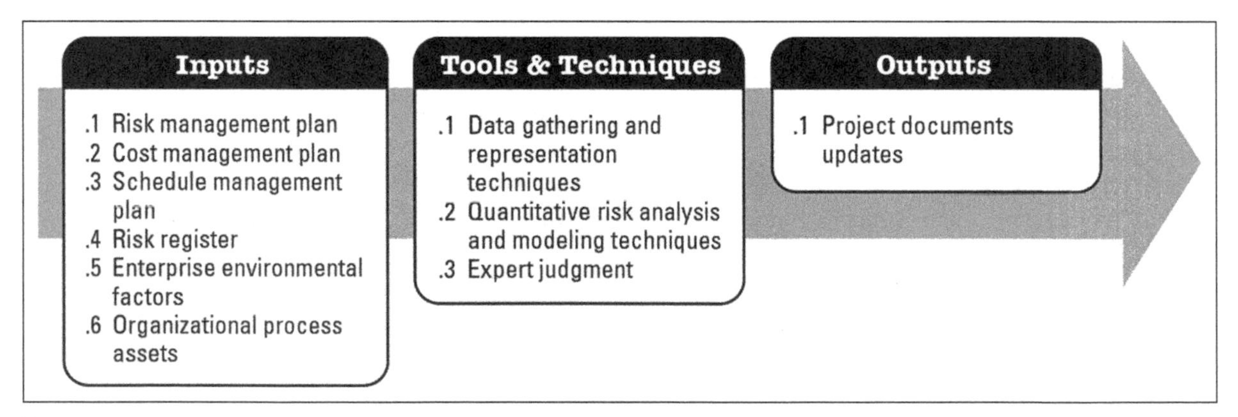

Figure 11-10
Perform Quantitative Risk Analysis: Inputs, Tools and Techniques, Outputs
Source: *PMBOK® Guide*—Fifth Edition

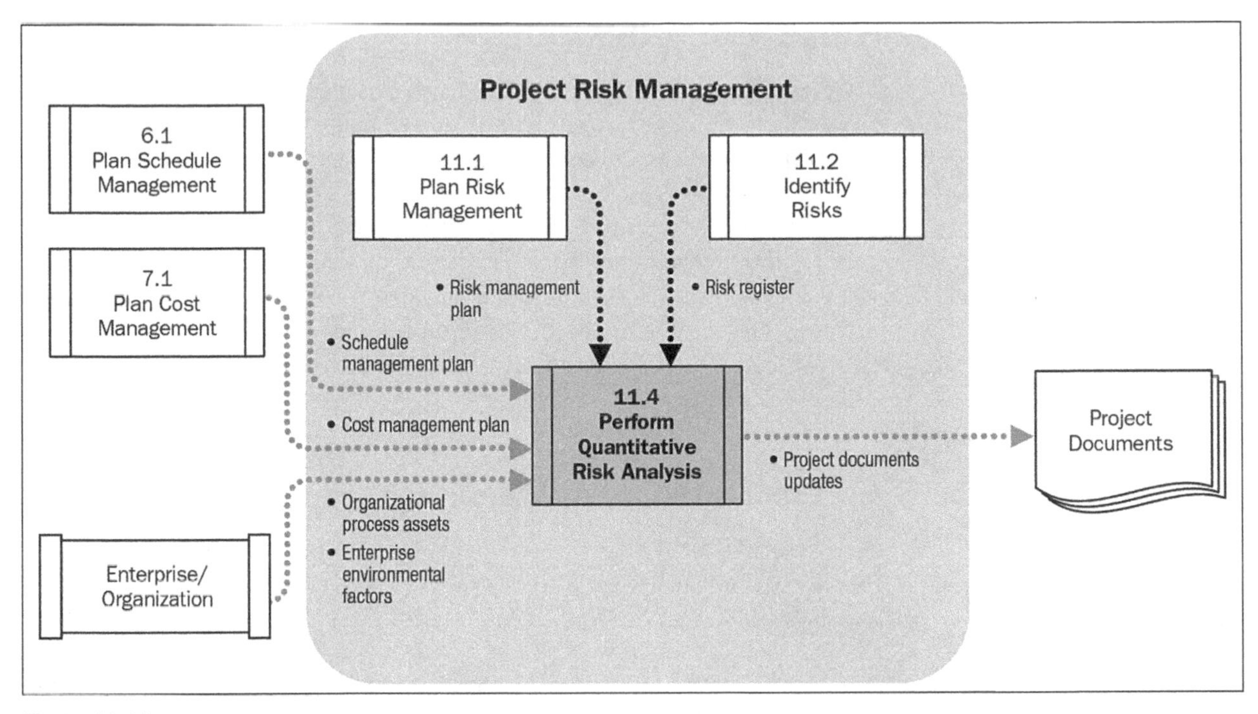

Figure 11-11
Perform Quantitative Risk Analysis Data Flow Diagram
Source: *PMBOK® Guide*—Fifth Edition

shows a data flow diagram for the Perform Quantitative Risk Analysis process.

> **Quantitative Opportunity Analysis**
>
> Projects don't generally require a quantitative analysis on opportunities

INPUTS

The risk management plan identifies the tools and techniques that will be used in performing quantitative risk analysis along with the people who will apply them.

The cost management plan and the schedule management plan have information that will be used in determining the overall cost and schedule risk for the project, such as the control limits, variance thresholds, and reserve methods.

The updated risk register contains the information from the Perform Qualitative Risk Analysis. Enterprise environmental factors include risk databases and any industry studies. *Organizational process assets* include policies and procedures for quantitative risk management as well as information from previous projects and risk database information.

TOOLS AND TECHNIQUES

Data-gathering techniques primarily consist of interviews. *Data-representation techniques* primarily consist of probability distributions. Interviews are used to take information on a specific risk and help quantify it. For example, if a risk was considered a high probability of having a medium schedule impact, you could interview stakeholders to understand what "high" meant in terms of probability and what "medium" impact meant in terms of the critical path, using, perhaps, float. At that point you might use a *quantitative analysis technique* such as expected monetary value to quantify the risk. Let's look at an example.

Expected Monetary Value

A risk with a 40 percent chance of having a $200,000 impact has an expected monetary value of $80,000 (.4 × $200,000 = $80,000).

Another use for interviewing is to begin to understand the uncertainty associated with risk events and estimates. The interviewer would ask the risk owner (or work package owner) questions about the extreme scenarios for the event, as well as the most likely occurrences of these events. This information is collected and can be used later for a modeling and simulation exercise. It can also be used to determine an expected value for an estimate. Here is an example:

Probability Distribution

Interviewer: Clara, if everything goes exceptionally well on this work package, what do you estimate it will cost?

Clara: Well, the best case scenario would be about $12,000.

Interviewer: Okay, what do you think is the most likely cost, knowing what you know about the environment, available resources, and so forth?

Clara: It will probably come in around $14,000. Most likely I will have the resources I need, but there may be a few glitches that we will have to work out.

Interviewer: Alright, now let's assume that the worst case scenario occurs; what would that look like?

Clara: Well, in the worst case scenario I would only have five out of the eight people promised, and the material would be late, causing a delay, which would mean overtime. And, if we could not resolve the technical issues as rapidly as we would like, then this work package could cost as much as $22,000!

This information can be used to create a probability distribution. The simplest form is a triangular distribution where each outcome is weighted equally, and then the sum is divided by 3 to get an expected outcome. In this case, the expected outcome would be $\frac{(12{,}000 + 14{,}000 + 22{,}000)}{3} = \$16{,}000$. You can also use a weighted formula where the most likely scenario is weighted higher than the best case and worst case scenarios, such as the PERT formula we looked at in the Estimate Activity Durations and Estimate Costs processes.

While the difference between $14,000 and $16,000 may be relatively small, this is only one work package. To fully understand the magnitude of the spread of possible outcomes you would do this for every outcome. This information would then be entered into a simulation software package where a Monte Carlo simulation could be run. A Monte Carlo simulation runs thousands of iterations of a project using a range of estimates within the range provided from interviews. The end result will show the likelihood of achieving various cost outcomes for the project. For instance, if you assume all work packages will come in at their most likely cost, a Monte Carlo simulation may indicate that you only have a 28 percent chance of success. This is because the expected value is the average, or a 50 percent likely outcome. So if the most likely result is less than the expected result, the most likely result will always have a less than 50 percent chance of being achieved. A Monte Carlo simulation can assist you by showing you the amount of budget you need to achieve 70 or 80 percent of the time, or any percent likelihood of success. You can see how having accurate information is important for this technique. This really is an example of "Garbage in, garbage out."

A decision tree is used in conjunction with expected monetary value to determine the expected value of a decision given uncertainty. Figure 11-12 is an example from the *PMBOK® Guide*. It models the decision of whether to build a new production plant or upgrade an existing one. It shows the uncertainty associated with the demand for the product and how that would play out financially with both scenarios. Keep in mind that the expected monetary value is not an actual value; it is a percent times a value. But it does highlight expected cost versus benefit overall given an uncertain event.

Expert judgment is especially useful in this process, particularly in knowing how to interview, set up models, and run simulations with software. It is also useful to have expertise available to help interpret the results of these processes.

OUTPUTS

Project documents that are updated include the risk register, which is updated with the prioritized list of risks based on the quantitative analysis. Information generated from this process can be used to determine the likelihood of achieving cost and schedule targets and establish the cost and schedule reserves needed to deliver the project successfully. As the project progresses and more information is available, these techniques can be repeated and you can look for trends in the results.

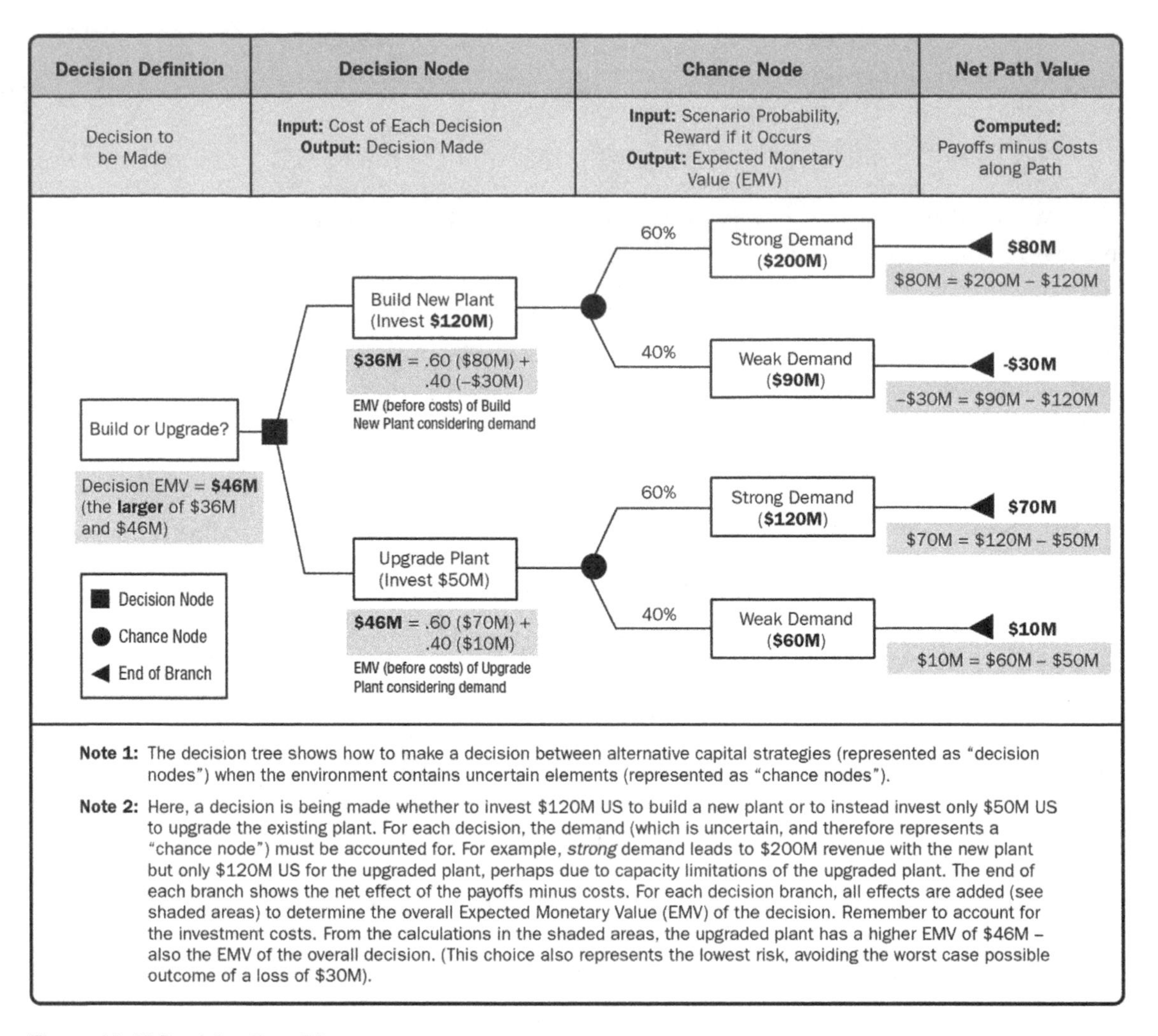

Figure 11-12 Decision Tree Diagram

Source: *PMBOK® Guide*—Fifth Edition

Plan Risk Responses

Plan Risk Responses is the process of developing options and actions to enhance opportunities and to reduce threats to project objectives. After risks have been prioritized in the analysis processes they are assigned to a responsible party to develop an appropriate response or responses. "Appropriate" means the response is cost and time effective given the project context. Many times there will be multiple options for responding to a risk, and some or all of those options will be employed.

Figure 11-13 shows the inputs, tools and techniques and outputs for the Plan Risk Responses process. Figure 11-14 shows a data flow diagram for the Plan Risk Responses process.

INPUTS

The risk management plan describes approaches for risk responses and the philosophy for contingency risk reserves. The risk register

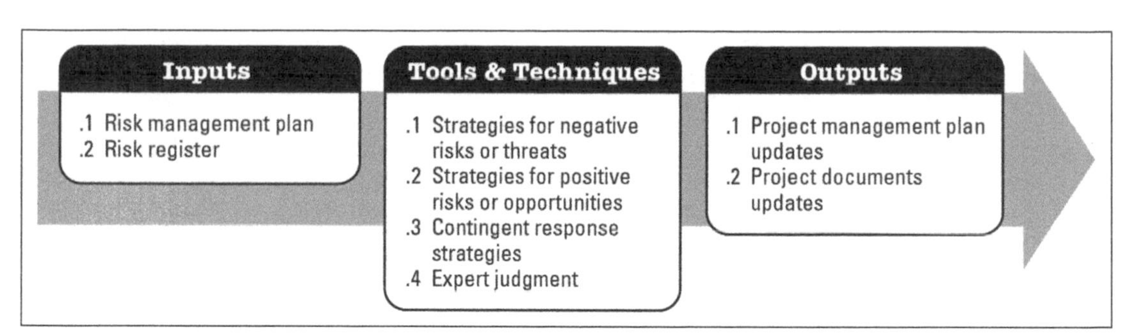

Figure 11-13
Plan Risk Responses: Inputs, Tools and Techniques, Outputs
Source: *PMBOK® Guide*—Fifth Edition

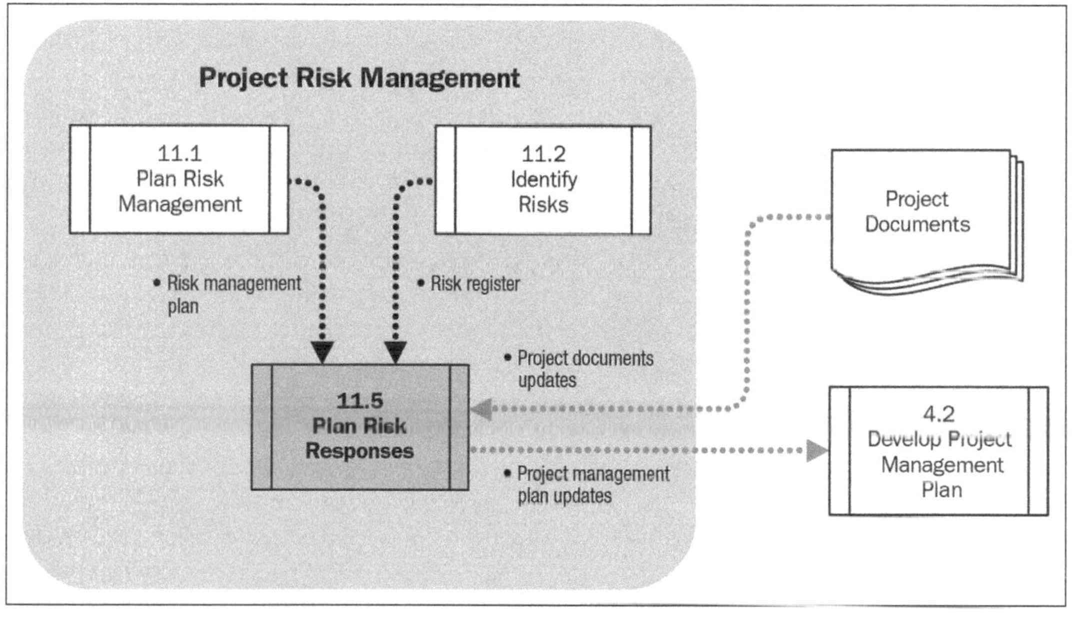

Figure 11-14
Plan Risk Responses Data Flow Diagram
Source: *PMBOK® Guide*—Fifth Edition

contains an updated list of risks that is prioritized by probability, impact, and urgency.

TOOLS AND TECHNIQUES

Strategies for negative risk events (threats) include:

- Avoid
- Transfer
- Mitigate
- Accept

Let's look at an explanation and then a couple of examples of each of these.

Avoid. Avoiding a risk entails changing the project management plan or project approach to eliminate the threat.

- If there are many risks that are categorized as overrunning the schedule you can avoid the risk by extending the schedule.
- If there are many risks associated with using an unproven technology or approach, you can find an approach that has worked before.

Transfer. Transferring a risk means you give the responsibility for managing the risk to someone who is better able to manage it. Transferring a risk usually involves a "premium" or payment for the organization or person who is taking on the risk.

- Construction contractors who do work for the government are required to get a completion bond as a guaranty the work will be completed. The contractor pays the premium to the bonding company. If the contractor defaults, the bond company pays to have the work completed. The government has transferred the risk of default to the bonding organization.
- An organization that is developing a new software in-house may have some risks associated with the availability or ability of their in-house resources to complete the work in a timely manner. They can choose to transfer the work to a contractor to complete the work.

Transferring work does not eliminate the operational impact of the risk. It can transfer the management of the risk or the financial impact of the risk. But if the event occurs, then the organization still bears the impact. In the second example above, if the contractor delivers late, the hiring organization still has the problem of a late delivery.

Mitigate. Mitigating a risk is reducing the probability and/or the impact of the threat. The most common mitigation strategies include defining requirements in more detail, building redundant processes, scaling up operations or models in a step-by-step fashion, and spending more time in up-front planning.

- A design team can reduce the probability of a design flaw by instituting a peer review process for all drawings.
- A backup generator can be used to reduce the impact of a power outage.

Accept. Accepting a risk means you cannot or will not develop a strategy to respond to it. This is used for low-probability and low-impact risks, or for those risks that cannot be managed.

- If there is a chance that a specific resource you have scheduled will not be available, but the work is routine and there are other resources available, you could accept the risk and deal with the situation if and when it arises.

- If one of your suppliers is in labor negotiations and their contract is set to expire, you cannot really impact that situation. You may accept it and develop a contingency plan.

Projects use time and cost contingency reserves to account for the risks they are accepting. Teams know that some, but not all, of the identified risks they are accepting will occur. Therefore, they build time into the schedule and funds into the budget to account for these.

For threats it is also useful to have a fallback plan that is implemented if all else fails. For example, if a new accounting system is being rolled out, and there is concern that if it doesn't work as planned you can have a fallback of using the old system until all the bugs are worked out of the new system.

Some risk responses introduce new risks. These are called secondary risks. A secondary risk is a risk that is identified as a result of a planned response. For example, if you transfer work to a contractor using a cost plus fixed fee contract, you have a new risk that the contractor could go over the target, causing your project to overrun.

Strategies for positive risks or opportunities include:

- Exploit
- Share
- Enhance
- Accept

Let's look at an explanation and then an example of each.

Exploit. Exploiting an opportunity is the equivalent of avoiding a threat. You are taking actions to ensure an opportunity is captured.

- If you are doing work on a contract that has an incentive fee for completing early, you may put your best workers on the contract and authorize overtime to ensure early completion and thereby maximize the incentive fee you will earn.

Share. Sharing an opportunity is the equivalent of transferring a threat. You work with another organization to help you capture the opportunity.

- You may decide to partner with another organization to respond to a bid in order to increase the likelihood of winning the business. By selecting an organization that has skills and strengths that offset some of your weaknesses you are in a better position to win.

Enhance. Enhancing an opportunity is the equivalent of mitigating a threat. You are taking actions to increase the probability and/or the impact of an opportunity.

- You may enhance your ability to save money by reusing information, code, material, or research from previous projects. While you may not be using the latest and greatest technology, you are increasing the likelihood of coming in under budget.

Accept. Accepting an opportunity is being willing to take advantage of the situation if the opportunity arises.

- There is a possibility that another project will be cancelled, freeing up some valuable resources. If this does happen you will work to get some of those resources assigned to your project, but you are not actively trying to do something that will cancel the project or reserve key resources if it is cancelled.

Contingent response strategies are employed if specific identified events occur. For example, if a component does not perform at a certain rate by a certain date, you may make the decision to bring in more resources, or use a different approach, or approve overtime. All of these are contingency responses or plans that will be deployed if the date arrives and the performance rate is below the trigger point you identified. Contingency plans should have a trigger point that indicates they should be deployed.

Expert judgment is used to help develop responses and select from multiple responses.

OUTPUTS

The project documents that are updated include the risk register, which is updated with the response plan, the person accountable for managing the risk and the response, and a revised analysis of the probability and impact of the risk with the response plan in place. Where the approach is to develop a contingency plan, the trigger is documented as well. The actions, schedule activities, and budgets associated with the risk responses are entered into the risk register and the risk data sheets if those are used. The residual, or remaining risk is indicated with the revised probability and impact assessment.

Where threats are transferred or opportunities are shared the team communicates and documents the risk-related contract decisions including the type of contracts used.

Risk responses, contingency plans, schedule, and cost reserve will cause a number of project management plan updates. The WBS, cost, and schedule baselines will need to be updated to incorporate changes in scope, budget, and activities.

The cost and schedule management plan are updated if there are changes to managing costs or duration estimates, contingency reserves, or roles and responsibilities for the schedule or budget.

Changes to the quality metrics, measurements, techniques, or processes are documented in the quality management plan. Changes in staffing, skill requirements, training, and project organizational structure are documented in the human resource management plan. Outcomes from the transfer or sharing of risks or opportunities are added to the procurement management plan.

Other project document updates include updates to requirements, technical documentation, assumptions, and stakeholder registers.

Chapter 12

Planning Procurement

TOPICS COVERED

Project Procurement Management

Plan Procurement Management

Project Procurement Management

Project Procurement Management includes the processes necessary to purchase or acquire products, services, or results needed from outside the project team.

Procurement management includes planning the procurement approach, selecting the contract type, developing bid documents, determining source selection criteria, negotiating a contract, and then managing the contractual relationship and performance throughout the life of the contract. For large projects, this can include many contracts and requires integrating multiple sellers with the work of the performing organization.

Procurements deal with contracts, which are legally binding. Contracts may be called subcontracts, purchase orders, agreements, or other various terms. Because they are legally binding and in many cases they are worth substantial amounts of money, the project needs legal, contracting, and purchasing expertise represented on the team. There are generally a whole set of policies, procedures, and processes associated with major procurements.

Some organizations "purchase" goods or services from other departments in the same organization. This is called intradivisional work. While many of the processes are the same, we will focus on external procurements. In some situations an organization may partner with another organization to win business. These are called teaming agreements. They are subject to the same processes as a procurement, but the decision to team with another firm is usually made at the very highest level of the organization.

WHAT'S IN A NAME

Agreements are also known as:

- Contracts
- Subcontracts
- Understandings
- Purchase orders

Buyers are also known as

- Clients
- Customers
- Prime contractors
- Acquiring organizations
- Government agencies
- Service requestors
- Purchasers

Sellers can be known as

- Bidders
- Vendors
- Sources
- Suppliers
- Subcontractors

Plan Procurement Management

Plan Procurement Management is the process of documenting project procurement decisions, specifying the approach, and identifying potential sellers. During this initial procurement process, a make or buy decision is made for the services and deliverables needed to meet project objectives. If all the work is to be done in-house, there is no need to conduct any further procurement processes. If a decision is made to purchase services or deliverables from outside the organization, then this process is used to determine how each step of the procurement will be managed and integrated with the in-house work. The type of contract is also determined during this process. Each process in Project Procurement Management is conducted for each procurement. For large projects that have numerous procurements integrating the contracting process into the schedule and coordinating the contracted deliverables with each other can be very complex.

Figure 12-1 shows the inputs, tools and techniques and outputs for the Plan Procurement Management process. Figure 12-2 shows a data flow diagram for the Plan Procurement Management process.

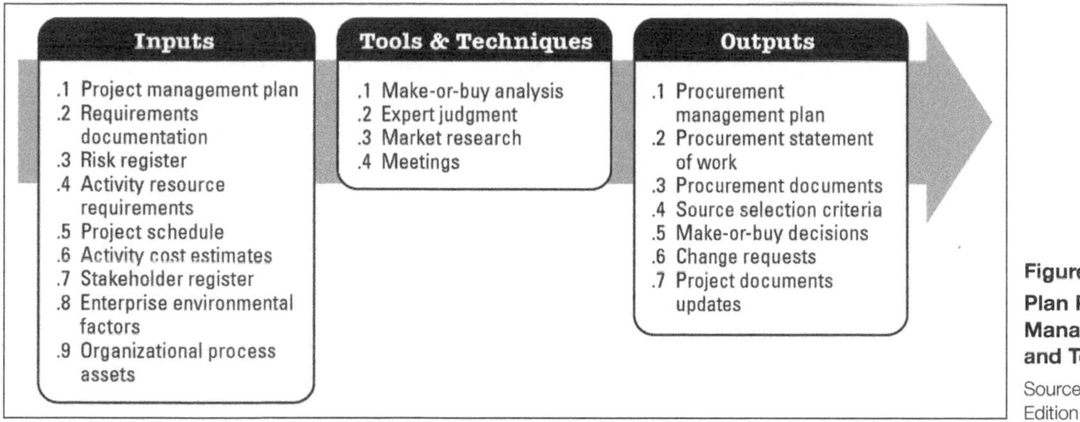

Figure 12-1

Plan Procurement Management: Inputs, Tools and Techniques, Outputs

Source: *PMBOK® Guide*—Fifth Edition

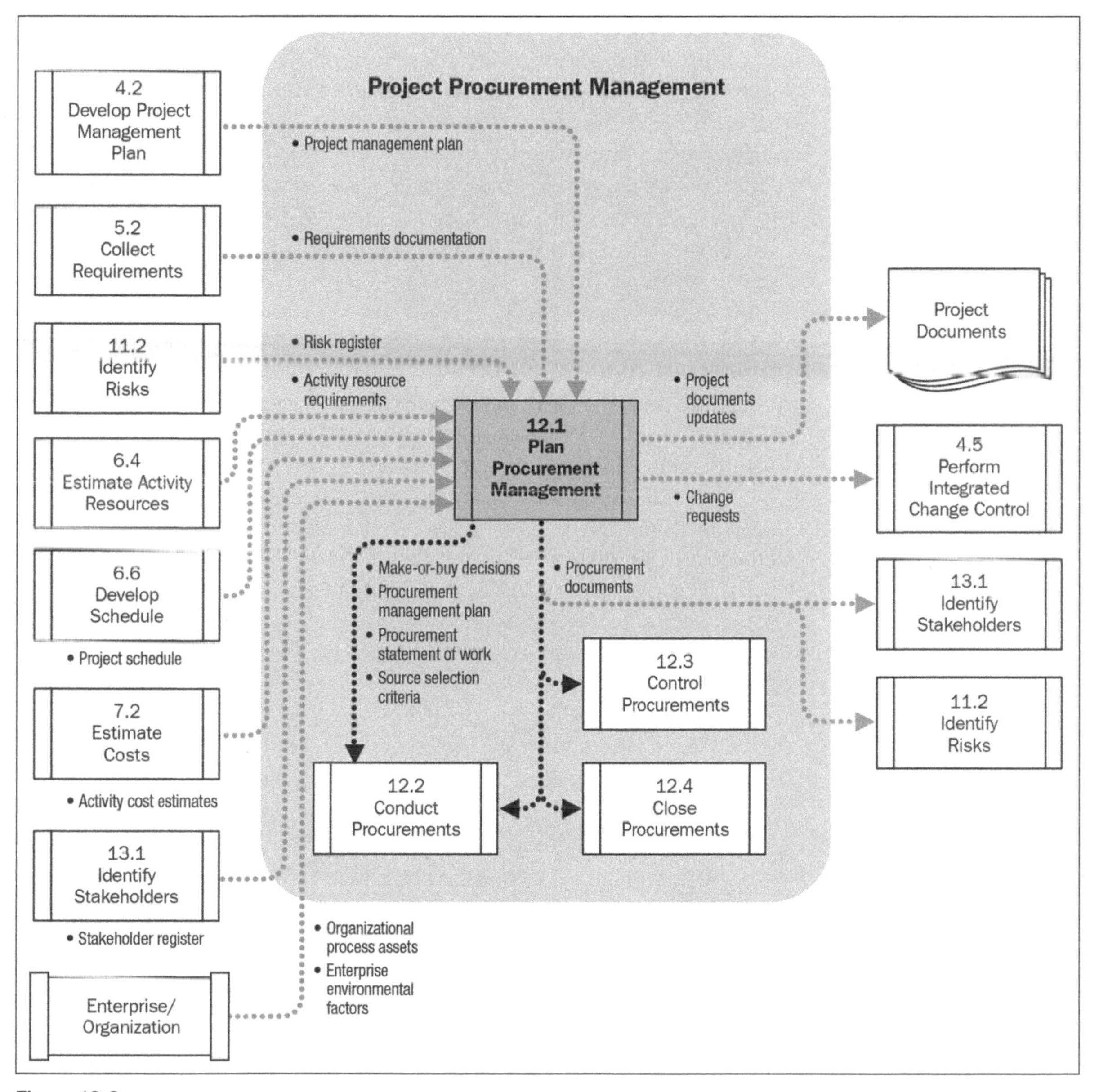

Figure 12-2

Plan Procurement Management Data Flow Diagram

Source: *PMBOK® Guide*—Fifth Edition

INPUTS

The *project management plan* contains the scope baseline. The scope baseline and the *requirements documentation* are reviewed to determine which deliverables and skills are available within the organization and which need to be acquired externally. There can be technical details in the scope baseline and requirements documentation that influence whether to purchase goods or services from outside the organization or perform the work in-house. Requirements documentation may also have information on certifications, permits, licenses, and so forth that must be obtained to perform the project. Such requirements can make the difference in whether a deliverable is developed inside the organization or externally.

In Chapter 11, Plan Risk Responses we mentioned that transferring a risk to a vendor is a risk response option. If that response is chosen, the risk register is an input to planning procurements.

Information about people, equipment, materials, and other resources are documented in the activity resource requirements. Depending on the type of materials needed, a purchase order may suffice rather than a more complex contract. The need for people with a particular skill set or licensure may necessitate looking outside the organization.

Activity cost estimates should be reviewed to determine the cost impacts of purchasing versus leasing goods. For example, a construction firm may opt to invest in purchasing a piece of heavy equipment if they think they will keep it in use a significant amount of time. However, if the current job requires a piece of equipment that will not be used more than once or twice the organization should look at renting or leasing the equipment. Activity cost estimates can also be used to determine the reasonableness of vendor bids.

The schedule shows the required delivery dates for project items and may influence the decision to purchase something rather than develop it in-house. The stakeholder register can be particularly useful if you have multiple divisions or organizations involved in the project. Many times a stakeholder organization will have a resource you will need for a project that you cannot get from inside your own division or organization.

The *enterprise environmental factors* can be very influential in this process. For example, the availability of material or labor and the number of suppliers in an area will have a significant impact on cost and schedule. If there are typical terms and conditions for a product, such as down payments, intellectual property rights, and warranty information, this information will need to be assessed and integrated into the project management plan and documents. Some municipalities have noise, work time restrictions, or environmental requirements that also have to be taken into consideration.

Organizational process assets include all the policies, procedures, and processes required for procurements. Information about previous seller performance is often maintained in the purchasing department. Prior performance should be reviewed to see how past performance influenced the success of previous projects. Some suppliers or vendors are pre-qualified for certain types of work. Having access to the prequalification information can save time when planning for procurements.

Many organizations have contract templates. Contract types fall into two main categories, fixed price and cost reimbursable. A third type, time and materials, is a hybrid of fixed price and cost reimbursable. The contract type is a big influence on the ultimate success of the procurement from a technical, cost, and schedule perspective. We'll discuss each type and some of the common fee structures associated with the various contract types.

Fixed-Price Contracts With all fixed-price contracts the scope of the procurement must be well defined. Any changes in scope require an amendment and price negotiations. With fixed-price contract, the seller is obligated to finish the work, regardless of the final price. There are three main types of fixed-price contracts: the firm-fixed-price, fixed-price-incentive-fee, and fixed-price-with-economic-price-adjustment. Let's look at each of the three types of fixed-price contracts in a bit more detail.

Firm-fixed-price (FFP) contracts are the most common type of fixed-price contract. The buyer and the seller agree on a price for the work. The price remains the same unless there is a scope change. All risk for cost growth is on the seller.

Fixed-price-incentive-fee (FPIF) contracts establish a price ceiling and build in an incentive fee (profit) for cost, schedule, or technical achievement. The term "fixed-price" can be misleading. When the buyer is incentivizing cost performance, the buyer and seller establish a cost target and a share ratio, such as 80/20, 70/30, or something similar. Cost performance below the target earns an incentive fee. Cost performance above the target means the seller gives back some of the fee. An example follows:

Target cost = $400,000

Price ceiling = $460,000

Target fee = $40,000

Share ratio = 80/20

If the actual cost comes in at $425,000 the incentive fee would be calculated as follows:

((Target cost – actual cost) × share ratio) + target fee

((400,000 – 425,000) ×.2) + 40,000 =

(−25,000 ×.2) + 40,000 = 35,000

Therefore, the total price would be $425,000 + $35,000 = $460,000.

In this case the price came in right at the ceiling. If the actual price had been more, the ceiling would have stayed at $460,000 and the seller would have started to lose some of their fee. The point at which the seller hits the dollar amount where he has to give back some of the fee is called the point of total assumption.

The following definitions are not from the *PMBOK® Guide*. They are merely descriptions to help you understand the way contract terminology is used.

Fee. Fee means profit in the contracting world. When a contract has a share ratio for an incentive fee, the first number is what the buyer keeps, the second number is what the seller keeps. Both numbers must total 100 percent. A 70/30 share ratio means that if the actual cost comes in under target by $20,000, the buyer keeps $14,000 of that and the seller gets $6,000.

Cost. Cost is the amount of funding it takes to complete the work.

Price. Price is the cost plus the fee.

Fixed-price-with-economic-price-adjustment (FP-EPA) contracts are used for long-term contracts to account for price fluctuations in labor, materials, and other costs. When negotiating the contract the buyer and seller agree on an objective financial index to use for price adjustments, for example, the consumer price index.

Cost-Reimbursable Contracts Cost-reimbursable contracts are used when the scope of work is not well defined, or subject to change. This is useful for research and development type of work. With this type of contract the buyer must reimburse the seller for legitimate costs associated with completing the work, plus a fee. The buyer and seller agree to a target cost up-front, and fees are calculated from that target cost. There are three main types of cost-reimbursable contracts: cost-plus-fixed-fee, cost-plus-incentive-fee, and cost-plus-award-fee. Let's look at each of the three types of cost-reimbursable contracts in some more detail.

Cost-plus-fixed-fee (CPFF) contracts reimburse the seller for all legitimate costs and provide a fixed fee, usually a percent of the target cost. The amount is fixed regardless of seller performance.

Cost-plus-incentive-fee (CPIF) contracts reimburse the seller for all legitimate costs, but the fee is based on performance. This type of contract operates much like the fixed-price-incentive-fee contract, but there is no price ceiling. Cost, schedule, and technical performance can be incentivized.

Cost-plus-award-fee (CPAF) contracts reimburse the seller for all legitimate costs, but the fee is subjective and based on a set of broad criteria. Usually there is an award board that takes into consideration the contractor performance when calculating the award fee.

Time and Material Contracts Time and materials (T&M) contracts are generally used for smaller contracts when there is not a firm scope of work, or when the work is for an indefinite period. Hourly labor rates and material rates are agreed to up-front (this is the fixed part of the contract), but the amount of time and material are subject to the needs of the job or the buyer. Generally, expenses are reimbursed at cost. Many times this type of contract is used until the scope of work can be well defined. A time and material contract may include a not-to-exceed amount.

TOOLS AND TECHNIQUES

When conducting a make-or-buy analysis the project manager will end up with three categories: those items that must be purchased, those items that must be developed in-house, and those items that can be either purchased or developed in-house. For the latter group, an analysis will be conducted on the best use of time and resources for the project before finalizing the make-or-buy decisions.

Some of the factors that influence the make-or-buy decision include:

- Availability of skill sets—if the organization has excess capacity or a needed skill set then keeping the work in-house is appropriate. However, if resources are already working at or beyond capacity the team should look at bringing in outside help.
- Availability of equipment—similar to above, equipment that is not fully utilized will influence the team to do work in-house versus outsourcing.
- Time constraints—sometimes a tight deadline will require the team to bring in supplemental staff or outsource work altogether.
- Cost—in some circumstances an outside vendor can do work less expensively than keeping the work in-house, in other situations the reverse is true.
- Core competencies—if the work is an organization's core competency, or a competency they want to develop, the work should stay in-house.
- Intellectual property—for some types of work the ownership of the intellectual property will be a deciding factor. For example, if a company outsources the development of some software, the vendor may quote a lower rate if they get to keep the intellectual property and reuse it or resell it.
- Sensitivity of information—there are some industries and some information that is too sensitive to contract out. Information that deals with security, salary, or of other personal nature, may need to stay in-house.
- Volume—if the volume requirement is very low for material, it is sometimes easier to purchase it.

For large procurements expert judgment in purchasing, legal, contracting, and technical disciplines is a necessity. You will use this information in developing a statement of work (SOW), terms and conditions, selecting a contract type, developing bid documents, and determining source selection criteria.

Market research is used to understand more about the market, the key players in the market, and vendor capabilities. Research can be done online, through associations and conferences, or by interviewing subject matter experts. Research is sometimes conducted via meetings to gather more detailed information, such as when interviewing subject matter experts. The team also uses meetings to develop content for the statement of work, procurement documents, and source selection criteria.

OUTPUTS

Based on the make-or-buy analysis, decisions are finalized regarding the products and services that will be acquired.

A procurement management plan is a part of the overall project management plan. It describes how all facets of the procurement will be conducted. For small projects with simple purchases, a paragraph in the project management plan will suffice. For complex procurements, or when a project has numerous procurements, or high-risk procurements, the plan should be more detailed and robust.

The procurement statement of work (SOW) is a narrative description of the work to be done. It describes the deliverables in sufficient detail to allow sellers to determine if they can complete the work, but it does not define how the work must be done. It includes all support and collateral work as well, such as training, documentation, performance reporting, and support. The SOW can be very simple, or very complex, depending on the needs of the project. It is the main part of the contract, therefore it needs to be clear, concise, and not open to interpretation.

Procurement documents are used to solicit proposals. There are many types of procurement documents. Some of the more common ones are:

- Request for Proposal (RFP)
- Request for Quotation (RFQ)
- Invitation for Bid (IFB)

Generally, a request for proposal emphasizes the desire for a proposed solution to a problem, whereas a bid or quotation is more focused on price. Procurement documents may be very simple or they can be quite involved. They should be reflective of the complexity of the work. The documents should be rigorous enough to ensure a valid comparison of responses, but also flexible enough to allow for seller suggestions.

When price is not the sole deciding factor the organization should define the source selection criteria and the weighting factors associated with each criteria. The *PMBOK® Guide* lists the following examples of selection criteria in addition to cost:

- Understanding of need
- Overall cost (includes operating and maintenance costs)
- Financial capacity
- Technical capability and approach

- Management approach
- Risk sharing
- Production capacity
- Business type (women- or minority-owned or small business)
- Past performance and references
- Warranty
- Intellectual property and proprietary rights

The team will identify the most relevant selection criteria and weight them relative to one another. Government contracts need to disclose this information, but commercial contracts do not.

Based on the make-or-buy decisions there may be components of the project management plan or the project documents that need to be updated. A change request is submitted to change these documents if they have already been baselined or finalized.

See Appendix A for a sample Procurement Management Plan

Procurement Management Plan Contents

Type of contract for each procurement

- Roles, responsibilities, and limits of decision-making authority on procurements
- Relevant policies, procedures, guidelines, and templates
- Strategy for managing and integrating multiple procurements
- Strategy for integrating procurements into the rest of the project (such as schedule and cost)
- Assumptions and constraints
- Strategy for handling long lead items
- Procurement milestones for each procurement
- Reporting requirements for sellers
- Required bonds, warranties, insurance, licenses, and permits
- Format for the statement of work
- A list of prequalified sellers
- Selection criteria and weighting
- Processes for managing contracts
- Process for procurement audits
- Risk management issues

Chapter 13

Planning Stakeholder Management

TOPICS COVERED

Project Stakeholder Management
Plan Stakeholder Management

Project Stakeholder Management

Project Stakeholder Management includes the processes required to identify all people or organizations impacted by the project, analyzing stakeholder expectations and impact on the project, and developing appropriate management strategies for effectively engaging stakeholders in project decisions and execution. Project stakeholders can be internal to the project, such as the project team, or external to the project, such as a regulatory agency. The ability to manage stakeholder expectations, involvement and engagement is a key factor in project success.

You will need to manage those stakeholders that have both a high degree of influence, and those with little influence, but high interest in the project outcome.

Communication is the main component in managing stakeholder expectations. To manage engagement you will also need skills in facilitation, conflict management and issue management.

Plan Stakeholder Management

Plan Stakeholder Management is the process of developing appropriate management strategies to effectively engage stakeholders in project decisions and execution based on the analysis of their needs, interests, and potential impacts. Once the project stakeholders are identified in the Identify Stakeholders process, a strategy for creating and maintaining productive relationships with the various stakeholders and stakeholder groups is developed.

Figure 13-1 shows the inputs, tools and techniques and outputs for the Plan Stakeholder Management process. Figure 13-2 shows a data flow diagram for the Plan Stakeholder Management process.

INPUTS

The stakeholder register lists all the stakeholders for the project. The project management plan contains various components that impact stakeholder management planning.

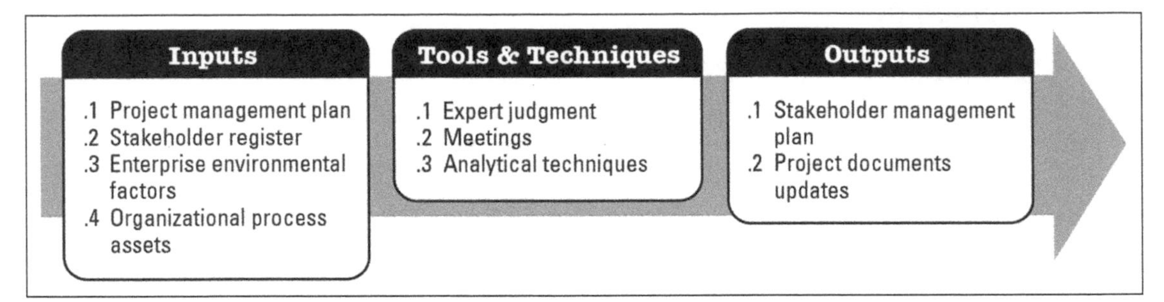

Figure 13-1
Plan Stakeholder Management: Inputs, Tools and Techniques, Outputs
Source: *PMBOK® Guide*—Fifth Edition

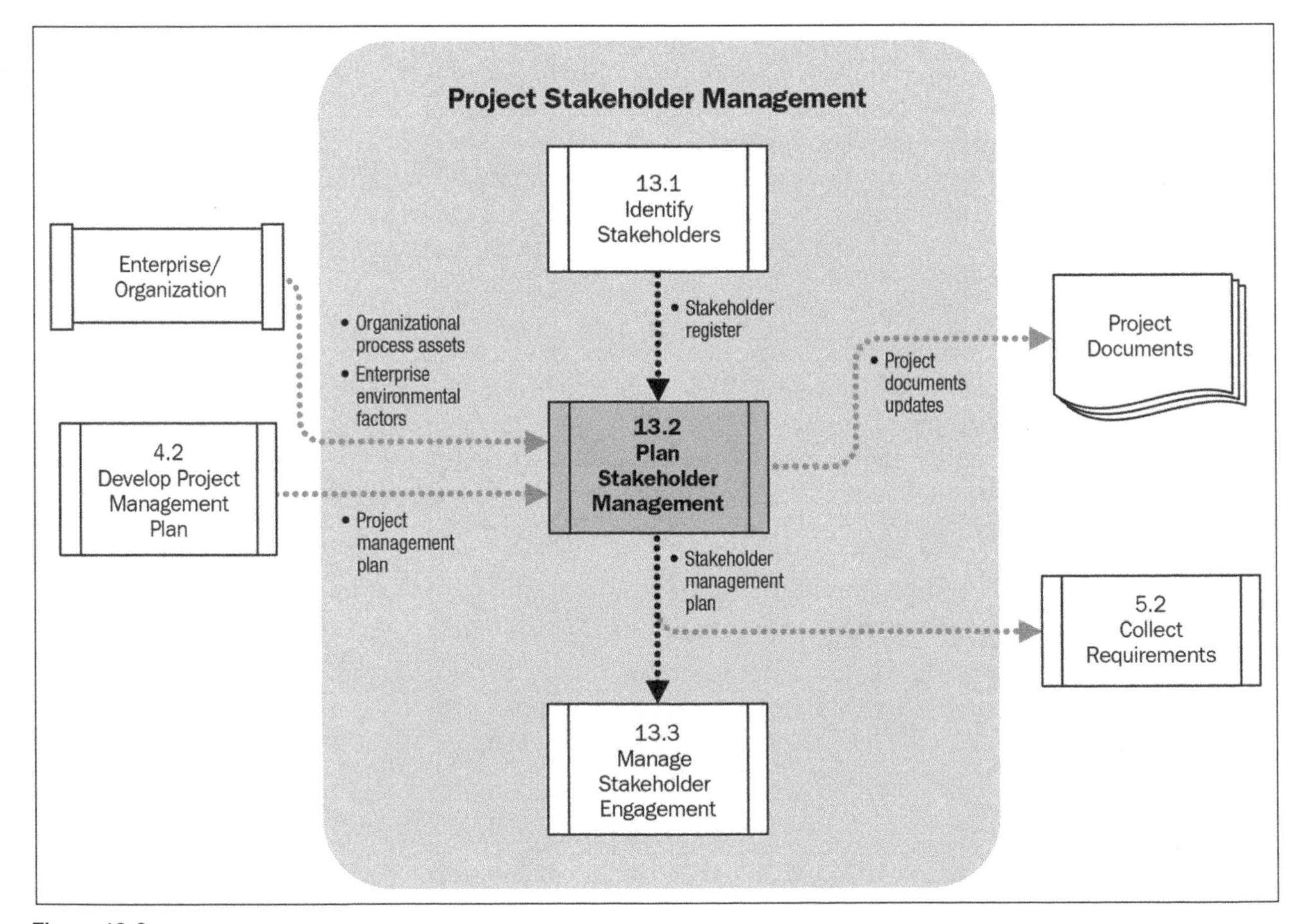

Figure 13-2
Plan Stakeholder Management Data Flow Diagram
Source: *PMBOK® Guide*—Fifth Edition

- The communication management plan defines the types, methods, frequency, and content of communications that will be delivered to stakeholders. It might outline newsletters, status reports, presentations, meetings, and the like, which are used to meet the communication needs of various stakeholders.
- The human resources plan contains the information on how the human resource requirements will be met, roles and responsibilities, and the reporting structure.
- The project approach describes how project execution will accomplish the project objectives. It has the lifecycle and may have the information needs by stakeholder broken out by lifecycle phase.

Any of the enterprise environmental factors impact how you plan to manage stakeholders. You need to know the company culture, the available systems, industry standards, personnel administration, and so on, in order to define a workable plan to manage stakeholders. Any organizational process assets, such as information on what has worked in the past to manage stakeholders, are also useful.

TOOLS AND TECHNIQUES

Meetings are a common way to call on expert judgment to categorize stakeholders and determine the best way to manage them. Meetings may entail discussion, brainstorming, problem solving, or analytical techniques. The *PMBOK® Guide*—Fifth Edition describes a stakeholder engagement assessment matrix, as shown in Figure 13-3, to categorize each of the stakeholders by their level of awareness and support for the project. The matrix is used to plot where the team wants the level of engagement for each stakeholder to be, and where it is currently. The five levels of engagement are:

Unaware. Unaware of both the project and the impacts.

Resistant. Aware, but not supportive of the project.

Neutral. Aware, but neither supportive nor against the project.

Supportive. Aware, and will support the project.

Leading. Aware of the project and actively engaged in supporting the project.

Stakeholder	Unaware	Resistant	Neutral	Supportive	Leading
Stakeholder 1	C			D	
Stakeholder 2			C	D	
Stakeholder 3				D C	

Figure 13-3
Stakeholder Engagement Assessment Matrix
Source: *PMBOK® Guide*—Fifth Edition

The C in this matrix indicates the current level of engagement and the D indicates the desired level of engagement. The gap in the levels of engagement will feed into the stakeholder management plan.

OUTPUTS

The stakeholder management plan describes how the team will move each stakeholder from the current level of engagement to the desired level of engagement. Part of the plan will include communications from the communications management plan. However, there are other elements in the stakeholder management plan including:

- Interrelationships among stakeholders
- Stakeholder needs by lifecycle phase
- Other strategies for engaging stakeholders effectively

Project documents that are updated include the stakeholder register and the project schedule. Be sensitive to the information in the stakeholder management plan. There may be information, particularly about unsupportive stakeholders, that could be potentially damaging or embarrassing.

Chapter 14

Executing the Project

TOPICS COVERED

Executing Process Group

Direct and Manage Project Work

Executing Process Group

The Executing Process Group consists of those processes performed to complete the work defined in the project management plan to satisfy the project specification.

The executing processes are where the majority of the project work is done and as such, the majority of the funds will be expended here. Because the executing processes are where the product, service, or result is developed, the processes are very dependent on the nature of the project, therefore much of the information in the *PMBOK® Guide* will be project-centric as opposed to product-centric for these processes.

As the team conducts the work on the project, performance may require change requests to align the work with the baseline plans. This can include corrective and preventive actions as well as defect repair.

Corrective Action. An intentional activity that realigns the performance of the project work with the project management plan.
Preventive Action. An intentional activity that ensures the future performance of the project work is aligned with the project management plan.
Defect Repair. An intentional activity whose purpose is to modify a nonconforming product or product component.

Elements of the project management plan and project documents are updated as the work is performed and completed.

A big part of conducting the project work is developing and managing the team and other stakeholders. Much of the project manager's time is spent communicating, managing issues, attending meetings, coordinating work, and resolving conflicts.

The scope, schedule, cost, and risk-executing processes are unified in the Direct and Manage Project Work process. The other knowledge areas have one or more executing processes.

Direct and Manage Project Work

Direct and Manage Project Work is the process of leading and performing the work defined in the project management plan and implementing approved changes to achieve the project's objectives. This process is where the bulk of the product work is focused. This is where all the plans are carried out to produce the project deliverables. The project manager is busy managing the work of the project team. Typical activities include:

- Implementing the project methodology and following the project lifecycle
- Staffing and managing the project team
- Acquiring and utilizing resource materials, supplies, and equipment
- Creating deliverables that meet project requirements
- Managing project communication
- Managing stakeholder engagement
- Generating information about project status
- Integrating approved changes into the project management plan and the project documents
- Managing risks and implementing responses
- Collecting lessons learned and continuously improving the team's processes
- Integrating work from all executing processes

Figure 14-1 shows the inputs, tools and techniques and outputs for the Direct and Manage Project Work process. Figure 14-2 shows a data flow diagram for the Direct and Manage Project Work process.

INPUTS

The project management plan and any approved change requests to either the project management plan, project documents, or product information is the foundation for doing the project work. The scope management plan, requirements management plan, schedule management plan, and the cost management plan are guiding components of the project management plan while executing the

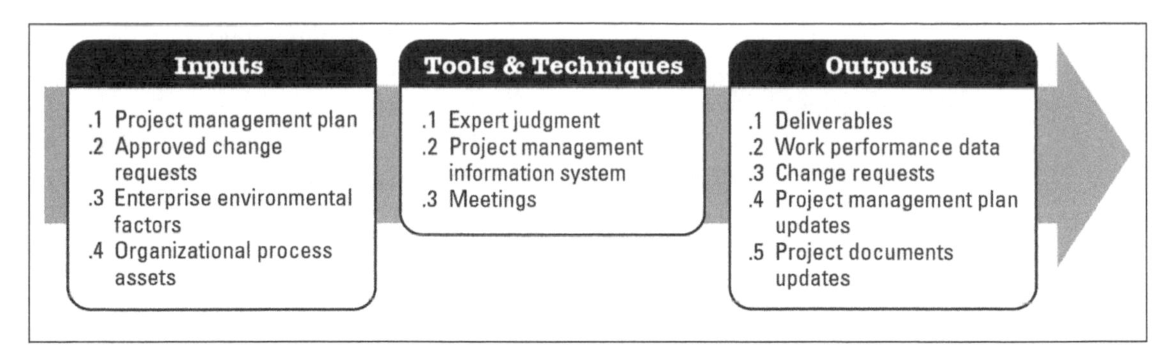

Figure 14-1
Direct and Manage Project Work: Inputs, Tools and Techniques, and Outputs
Source: *PMBOK® Guide*—Fifth Edition

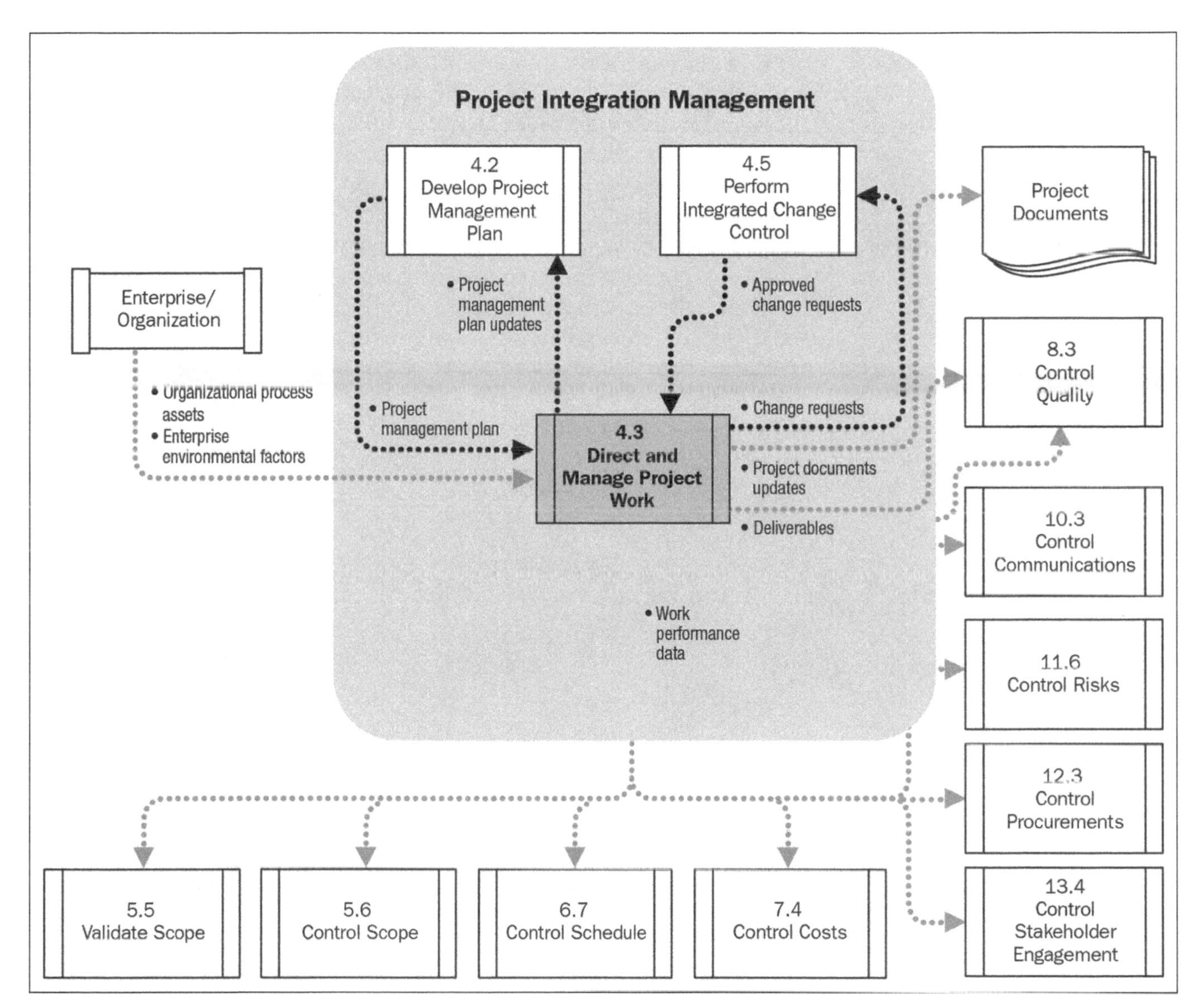

Figure 14-2
Direct and Manage Project Work Data Flow Diagram
Source: *PMBOK® Guide*—Fifth Edition

project work. These subsidiary management plans give direction and provide the roadmap for the work that needs to be done and the way it should be conducted.

It is in this process that enterprise environmental factors will have the greatest influence on the project. For example, the company culture, the information management systems, HR policies, and the organization's infrastructure all influence and constrain the way the project work is accomplished.

Many of the organizational process assets are used when performing the project work. Examples include work authorization guidelines, defect management processes, standards and guidelines, policies and procedures, and information from previous projects.

TOOLS AND TECHNIQUES

Expert judgment on the product side comes from the people doing the work—the people with the skills and abilities to produce the deliverables. Expert judgment on the project side comes from the project manager and the project management team. Project expert judgment is used to provide guidance and oversight to the product work.

The project management information system is used as a tool here to enter, store, manage, and analyze project information. An example is the scheduling and budgeting systems. While the systems themselves are enterprise environmental factors (as an input) the data in them is used as a tool in performing the project.

Meetings are an important part of project execution. Typical project meetings include:

- Weekly status meetings
- Review meetings
- Risk management meetings
- Meetings with stakeholders
- Brainstorming or problem solving sessions

OUTPUTS

Finally . . . a deliverable! A deliverable is a unique and verifiable product, result, or capability to perform a service. In other words, it is the final result, or a discrete component of the result of the project. Work performance data is generated as a result of this process. The data includes raw data on the work that has been accomplished thus far, including deliverable status, activity start and finish dates, and costs incurred to date.

Based on how the work is progressing, there may be change requests for either the product or the project. Product changes will show up in the requirements documents, the WBS, and even the quality metrics. Project changes can include schedule, budget, resource, and policy changes, just to name a few. The nature of the changes can include a different way of doing something; or it can include corrective action to stay aligned with the project plan; preventive action to reduce the likelihood of going off target or incurring a risk, or a defect repair, if there is an error in the product or process.

Change requests will need to be formally documented and go through the Perform Integrated Change Control process. Generally, a change request form is used to document the proposed change and the reason for the change. Refer to the following list for common elements in a change request form. Many times the change request form also has an area to record the disposition of the change request and the justification.

CHANGE REQUEST FORM CONTENTS

Change category

Detailed description of the proposed change

Justification

Impacts of change on scope, quality, requirements, schedule, and cost

As the work on the project progresses, the project management plan will be updated along with various project documents. Any of the subsidiary plans in the project management plan may be updated to either indicate progress or to implement revised approaches to the project. Much of the information in the project documents is expected to be dynamic, such as the assumption log, decision log, issue log, and the like. This information is updated throughout the project.

Chapter 15

Executing Quality Management

TOPIC COVERED

Perform Quality Assurance

Perform Quality Assurance

Perform Quality Assurance is the process of auditing the quality requirements and the results from quality control measurements to ensure appropriate quality standards and operational definitions are used.

There are several goals for the Perform Quality Assurance process:

- Provide confidence that the deliverables will meet the requirements and stakeholder expectations
- Ensure the project is following the quality policies, procedures, and plans as documented in the quality management plan
- Determine if the identified quality policies, procedures, plans, and metrics are sufficient to meet the product and project objectives
- Improve processes as appropriate to improve overall project performance

Figure 15-1 shows the inputs, tools and techniques and outputs for the Perform Quality Assurance process. Figure 15-2 shows a data flow diagram for the Perform Quality Assurance process.

Performing quality assurance is really the quality management part of the project. The planning process determines the processes and metrics that will be used, the control process measures against the metrics, and quality assurance oversees the overall quality management and improvement process to ensure optimum performance.

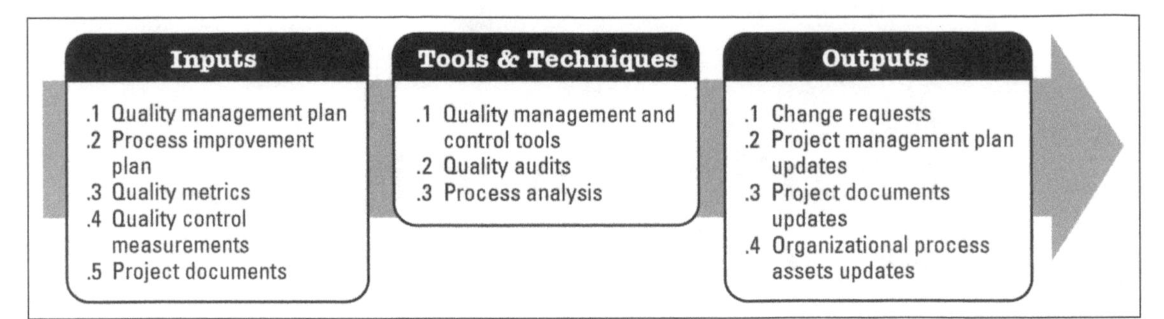

Figure 15-1
Perform Quality Assurance: Inputs, Tools and Techniques, and Outputs
Source: *PMBOK® Guide*—Fifth Edition

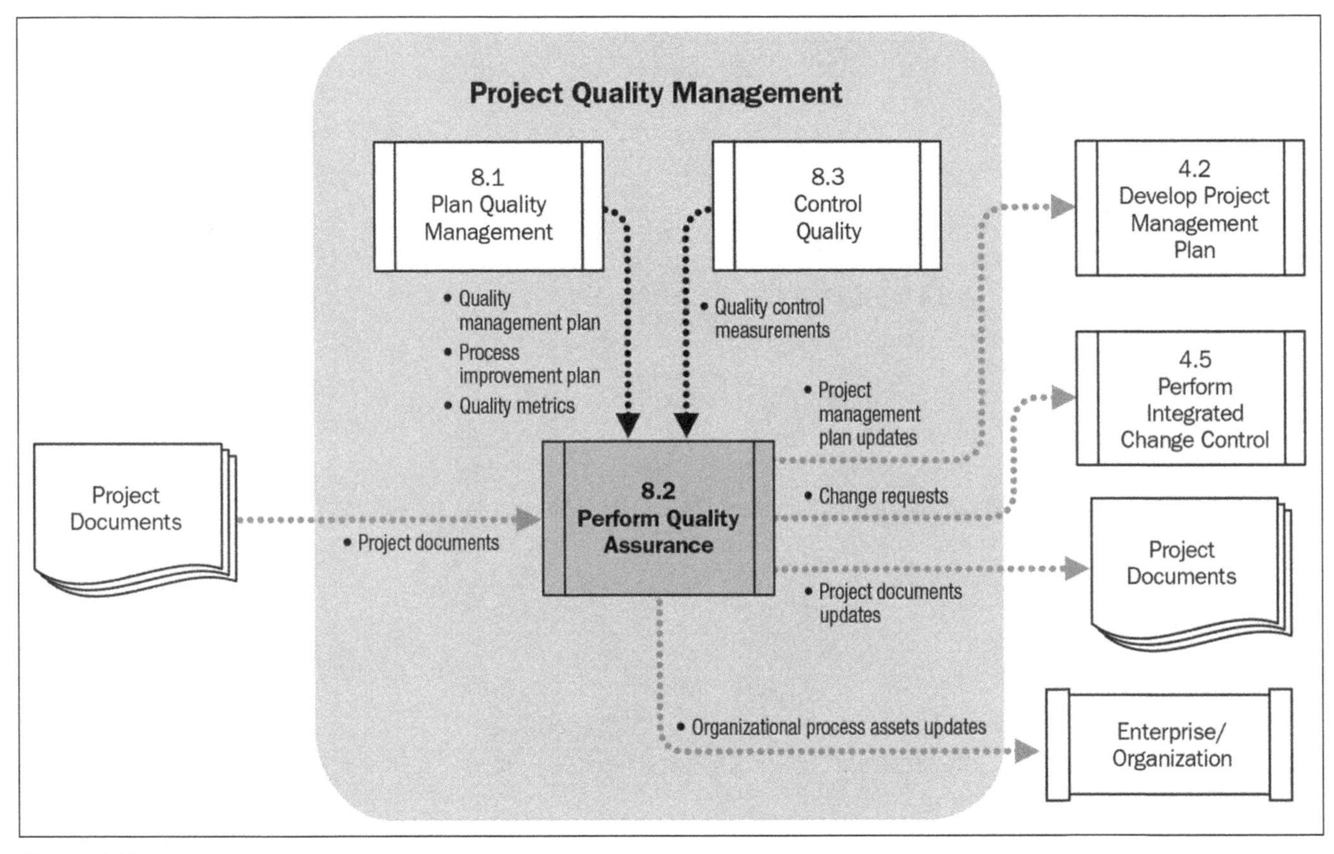

Figure 15-2
Perform Quality Assurance Data Flow Diagram
Source: *PMBOK® Guide*—Fifth Edition

INPUTS

The quality management plan and the process improvement plan contain approach, processes, policies, and procedures that will be used to ensure quality and, if appropriate, process improvement activities.

Quality metrics include the measurements that are used to measure quality for both the project and the product. They include measurements for performance, tolerance, acceptance, and the tools and techniques that are used to measure them.

Quality control measurements are the results of applying the quality metrics to the deliverables and project performance as outlined in the quality management plan.

Project documents can include any project document. You might expect to reference requirements documentation, the stakeholder register, roles and responsibilities, and work performance information (status reports in particular) while performing quality assurance activities.

TOOLS AND TECHNIQUES

A quality audit is the main technique used to determine the effectiveness of the quality management plan. It compares the project and organizational quality policies, processes, and procedures to the actual actions on the project. The audit determines whether the quality management plan is being followed and, if it is being followed, whether or not it is effective.

In some cases, the quality plan is being followed, but the results are not acceptable. In this situation the auditor would look at the work performance information, the metrics, and the quality control measurements to determine whether or not the metrics are appropriate and if the measurement methods are appropriate, whether or not they are being conducted correctly.

The quality audit should document both areas for improvement in the processes as well as those areas that are performing well. It is not the intent of the audit to only highlight negative results—both good practices and those that need improvement should be documented.

The tools and techniques that are used in the Plan Quality Management and the Perform Quality Control processes may be used in this process to determine the performance of the quality processes and to determine if they are the appropriate techniques to use to achieve the desired results. The quality planning and control tools include:

- Control Charts
- Benchmarking
- Design of Experiments
- Statistical Sampling
- Flowcharts
- Histograms
- Cause and Effect Diagrams
- Pareto Charts
- Run Charts
- Scatter Diagrams

Figure 15-3 shows a sample of the seven quality management and control tools.

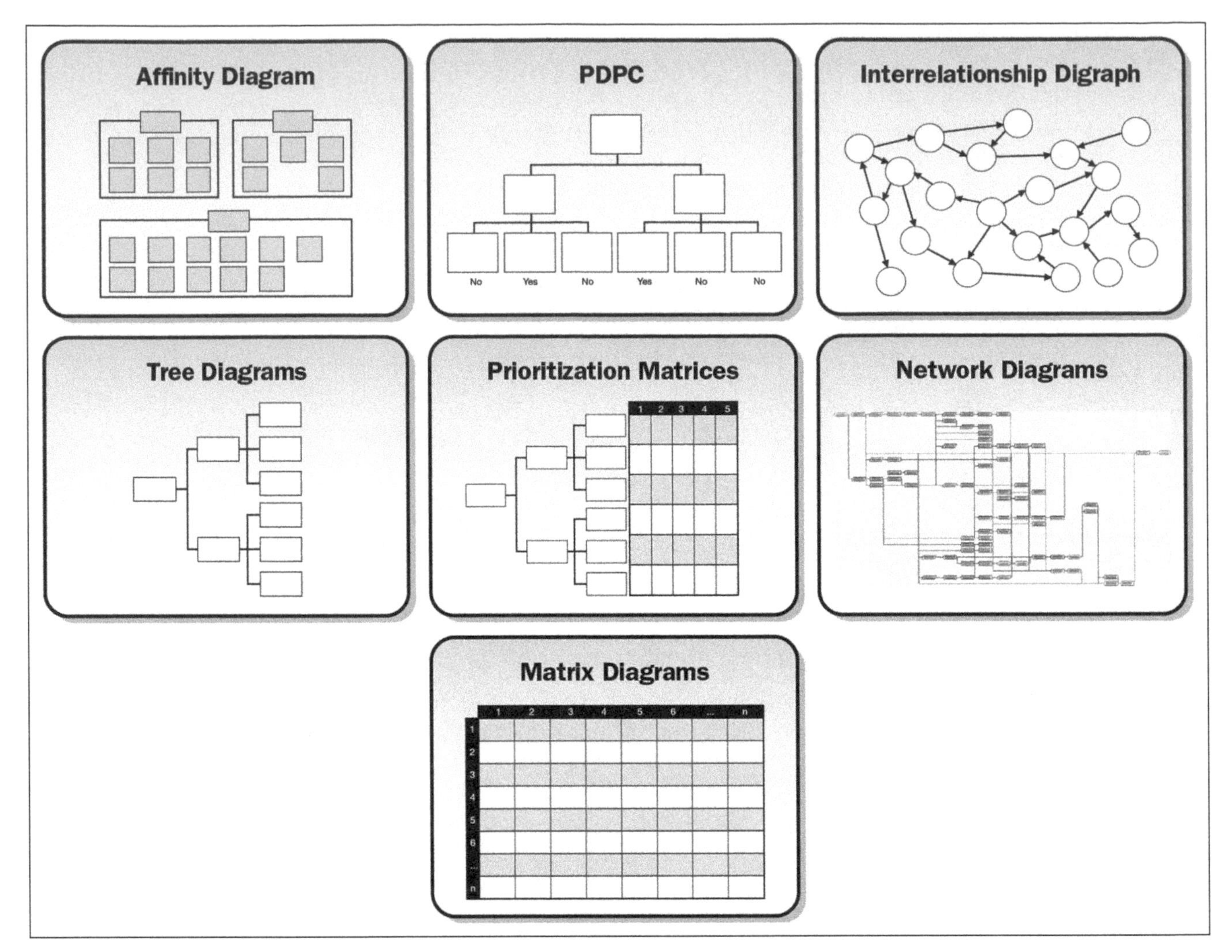

Figure 15-3
Storyboard Illustrating the Seven Quality Management and Control Tools
Source: *PMBOK® Guide*—Fifith edition

The *PMBOK® Guide* – Fifth Edition lists additional quality management and control tools. The tools shown in Figure 15-3 are most frequently used on a Six Sigma project or on highly technical projects.

Additional Quality Management and Control Tools

Additional quality management and control tools include:

- Affinity diagrams
- Process decision program charts
- Interrelationship diagraphs
- Tree diagrams
- Prioritization matrices
- Activity network diagrams
- Matrix diagrams

Process analysis is used when looking for ways to improve processes. This can include looking at bottlenecks in a process, redundancies, gaps, and any other problems. Many times a root cause analysis is conducted to determine how problems start and the factors that contribute to them.

OUTPUTS

Depending on the results of the audit, the team may have change requests that would go through the Perform Integrated Change Control process. Change requests can include corrective actions, preventive actions, and defect repairs. These may need to be integrated into the project management plan as updates to the schedule, quality management plan, or other components of the project management plan. Project documents that would be updated include process documentation, requirements, and training plans. In addition, there may be a need to change or update some of their organizational process assets, such as quality management processes, policies, procedures, or standards.

Chapter 16

Executing Human Resource Management

TOPICS COVERED

Acquire Project Team

Develop Project Team

Manage Project Team

Acquire Project Team

Acquire Project Team is the process of confirming human resource availability and obtaining the team necessary to complete project activities. As a project transitions from planning to conducting the work and developing a product, the project management team begins to bring on project staff. On small projects the planning staff and the project team may be the same. But consider when you have hundreds of people who will be working over a period of several years. The staffing management plan may say that you need to have 25 engineers as the project begins. It can be very challenging to move from 0 to 25 engineers in a week. Often, projects get behind from the very beginning because there is a delay in the time when staff is scheduled and needed and when they are actually available.

Sometimes less qualified staff will be brought onto the project, or overtime will be approved to maintain schedule performance. Of course, this can have an impact on quality and cost performance. These are some of the challenges involved in acquiring the project team.

Figure 16-1 shows the inputs, tools and techniques and outputs for the Acquire Project Team process. Figure 16-2 shows a data flow diagram for the Acquire Project Team process.

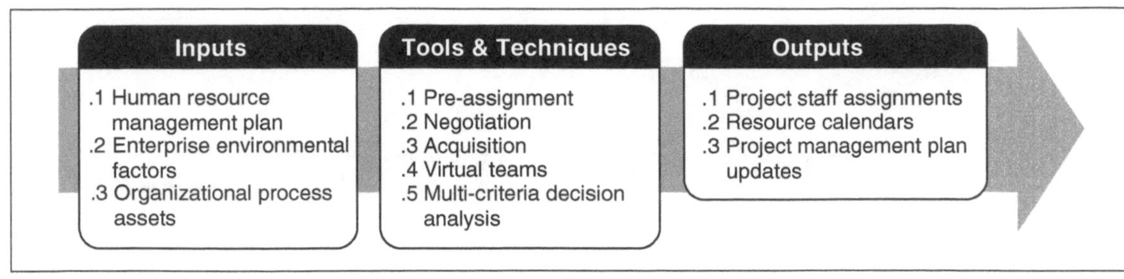

Figure 16-1
Acquire Project Team: Inputs, Tools and Techniques, Outputs
Source: *PMBOK® Guide*—Fifth Edition

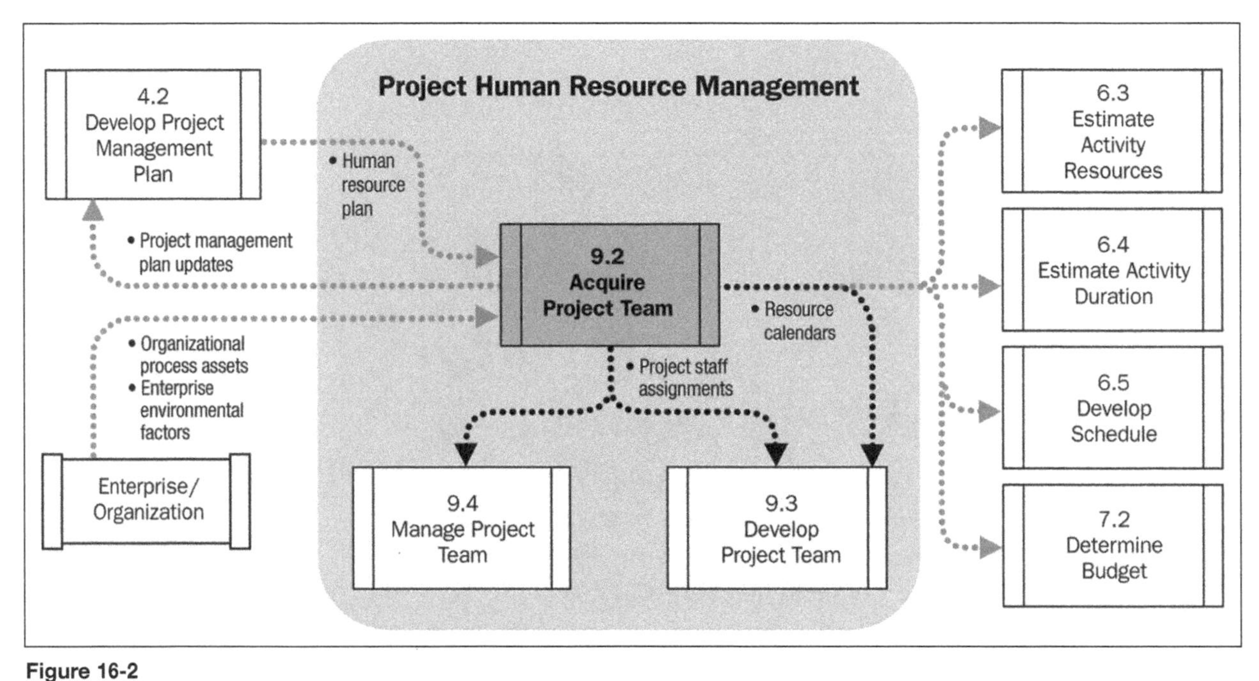

Figure 16-2
Acquire Project Team Data Flow Diagram
Source: *PMBOK® Guide*—Fifth Edition

INPUTS

The human resource management plan describes how the project will be staffed, the skills and competencies needed, and how the staff will be acquired. It can also address how new staff will be initiated into the project and any steps necessary to bring them on board. For example, if the staff is from an external source, how does the project manager go about getting them access to the organization's information system? Do they need a badge? Where will they be located? This information should be defined when planning the project so the proper mechanisms are in place when people start coming on board.

Enterprise environmental factors include the organization's existing staff and skill sets, the availability of skills in the marketplace, the organization's structure (whether it is a functional, matrix, or project-centric organization) and policies regarding bringing in external workforce. Another aspect to keep in mind is geography. If the project will be completed onsite in one location the logistics are much easier. However, if there are multiple locations throughout the country, or even throughout the world, staffing the project becomes more complicated. In this case you will have to consider how the team will interact virtually, if everyone needs to be in some kind of office or, if people can work from home; how often to meet face-to-face versus using electronic meetings; how time differences impact the work; and so forth.

Organizational process assets that can help in these situations include procedures for bringing outside staff on board, such as getting contractor badges, laptops, cube space, and so on. Information on how past projects have managed the process can help you avoid making mistakes that someone else has already made.

TOOLS AND TECHNIQUES

In some situations, such as when an organization wins a contract to work on a project, resources are promised as part of the bid response. This is a case of pre-assignment. Another situation that necessitates pre-assignment is when there is work that has to be done and there are limited resources that have the skills to perform the work. This usually introduces risk into the project because those limited resources are generally in high demand. If your project is not the most important one to the organization, you may experience a schedule slip if the limited resource is assigned to a more important project.

In many cases, such as in a matrix organization, you will have to negotiate for the best resources for your project. Most organizations have more projects than they do available resources. So you will be competing for resources with other project managers and functional managers. Many times the "best" resources are already assigned to one or more projects.

Acquisition is used to fill in the gaps in skills and availability. In certain instances the decision is made to subcontract work to another organization. In these cases you need to make sure all the contracts and agreements are in place through the procurement processes. In other cases you are bringing on labor for a specified period of time. In either instance, you will want to ensure you have the correct materials, equipment, site access, and reporting mechanisms in place for the work and interface to go smoothly.

For projects that have multiple locations, managing virtual teams is a necessity. Even for projects in one location, you may have people who work at home, so having processes established for virtual communication is important.

Multicriteria decision analysis, as used as a tool for acquiring team members, analyzes multiple variables to determine the best

GETTING THE RIGHT RESOURCES

Many times the highest skilled resource is not the best resource for your project. Consider the following scenarios:

1. You have some work to do that is not terribly complex; in fact, it is fairly standard work that almost anyone who has basic skills in a particular role can perform.
2. You have work that is not on the critical path and has quite a bit of float.

In both of these instances using someone who is a junior level or mid-level employee as opposed to an expert would be a better option. In the case where the work is fairly standard, the expert would be bored, and there are probably other places where their expertise could be better used. In the case where there is a lot of float, why not use the opportunity to train someone and develop their skills? Maybe you use a more skilled person to review their work, but let them do the majority of the effort. They will end up with better skills and probably cost less than the expert!

VIRTUAL VERSUS IN PERSON

It is a good idea to have some type of regular face-to-face interaction, even if it is only a few times a year. Nothing can replace the relationships you build when you can sit across from someone and solve a problem or discuss an issue. Telephone, web conferencing, emails, and faxes are good for addressing routine situations, but to build a team and to solve complex problems, being in a room together is best.

Table 16-1 Multicriteria Decision Analysis for Resources

	Weight	Myra	Mary	Bob	Dan	Barbara
Availability	40					
Cost	15					
Experience	15					
Skills	30					

resource to fill a position. The team identifies the variables they need to evaluate and establishes a weighting system based on the importance of the variable. Resources are then ranked according to the scoring system. Table 16-1 shows a multicriteria decision analysis using availability, cost, experience, and skills as the variables being evaluated.

OUTPUTS

Project staff assignments document how the project is actually staffed. The staffing management plan identifies the needs and the staff assignments show how those needs will be met and by whom. Once the positions are filled you should incorporate each person's availability into a resource calendar. Some people may work on the project part-time, others may work 80 hours in nine days instead of ten. Different organizations may have different work holidays, or different work shifts. Having the actual availability will make the scheduling process more accurate.

The project management plan may be updated based on staff assignments—in particular, the human resource management plan and the schedule. The human resource plan may need to be updated to include more or less training, fewer or more resources, virtual team logistics, and other aspects that will need to be adjusted based on the actual committed resources. The schedule may need to be adjusted based on availability as well.

Develop Project Team

Develop Project Team is the process of improving competencies, team interaction, and overall team environment to enhance project performance. We will address each of these aspects of developing the project team in this section. This process is separate from the process on managing the project team because the focus and techniques for developing a team are different than for managing a team. In developing the project team, you are looking to develop individual skills and competencies, as well as a functional team environment. When we look at managing the project team, we will look at ways to facilitate the team in delivering the project results.

There are some similarities in that both processes rely heavily on interpersonal skills.

Figure 16-3 shows the inputs, tools and techniques and outputs for the Develop Project Team process. Figure 16-4 shows a data flow diagram for the Develop Project Team process.

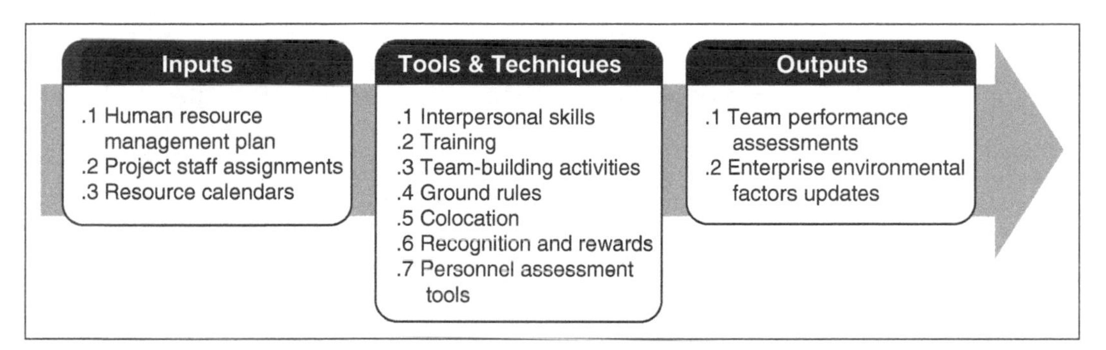

Figure 16-3
Develop Project Team: Inputs, Tools and Techniques, Outputs
Source: *PMBOK® Guide*—Fifth Edition

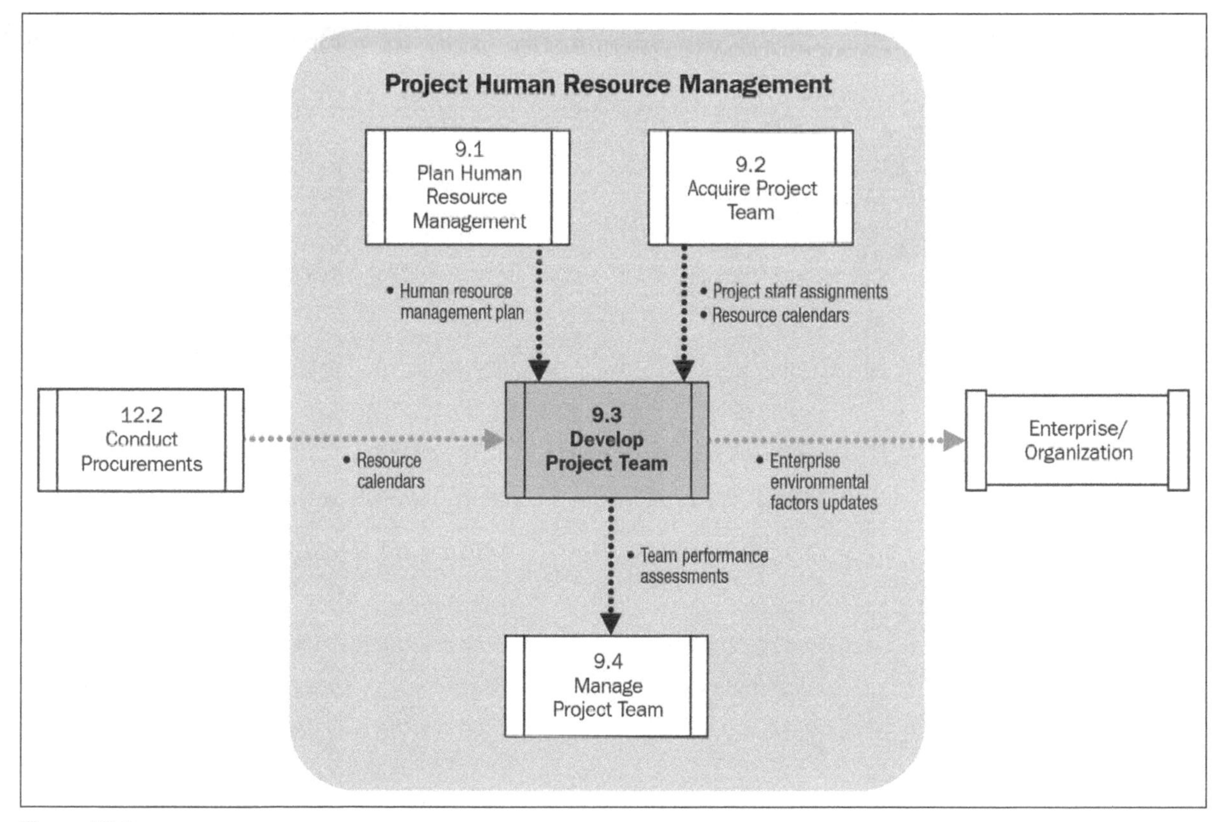

Figure 16-4
Develop Project Team Data Flow Diagram
Source: *PMBOK® Guide*—Fifth Edition

INPUTS

The project staff assignments and the resource calendars that are outputs from acquiring the team feed into developing the team as they indicate who will be participating and when they will be involved. The human resource management plan has information regarding recognition and rewards, team building plans, training needs, and organization-specific information on developing the team.

TOOLS AND TECHNIQUES

The interpersonal skills the project manager uses most in this process are leadership, motivation, and team building. Let's take a look at each of these a little closer.

Leadership is the ability to articulate the vision of the project, enroll others in that vision, and lead them in the endeavor. Effective leadership lets the team know where they are going and that the project manager can help them get there. Another key aspect is that the team members need to trust the project manager. They need to trust the project manager's skills and abilities, but also that the project manager will empower them, remove barriers, and do everything in their ability to make sure the team "wins." Leadership also includes making decisions, course corrections, and providing feedback on team performance.

Motivation is incentivizing people to perform certain actions. Not everyone is motivated by the same things. Some people are motivated by a challenging work environment, recognition, and contribution, others are motivated by contributing to a team or to a cause they believe in. There are lots of theories on motivation; some of the more common ones are highlighted in the following features.

Herzberg's Theory of Motivation

Frederick Herzberg believed there were two aspects to the work environment: hygiene and motivation. He postulated that hygiene factors do not motivate a worker to perform. However, the way they are implemented, or not implemented, may lead to employee dissatisfaction. Motivation factors, such as recognition, lead to higher individual performance. Table 16-2 shows examples of hygiene and motivation factors.

Table 16-2 Hygiene and Motivation Factors

Hygiene Factors	Motivation Factors
Policies	Achievement
Administration	Recognition
Working conditions	Growth
Salary	Advancement
Status	Interest in the job
Supervision	Job challenge
Security	

Theory X and Theory Y

Theory X states that management believes that workers will do as little as possible to get by, and thus need a great deal of direction. Theory Y states that management believes that workers are interested in doing their best and, given the freedom, will perform well. Table 16-3 compares the two.

Table 16-3 Theory X and Theory Y

Theory X	Theory Y
The average worker has an inherent dislike of work and will avoid it if possible.	The average worker wants to be active and finds the physical and mental effort on the job to be satisfying.
Because of their dislike for work, most people must be controlled before they will work hard enough.	The greatest results come from willing participation, which will tend to produce self-direction toward goals without coercion or control.
The average worker prefers to be directed and dislikes responsibility.	The average worker seeks opportunity for personal improvement and self-respect.
The average worker is not ambitious, and desires security above everything.	Imagination, creativity, and ingenuity can be used to solve work problems by a large number of employees.

There are many additional theories of motivation, but these are two of the better known ones.

Team building is getting individuals to work together toward a common goal. Trust and respect are important factors in team building. The intended outcome is for the team to develop a synergy and a relationship where problem solving and project work are performed more effectively and efficiently by the team members. Team building takes time. Some project managers include team building activities as part of their project meetings, or they build in time and budget for offsite activities. To form an effective team, people need to know the other team members and understand how to work well with them. Bruce Tuckman, a psychologist who conducted research in group dynamics, developed a ladder that describes the various stages of team building.

Stages of Team Building

Phase	Description
Forming	The team first comes together. They get to know each other's name, position on the team, department, and other pertinent background information. This may occur in the kickoff meeting.

Phase	Description
Storming	Team members jockey for position on the team. This phase is where people's personalities, strengths, and weaknesses start to come out. There may be some conflict or struggle as people figure out how to work together. Storming may go on for some time, or may pass relatively quickly.
Norming	The team starts to function as a team. At this point, everyone knows their place on the team and how they relate to and interface with all the other members. They are starting to work together. There may be some bumps in the road, but these are resolved quickly and the team moves into action.
Performing	The team becomes operationally efficient. This is the mature team stage. Teams that have been together for a while are able to develop a synergy. By working together they accomplish more than other teams, and produce a high-quality product.
Adjourning	The team completes the work and moves on to other projects. If the team has formed a good relationship some members may be sad about leaving the team.

Training is usually used to develop individual technical skills; though in some cases, team members can benefit from training in interpersonal and leadership skills as well.

Ground rules are useful to establish in the beginning of the project to determine expectations on how people will interact with each other. Some key items to include are how decisions will be made, how the team will communicate with each other, and how conflicts will be addressed. Some teams use a Team Operating Agreement to document their team approach. See Appendix A for an example of a Team Operating Agreement.

Co-location can help build team camaraderie by having all the team members in the same location. Co-location facilitates communication, problem and issue resolution, and relationship building. If it is not feasible to have the team all in one place, sometimes having a team meeting room where the team has their status meetings and stores project documentation is a good substitute.

Recognition and rewards should be based on what motivates people. Remember we talked about motivating factors earlier? Some people will be thrilled with being mentioned in a company newsletter as doing a great job; other people would rather have the opportunity to have more responsibility. Be careful to reward behavior you want to see more of, such as meeting a milestone because the work was well planned and executed, not because heroics and overtime were needed to meet it because the work wasn't well planned.

Personnel assessment tools can be used for self-assessment or for the team and project manager to understand each other better. Assessment tools are available to identify leadership strengths and weaknesses, emotional intelligence, personality traits, management styles, and almost anything else you want to know about your fellow team members. Working on and discussing the outputs of individual assessments can provide insights that help the team work better together and help individual team members understand their own strengths and weaknesses.

OUTPUTS

Team performance assessments are used to assess the performance of the team as a whole. Team performance assessments can be useful at the end of a project phase to plan how to improve performance in future phases. Whether the project manager provides individual team member performance evaluations will depend on the organizational structure and policies. Even if the organization does not seek formal input by project managers on a team member's performance, the project manager can write an informal assessment and give it to the functional manager and copy it to the Human Resources department. This is a good practice if a team member has done an exceptionally good job on the project and you want to make sure their performance is recognized.

Enterprise environmental factors include updating training and personnel records.

Manage Project Team

Manage Project Team is the process of tracking team member performance, providing feedback, resolving issues, and managing team changes to optimize project performance. Managing the project team effectively requires a blend of project skills and interpersonal skills.

Figure 16-5 shows the inputs, tools and techniques and outputs for the Manage Project Team process. Figure 16-6 shows a data flow diagram for the Manage Project Team process.

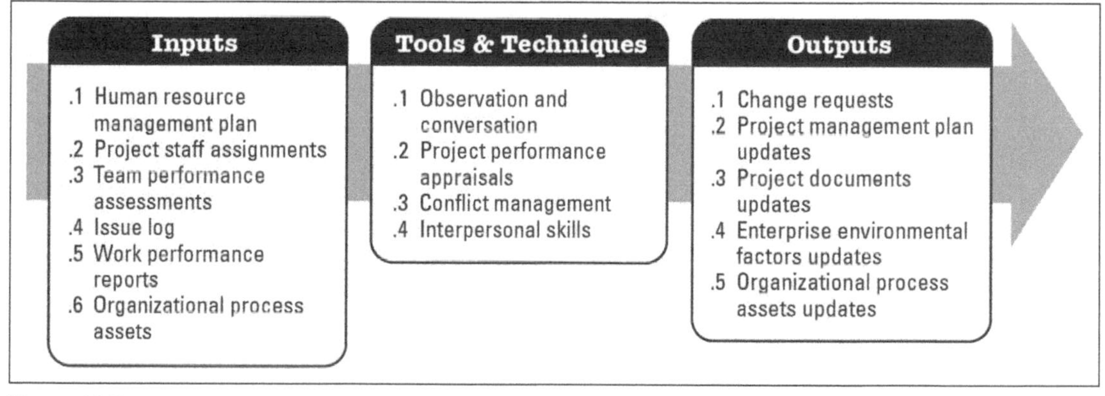

Figure 16-5

Manage Project Team: Inputs, Tools and Techniques, Outputs

Source: *PMBOK® Guide*—Fifth Edition

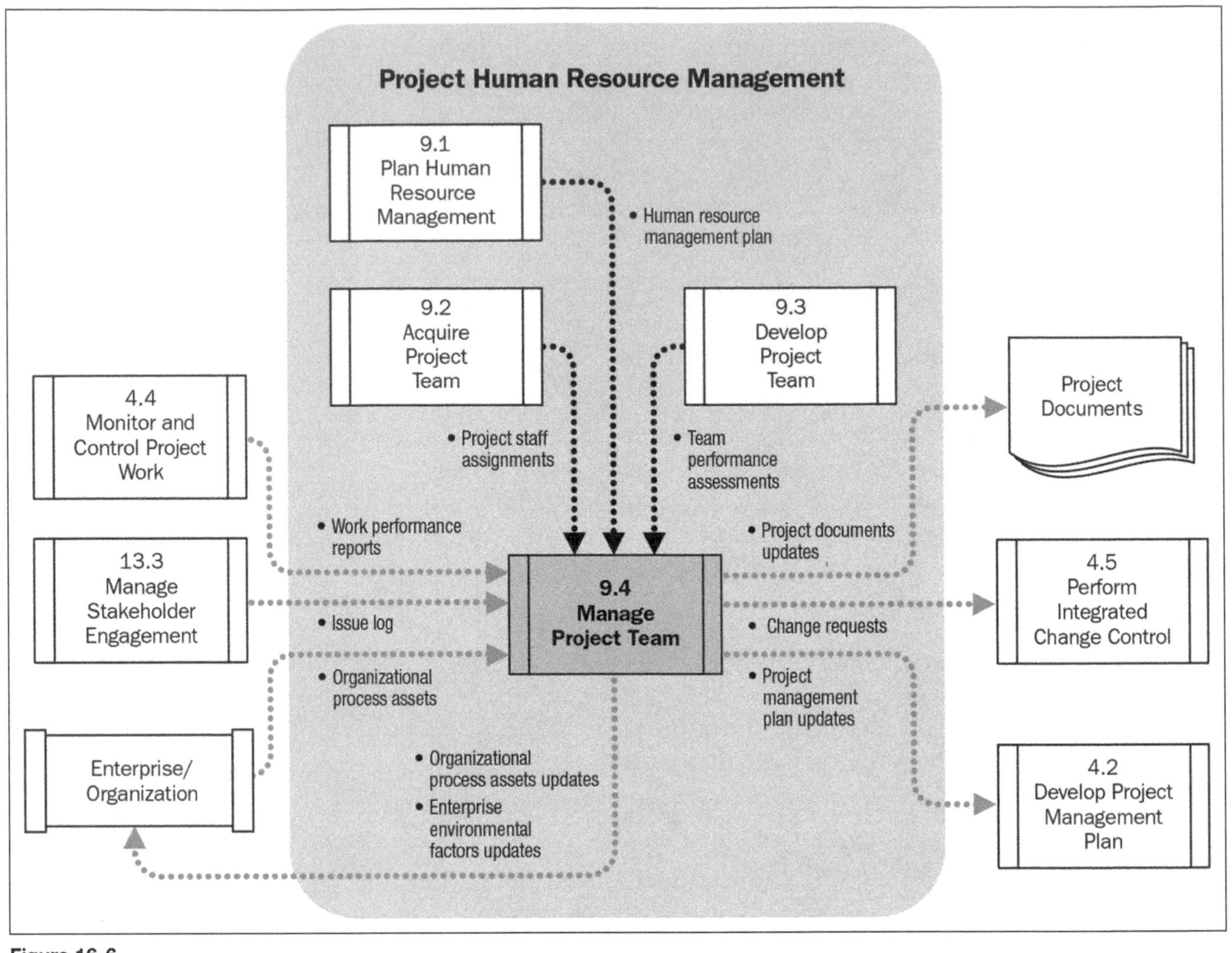

Figure 16-6
Manage Project Team Data Flow Diagram
Source: *PMBOK® Guide*—Fifth Edition

INPUTS

Project staff assignments list the people who are assigned to the project team. The human resource management plan defines roles and responsibilities, the project organization, and the staffing management plan. Together these two documents give an overview of the staffing profile for the team.

Team performance assessments provide information on how well the team is performing collectively. The work performance reports provide insight into project performance for scope, schedule, cost, and quality. Both the assessments and the performance reports help the project manager determine where he or she should focus his or her attention to improve performance at the team level and the individual resource level.

An issue log is used to document situations that need attention on the project. This can include a risk event that has occurred, a decision that has to be made, or a situation that must be resolved before the project can move forward. In general, an issue is some event that needs input or participation by one or more stakeholders to resolve. The issue log helps track this information along with the person responsible for taking action about the issue and the target date for resolution.

Organizational process assets can include the standard policies and procedures, but it can also include organization-specific benefits such as company logo wear, bonus structures, team newsletters, and other such assets.

TOOLS AND TECHNIQUES

Observation and conversation are informal ways of staying in touch with the project team. They will provide the project manager information about how the project is going that is not visible in performance reports. Information about team morale, challenges, working conditions, relationship issues, and the like are more obvious when you are walking the halls, having a conversation in a team member's cube, or just chatting informally.

Depending on the length, complexity, and organizational structure the project manager may be expected to conduct project performance appraisals on team members. Performance appraisals are a more formal approach than the performance assessments mentioned in the Develop Project Team process. A performance appraisal will generally include areas for improvement, goals for development, and feedback on performance.

Conflict is natural, especially on projects. Common sources of conflict include:

- Schedules
- Resources
- Technical approach
- Priorities

Some people get uncomfortable around conflict. Handled appropriately and respectfully, however, conflict can lead to better alternatives and solutions. There are a number of methods you can use to manage conflict depending on the urgency, the relative power of the people involved, and the importance of maintaining a good working relationship. Let's look at some of the methods and the optimum situations for using them in Table 16-4.

Table 16-4 Conflict Management Techniques.

Technique	Situation
Problem solving Treating the conflict as a problem and working together to resolve it	• If there is time and trust • When the objective is to learn • If you have confidence in the other's ability • When you need a win-win

(continued)

Table 16-4 (continued)

Technique	Situation
Collaborating Incorporating multiple viewpoints and coming to consensus on a solution	• If you want to maintain and build good relationships • When you need buy-in to the solution • If the situation is complex • When you need the best possible resolution
Compromise Finding a way for each party to come away with something of value	• When both parties need to win • When you can't win • Equal relationship • To maintain relationship • When the stakes are moderate • To avoid a fight
Smoothing or accommodating Emphasize areas of agreement instead of areas of disagreement	• To reach an overarching goal • To maintain harmony • When any solution will be adequate • When you will lose anyway • To create goodwill
Forcing Win-lose; imposing the resolution on the other party	• When you are right • In a do-or-die situation • High stakes • To gain power • If relationship is not important • When time is of the essence
Withdrawing or avoiding Retreating from the situation; sometimes used as a cooling-off period	• When you can't win • When the stakes are low • To preserve neutrality or reputation • If the problem will go away

Here is an example of using various types of conflict management.

Conflict Management Techniques

Problem Solving. Joe needs Ben to work an extra week on his project, but Ben is supposed to be moving over to Jennifer's project. Joe, Ben, and Jennifer come up with several options and end up using Ben for two days on Joe's project and, while he is there, he trains someone to perform his role for the rest of the week. Jennifer is able to postpone the start of her activities without impacting the critical path.

Collaborating. A customer has asked for a new feature on a project, after the design has been approved. The customer is willing to pay for the change, the conflict is in finding the best approach to incorporate the new scope. Several team members get together to resolve the conflict. The project manager leads the group in identifying the solution and decision criteria. Then the project manager acts as a facilitator during the brainstorming session, and records all the ideas. When everyone has exhausted their ideas the team weighs the solutions against the decision criteria and comes up with the best solution to the situation.

Compromising. A resource's boss tells you he needs your best engineer on another project starting next week. You push back, stating that this is a critical time in the project and that the engineer is on the critical path. The resource's boss agrees to give you one engineer full-time and another half-time. Both are as qualified as the originally assigned engineer. You agree because you can meet your needs, even though it will take some ramp-up time to train the new people on the particulars of your project.

Smoothing. The project sponsor states that she wants you to take on training a new project manager for the company. You don't really have the time to do this, but because this will ultimately benefit the company, and you can't really tell the sponsor no, you agree.

Forcing. The quality assurance director informs you they are going to do a project audit on your project next week. You tell them that next week is really hectic and suggest an alternative time three weeks in the future. The quality assurance director reminds you that she reports directly to the CEO and tells you to be ready first thing Monday morning.

Withdrawal. You are having a forecasting meeting. Pete and Maggie are both senior cost estimators on your project. They have come up with independent calculations for the estimate at completion and they are 15 percent different. They are currently engaged in a heated argument on whether the Monte Carlo analysis is more accurate than a bottom-up analysis. In the meantime, the rest of the meeting attendees are starting to get very uncomfortable with the escalating conflict. You suggest that you table the discussion until a later time when you can look at the underlying assumptions and everyone will have some time to get some perspective on the situation.

Interpersonal skills are key to successfully managing the team. We will discuss listening and communication skills in Chapter 17, Executing Communications Management, though they are certainly needed in managing team members as well. In this chapter, we will focus on influencing, decision making, and cultural awareness. We discussed leadership in the last process, and while leadership is necessary to manage the team, we will not repeat the information here.

Influencing is finding ways to compel others to take a specific action or behave in a certain way. Many times project managers don't have a great deal of position power, and so we must use our ability to influence people to achieve our objectives. A key part of influencing people is gaining their trust. Another aspect is identifying the benefits to others of helping you achieve the project outcomes.

Decision-making techniques use various models. Some involve consensus, some involve majority or plurality, and some involve command techniques. On projects, gaining consensus or at least consulting with stakeholders usually yields better results, even if the project manager ends up making the final decision. A process that helps improve decision outcomes is getting a clear definition of the problem or decision that needs to be made, then defining the decision-making criteria that will be applied against the alternative options. Engaging the team in providing alternatives and then applying the decision-making criteria can help attain buy-in for the ultimate decision.

Cultural awareness means not only awareness of global diversity, but also of the culture of the organizational environment. Knowing who has power and how to influence them is useful information in managing a project.

OUTPUTS

In the event that the planned resources are not available or are changed in the midst of the project, it may be appropriate to record that via the change control process by filling out a change request. A change to staffing will most likely impact cost, schedule, or quality and the impact should be analyzed.

The human resource management plan, schedule, budget, or other elements of the project management plan may need to be updated. Project documents that may be updated include the issue log, roles and responsibilities, or staffing assignments.

Enterprise environmental factor updates can include input into performance appraisals or human resource skills databases. Updates to organizational process assets might include lessons learned, and techniques or processes used to effectively manage the team. For example, if you developed an effective decision-making process, or a good conflict resolution process, this can be shared with other projects via lessons learned or by recording it in the project documentation.

Chapter 17

Executing Communications Management

TOPIC COVERED

Manage Communications

Manage Communications

Manage Communications is the process of creating, collecting, distributing, storing, retrieving, and the ultimate disposition in accordance with the communications management plan. The main method the project manager uses to accomplish everything in the project is communication. Thus, managing communications occurs from the start of the project to the finish. Communication takes place verbally, in written form, or electronically. It can be formal or informal, planned or spontaneous, intentional or unintentional. The Manage Communications process looks at means and methods of communicating project information. It is the implementation of the communications management plan.

Figure 17-1 shows the inputs, tools and techniques and outputs for the Manage Communications process. Figure 17-2 shows a data flow diagram for the Manage Communications process.

INPUTS

The *communications management plan* identifies the stakeholders who need information, when they need it, and the method that will be used to communicate.

Work performance reports include information on past performance (what has been accomplished), current status (what is being worked on currently), and forecast information (what is expected to happen in the future). Performance reports are the most common formal information distributed to project stakeholders.

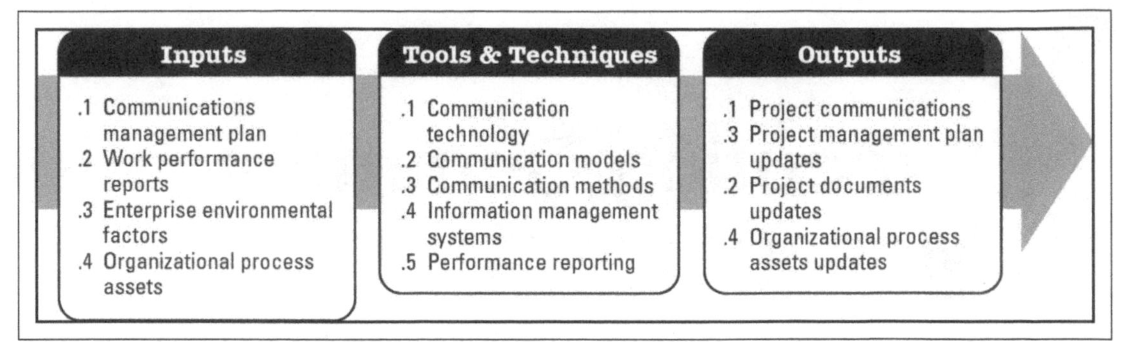

Figure 17-1
Manage Communications: Inputs, Tools and Techniques, Outputs
Source: *PMBOK® Guide*—Fifth Edition

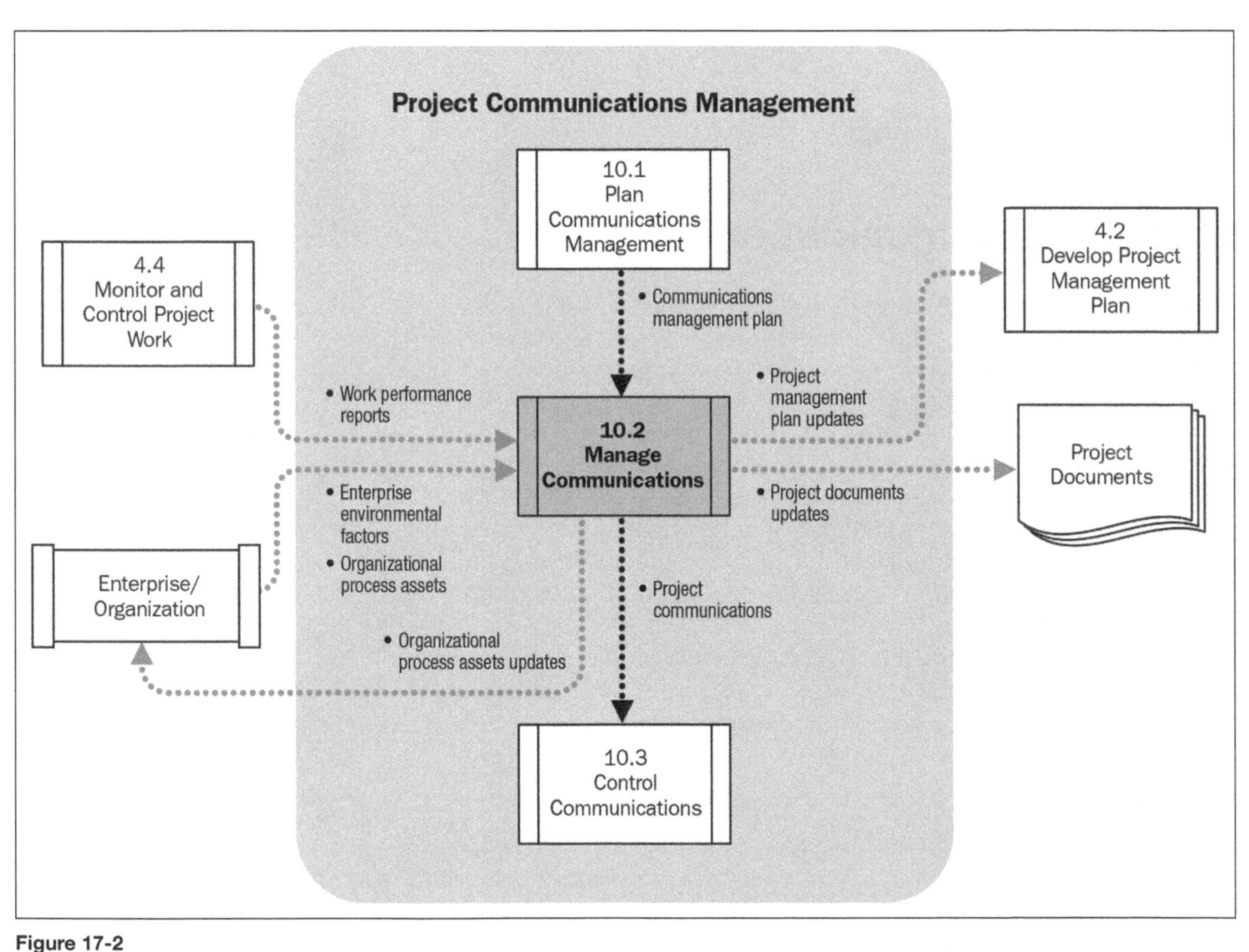

Figure 17-2
Manage Communications Data Flow Diagram
Source: *PMBOK® Guide*—Fifth Edition

Communications management is influenced by enterprise environmental factors such as information management systems, culture, industry standards, and the proprietary nature of the work.

Organizational process assets contain policies and procedures as well as templates for information distribution. They may contain guidelines on information that can be shared with parties external to the project and the organization.

TOOLS AND TECHNIQUES

The communication technology, communication model, and communication methods that were identified in the communications management plan are used to manage the communications for the project.

The information management systems that will be used influence whether information will be pushed out to stakeholders or pulled down from a central location. With the speed of technology advancements there are many more distribution methods using information management systems than ever before. It is common to have web conferences, video conferences, and to talk via Instant Messenger (IM) and texting. Some people keep all their day-to-day business on a tablet and sync it with the main system every night. Thus when planning and managing communications there are a wide variety of options you can consider to communicate effectively.

Performance reporting is about collecting and distributing project performance information. Reports can include status reports, progress reports, and forecasts. A performance report can be as simple as a one-page update or it can include complex electronic dashboards that pull information from multiple information systems.

OUTPUTS

The main outputs of the Manage Communications process are the project communications themselves. This includes reports, memos, meetings, presentations, and all other types of formal and informal communications.

Throughout the project the project documents and the project management plan will be updated based on the various types of communications.

All of the formal and much of the informal communications become organizational process asset updates. In other words, they become part of the project records. Status reports, presentations, lessons learned, meeting minutes, decision logs, letters, and most other written project information will go into the project documentation and eventually be archived at project closure.

Chapter 18

Executing Procurement Management

TOPIC COVERED

Conduct Procurements

Conduct Procurements

Conduct Procurements is the process of obtaining seller responses, selecting a seller, and awarding a contract. You may notice that some of the inputs and tools and techniques from the Conduct Procurements process should occur in sequence as opposed to simultaneously. For example, in the inputs you have procurement documents and seller proposals. In the tools and techniques you have a bidder conference and proposal evaluation techniques. You will need the input of the qualified seller list and the procurement documents before you have the bidder conference. After you receive the seller proposals (another input), you will apply proposal evaluation techniques. If the topic of procurements is new to you, this may be confusing. I have tried to make it as logical as possible to help you understand all the events that happen when conducting procurement and the order in which they occur.

Figure 18-1 shows the inputs, tools and techniques and outputs for the Conduct Procurements process. Figure 18-2 shows a data flow diagram for the Conduct Procurements process.

INPUTS

The *procurement management plan* describes how the procurement process will be conducted. Of particular interest in this process are the roles and responsibilities for distributing the procurement documents, managing the bidder conference, participating on the source selection committee, and the contract type that will be used.

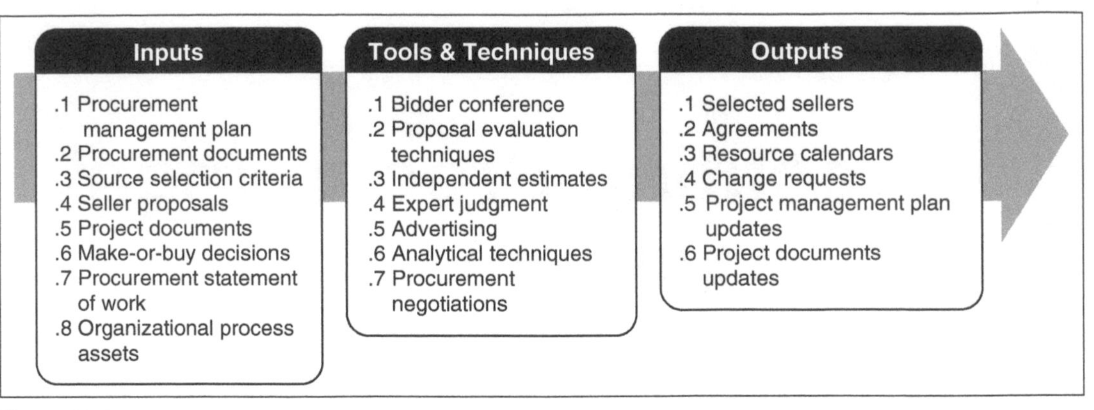

Figure 18-1
Conduct Procurements: Inputs, Tools and Techniques, Outputs
Source: *PMBOK® Guide*—Fifth Edition

You will recall the procurement documents can take the form of a Request for Proposal (RFP), an Invitation for Bid (IFB), or a Request for Quotation (RFQ). There are other types of bid documents, but these are the most common. The bid documents contain the *procurement statement of work*, which documents the goals, requirements, deliverables, and outcomes for the procurement.

The most common project documents are from the Plan Risk Responses process and include the risk register with documentation of the risk-related decisions that are transferred via a contract. For example, if there is a risk that the organization will not have enough people to work on a particular activity, the risk response may be to either outsource the work or bring in temporary outside help.

The make-or-buy decisions will provide a list of all the items that will need to be purchased for the project. Along with the list of purchases the team will have determined the source selection criteria they will use to select the winner.

The organizational process assets used in this process can include qualified seller lists, or if such a list does not exist, it can include history from working with various vendors in the past and information on how well they performed.

The final input is seller proposals. These will be an input after some of the tools and techniques have been applied. The seller proposals are responses to the bid documents. The remaining tools and techniques will be applied to the proposals.

TOOLS AND TECHNIQUES

If the organization does not have qualified seller lists or teaming agreements, sometimes they will need to do an Internet search or place advertising to generate bidders. For government contracts there are rules in place that mandate advertising bid opportunities.

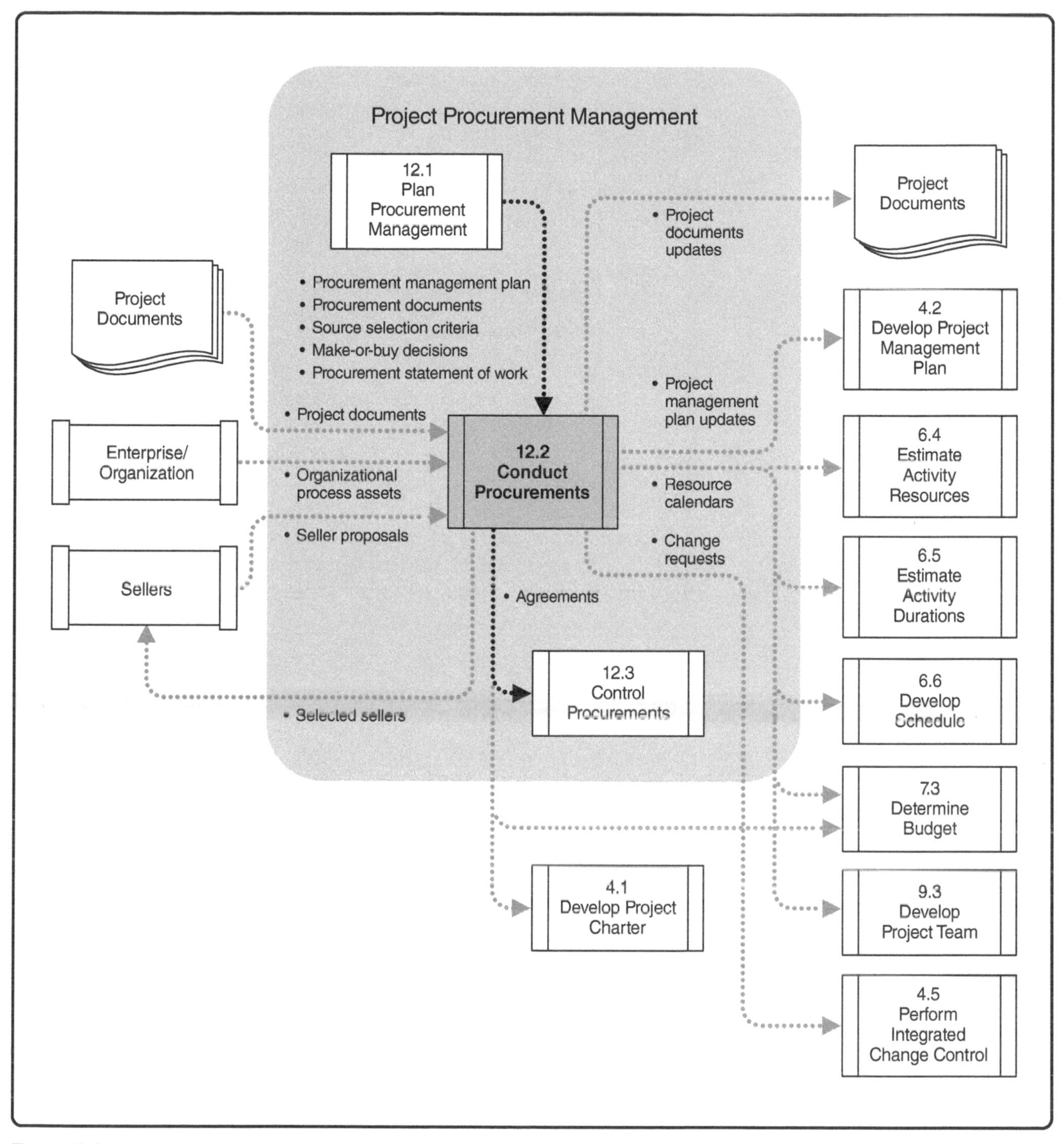

Figure 18-2
Conduct Procurements Data Flow Diagram
Source: *PMBOK® Guide*—Fifth Edition

In many cases the purchasing organization will have bidder conferences. Bidder conferences can be in person, over the phone, or in a virtual environment. The purpose is to ensure that all bidders have complete information, a clear understanding of the contractual and technical requirements, and that everyone is on equal footing.

Once the proposals have been received they undergo a proposal evaluation by an evaluation committee. The evaluation committee uses their expert judgment in applying the source selection criteria.

Bidder Conferences

Many times construction conferences are held at the site where the work will take place. Other times, the bid conferences are held at the buyer's offices or at a neutral location, such as a hotel conference room.

It is a common practice to have someone taking notes of the proceedings and documenting all comments or questions. The responses are then sent out to all attendees, posted on a website, or incorporated into an amendment to the bid documents.

If the conference is virtual or over the telephone, a recording of the conference may be maintained until the contract has been awarded.

Subject matter expertise can consist of technical experts, legal representatives, contracting professionals, the project manager, and other areas as appropriate. Expert judgment is also applied via analytical techniques that are used to evaluate the proposals to determine if they are realistic from a scope, schedule, cost, resource, and risk perspective.

For large procurements, an independent estimate may be developed either in-house or by an outside third party. The independent estimate is compared to the submitted bids. If there are significant differences, then either the bid documents were vague, or the bidder did not understand the scope of the work.

Once one or more finalists are identified, the team conducts procurement negotiations. Negotiations can include technical information, contract terms and conditions, payment terms and timing, reporting and project support efforts, schedule and, of course, price and fee structure.

NEGOTIATED CONTRACT ELEMENTS

- Statement of work
- Schedule
- Period of performance
- Place of performance and delivery
- Roles and responsibilities
- Reporting requirements
- Price, cost, and fee details
- Inspection and acceptance criteria
- Product support, liability, and warranty information
- Change management
- Termination and dispute resolution

OUTPUTS

After all the negotiations have been finalized a seller is selected and the agreement is formalized. The agreement can be a simple purchase order or a very complex contract with different layers of fees, very detailed statements of work, performance criteria, reporting requirements, and so forth. The callout list shows some of the elements that are commonly included in more robust contracts.

Resource calendars from the Estimate Activity Resources process are updated with the information from the procurement. There may be several other project documents, project changes, or project management plan components that need to be updated based on the outcome of the procurement. If documents are under configuration control or are baselined, then the changes will have to go through the Perform Integrated Change Control process.

Chapter 19

Executing Stakeholder Management

TOPIC COVERED

Manage Stakeholder Engagement

Manage Stakeholder Engagement

Manage Stakeholder Engagement is the process of communicating and working with stakeholders to meet their needs/expectations, address issues as they occur, and foster appropriate stakeholder engagement in project activities throughout the project life cycle. Managing stakeholder engagement entails carrying out the stakeholder management plan that was developed in the Plan Stakeholder Management process. The Manage Stakeholder Engagement process seeks to influence and manage expectations and the engagement of groups of stakeholders and individual stakeholders. For example, it might seek to influence end users of the product, or it might look to influence the customer.

The desired outcomes are:

- Increased support and decreased resistance from stakeholders
- Realistic expectations for the product and project
- Addressing pending and existing issues
- Proactive communication

Figure 19-1 shows the inputs, tools and techniques and outputs for the Manage Stakeholder Engagement process. Figure 19-2 shows a data flow diagram for the Manage Stakeholder Engagement process.

INPUTS

The stakeholder management plan defines the selected approach for maximizing the influence of supporting stakeholders and minimizing

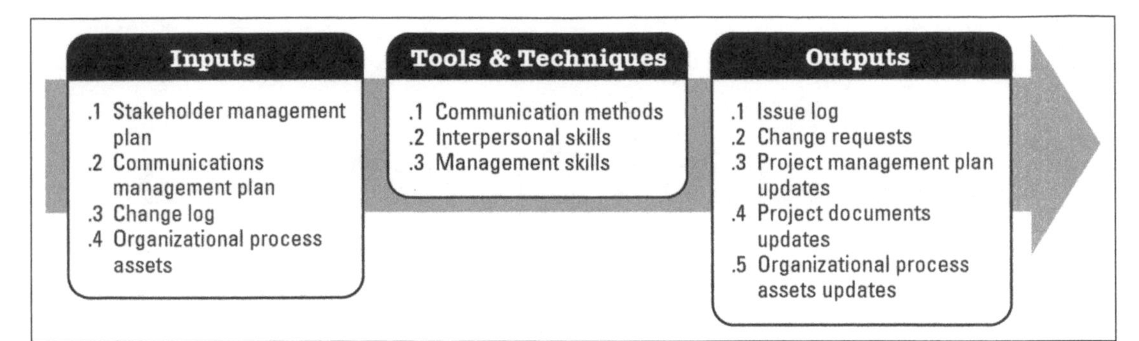

Figure 19-1
Manage Stakeholder Engagement: Inputs, Tools and Techniques, Outputs
Source: *PMBOK® Guide*—Fifth Edition

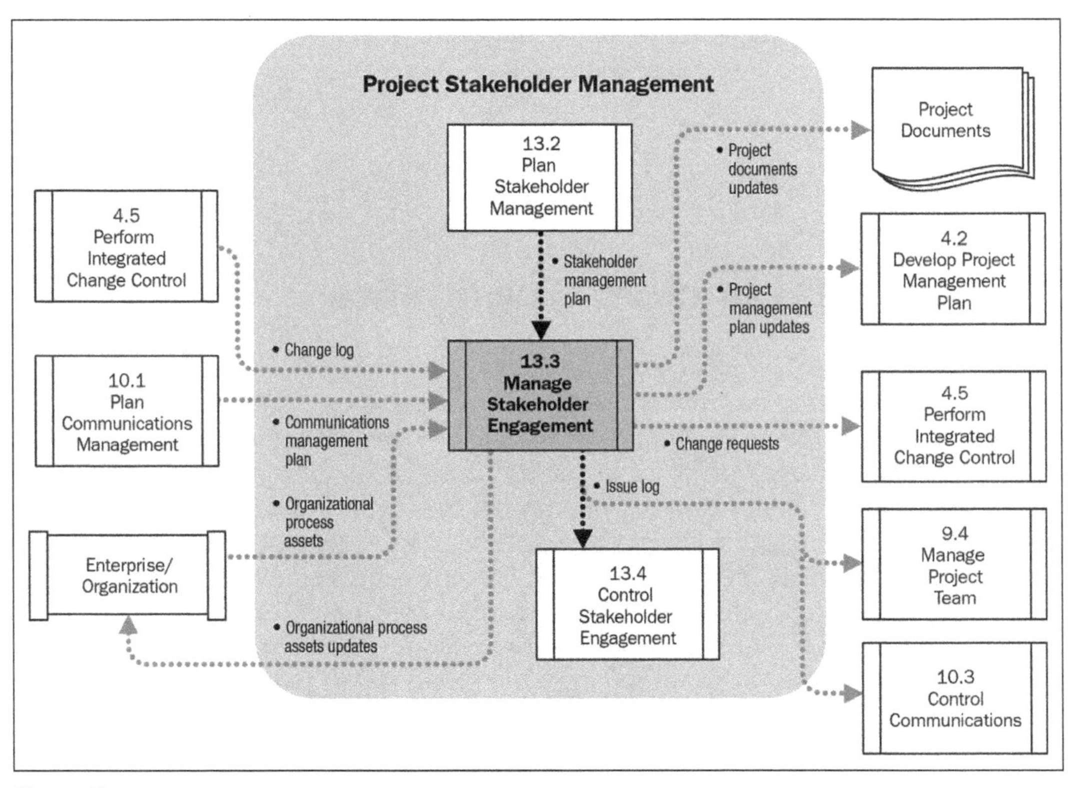

Figure 19-2
Manage Stakeholder Engagement Data Flow Diagram
Source: *PMBOK® Guide*—Fifth Edition

the influence of negative stakeholders. The communications management plan defines the types, methods, frequency, and content of communications that will be delivered to stakeholders. It might outline newsletters, status reports, presentations, meetings, and the like, which are used to meet the communication needs of various stakeholders.

The Change Log contains information on changes and their impacts. Many times changes require more than just communication, they require meeting with various stakeholders to explain the impacts of the change and the reasons for the change. Stakeholders may be called on to help re-plan based on the changes as well.

Organizational process assets used in this process include policies, procedures, and templates for issue management, change management, and communication. Information from previous projects may come in handy as well.

CHANGE LOG CONTENTS

The Change Log contains at least:

- Change ID
- Category of change
- Description of the change
- Who submitted the change request
- Submission date
- Status
- Final disposition

TOOLS AND TECHNIQUES

The communications management plan defines the communication methods that will be used to communicate with stakeholders. This can include meetings, memos, Web meetings, phone calls, presentations, and so forth. Some of the communications will be formal, such as presentations and status reports; and others will be more informal, such as brainstorming, conversations, and problem solving sessions. The project manager will discern which method is most appropriate for communicating with stakeholders:

Push Communication. Deliberately sending out information to specific stakeholders. For example, email, memos, presentations, and the like.

Pull Communication. Making information available to stakeholders to access as needed. Examples include Web sites, intranet, knowledge repositories, and so forth.

Interactive Communication. A discussion between two or more people. This includes meetings, phone calls, and video conferences.

As a project manager, this is a critical area for interpersonal skills, particularly building trust, resolving conflict, active listening, and overcoming resistance to change. Let's look at each of these a bit more closely:

Building trust. By listening, as described above, and then following through on your commitments, keeping communication lines open, and being transparent with decision making you will start to establish trust with your team members, customers, and other stakeholders.

Resolving conflict. We discussed conflict resolution in the human resource management chapter. Suffice it to say that it isn't only team members who have conflicts. The same techniques you use to keep your team functioning are useful in managing other stakeholders as well.

Active listening. Understanding what your stakeholders expect, want, and need is paramount in achieving their support for the project. Telling stakeholders what they need and why they need it is not a very successful approach. Developing the ability to listen, summarize what you have heard, and indicate how you will achieve that goes a long way in creating satisfied stakeholders.

Overcoming resistance to change. Many times stakeholders are comfortable with the status quo. People often don't want to change their habits if that change can be reasonably avoided. In order to overcome that resistance you need to help stakeholders see the benefit to the change both for them and for the organization. Brainstorming ways to make the change as smooth as possible also helps reduce resistance to the change.

Management skills comprise a large body of knowledge and are a big part of a project manager's day-to-day activities. The *PMBOK® Guide* lists four skills that we will explore in a bit more detail:

1. **Facilitating consensus toward project objectives.** Different stakeholders have different objectives. It can be challenging to get all stakeholders to agree to a defined set of objectives for the project. Using techniques such as the nominal group technique (discussed in Chapter 5) can help the group come to an agreement on the prioritization of objectives. You can use a voting process where the majority or plurality (the biggest block of votes, but not necessarily the majority) opinion is carried forward. If your team has a team operating agreement you can apply the decision-making criteria outlined in that document.
2. **Influencing people to support the project.** Influencing people can be as simple as asking for support or it can entail negotiations, overcoming resistance to change and taking time to help people identify the benefits of the project.
3. **Negotiate agreements to satisfy project needs.** The key to successful negotiations is for both parties to feel they came away with something of value. You must be able to analyze what the other party needs, and help them achieve their needs while making sure the project needs are being met. To keep the relationship professional, make sure you focus on the current issues, not the past and not the personalities.
4. **Modify organizational behavior to accept the project outcomes.** Modifying organizational behavior includes managing stakeholders' behavior as well as ensuring there are processes in place to facilitate the deployment and operations of the new product or service.

OUTPUTS

In response to interactions with stakeholders there may be some change requests that go through the integrated change control process.

The project management plan may be updated based on the change requests, issue management, and changes to the communications management plan and the stakeholder management plan. Project documents that are updated include the stakeholder register, issue log, and the change log.

Organizational process assets that are commonly updated as a result of this process include documentation to and from stakeholders, reports and presentations, and lessons learned.

Chapter 20

Monitoring and Controlling the Project

TOPICS COVERED

Monitoring and Controlling Process Group

Monitor and Control Project Work

Perform Integrated Change Control

Monitoring and Controlling Process Group

The Monitoring and Controlling Process Group consists of those processes required to track, review, and regulate the progress and performance of the project; identify any areas in which changes to the plan are required; and initiate the corresponding changes. In a number of the monitoring and control processes you will see similar inputs: the project management plan and work performance data. The outputs will include work performance information, updates, and change requests. Look for this pattern as you go through the various processes.

Another pattern you may notice is how the work performance data from the Direct and Manage Work process is an input to nine of the control processes. From the control processes the work performance data becomes an output as work performance information. The work performance information becomes an input to Monitor and Control Project Work. Through that process it becomes an output as work performance reports. Figure 20-1 demonstrates this progression of data to information to reports.

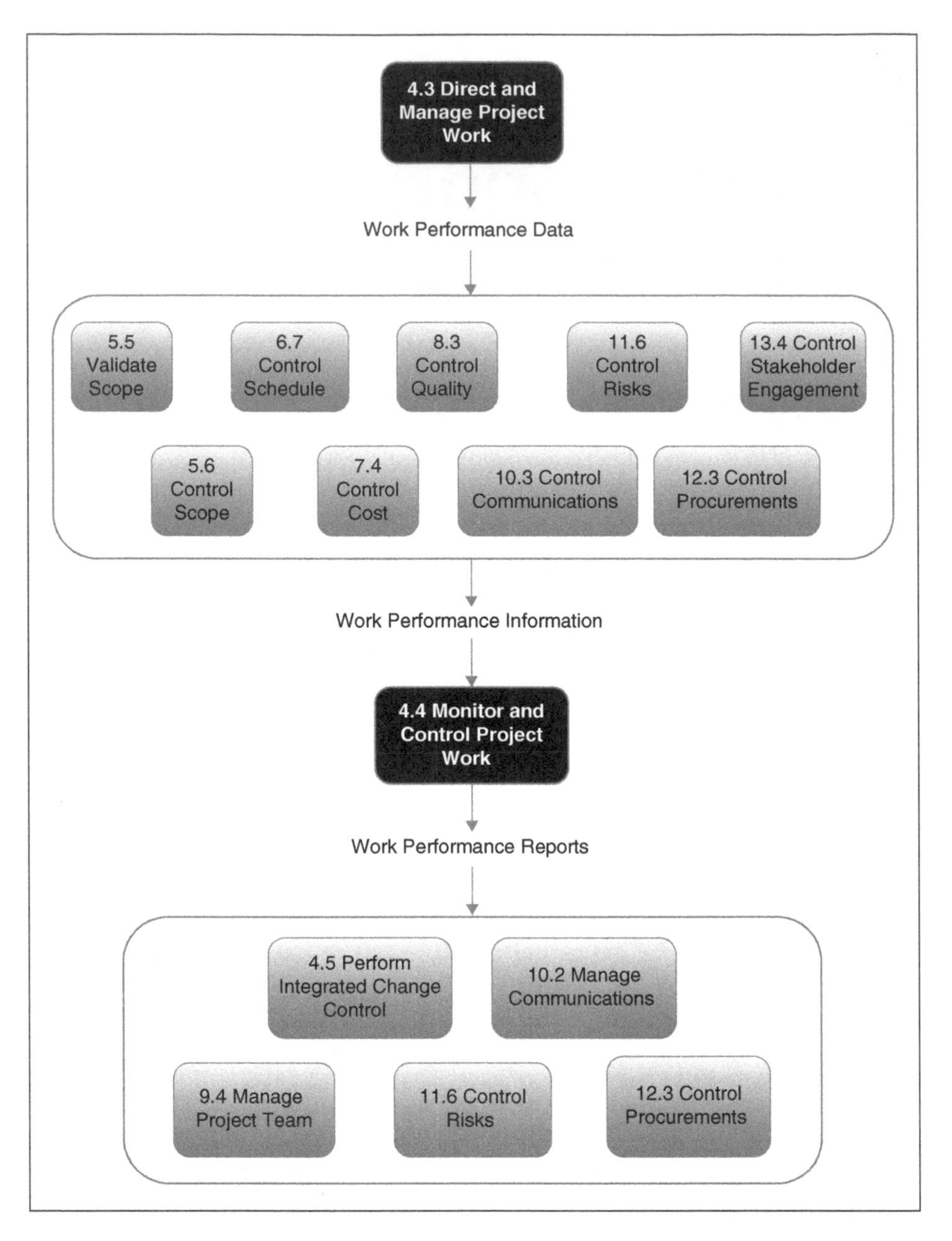

Figure 20-1
Work Performance Data, Work Performance Information, and Work Performance Reports

Monitor and Control Project Work

Monitor and Control Project Work is the process of tracking, reviewing, and reporting the progress to meet the performance objectives defined in the project management plan. The Monitor and Control Project Work process is concerned with the status of the project overall. Some of the activities include:

- Comparing planned results with actual results
- Determining if action is needed, and what the right action is

- Reporting performance
- Ensuring deliverables are correct
- Assessing overall project performance
- Providing forecasts
- Ensuring approved changes are implemented correctly

Monitoring the project work consists of collecting information about the project, compiling and analyzing the data, communicating the status, and assessing measurements and trends. *Controlling* the work consists of developing and implementing corrective and preventive actions consistent with the needs of the project.

Figure 20-2 shows the inputs, tools and techniques and outputs for the Monitor and Control Project Work process. Figure 20-3 shows a data flow diagram for the Monitor and Control Project Work process.

INPUTS

The project management plan contains all the subsidiary management plans that define how each knowledge area of the project will be managed. The project management plan also contains the scope, schedule and cost baselines, and defines the thresholds for variances for each of the project objectives. The baselines and variance thresholds are compared to the work performance information that comes from the various other controlling processes.

The schedule and cost forecasts project what will happen with the schedule and cost performance in the future if current performance trends continue. Forecast information may include calculated forecasts such as an estimate at completion (EAC) for cost, or it may include a bottom-up re-forecast.

Changes that were approved through the Perform Integrated Change Control process are implemented in the Direct and Manage Project Work process. In the Monitor and Control Project Work process the team validates that the changes were implemented properly.

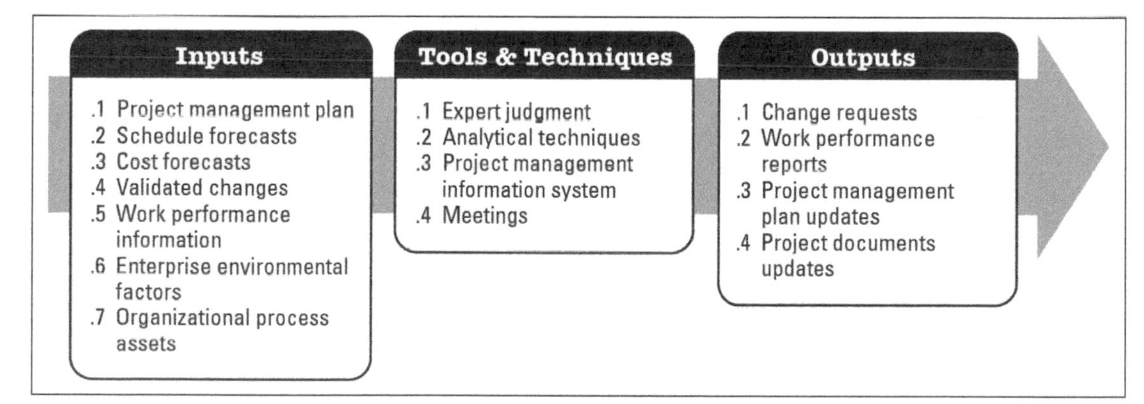

Figure 20-2

Monitor and Control Project Work: Inputs, Tools and Techniques, Outputs

Source: *PMBOK® Guide*—Fifth Edition

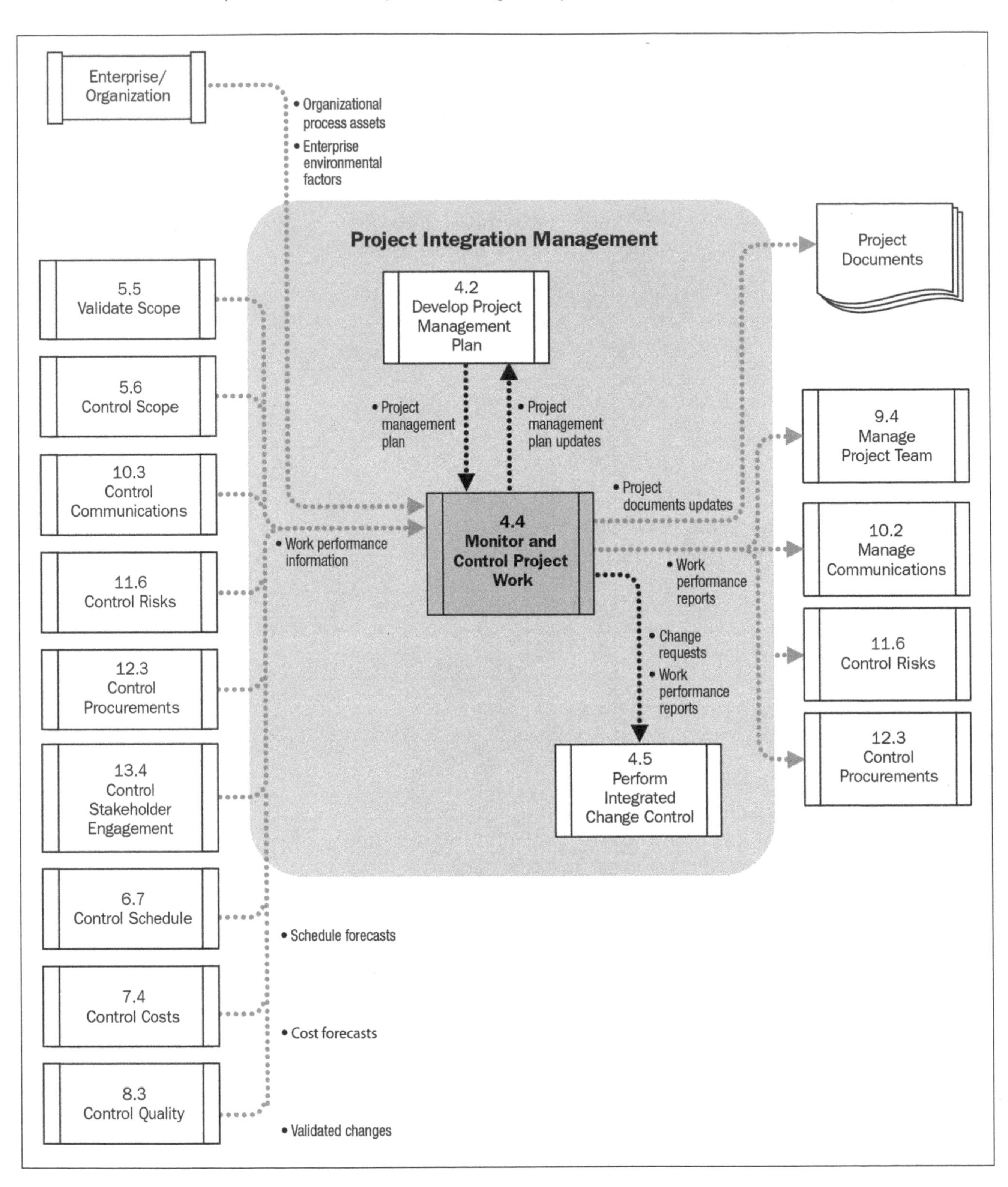

Figure 20-3
Monitor and Control Project Work Data Flow Diagram
Source: *PMBOK® Guide*—Fifth Edition

Industry standards, benchmarks, regulations, and the organization's information management systems are the enterprise environmental factors that shape how this process is conducted. The organizational process assets include guidelines and policies for financial controls, risk management, and communication policies.

TOOLS AND TECHNIQUES

Based on comparing the planned to actual results, the team members will apply expert judgment and analytical techniques to determine the best response and the appropriate actions.

Analytical techniques often require the project management information system to run regression analysis and root cause analysis to determine which variables are causing unacceptable variances or unfavorable trends. Another common analysis technique is to compare the amount of reserve expended to the amount of work performed. If the project is 35 percent complete, but 60 percent of the reserve has been used, this indicates an unfavorable position.

When performance is trending poorly the project manager may call a team meeting to brainstorm causes and solutions to the poor performance.

OUTPUTS

The information from the various knowledge areas is analyzed and compiled into a work performance report. Reports can include information on what has been accomplished in the previous time periods (progress reports), information on where the project is currently (status reports), or information on projected accomplishments (forecasts). Additional report contents can include information on current risks and issues and other relevant data. Performance reports organize and summarize the status and forecast information into a comprehensive and logical format. The reports can include tables, charts, and written explanations. Work performance reports can be relatively simple, or can include automated reports and dashboard views of status.

If the results are not acceptable the team will put forth a change request for corrective or preventive action, or even a defect repair. Depending on the issues uncovered, any of the subsidiary plans or baselines from the project management plan may need to be updated. Forecasts, issues, or risks are project documents that may need updating depending on the situation.

PERFORMANCE REPORT INFORMATION

- Past Performance
- Work Completed During the Past Period
- Work to Be Completed in the Next Period
- Measurements and Results from the Variance Analysis
- Information on Current Risks, Issues, and Pending Changes
- Forecasted Completion Dates
- Forecasted Costs

Perform Integrated Change Control

Perform Integrated Change Control is the process of reviewing all change requests; approving changes and managing changes to deliverables, organizational process assets, project documents, and the project management plan; and communicating their disposition. Change control is performed throughout the project. Change control applies to deliverables, documents, and project management plan components.

By performing change control diligently you can protect the project from scope creep and from finding yourself at the end of your funds, but with work left to do. A change control board (CCB) with a formal process for initiating, evaluating, and communicating changes will help to keep the change process functioning smoothly.

The Perform Integrated Change Control process is where most of the information on configuration management is integrated. On large projects or projects with a lot of parts, documents, or deliverables, it is important to develop a system to identify and name the product components and the project management documents. A rigorous process of tracking and maintaining an auditable trail of pieces, parts, components, design drawings, and project documents can help keep a complex project from getting out of control.

In some organizations the change control system is considered a subset of the configuration control system. In this case the configuration management system defines how changes will be identified and managed and the change management system carries out the actual daily operations of managing the change.

In other organizations the change management system defines the change control processes and the configuration system is a subset that helps define, identify, organize, and manage the individual items.

As in all things in project management, how it is set up depends on the organization and the project. Let's look at some of the definitions used in configuration management.

Configuration Management System. A subsystem of the overall project management system. It is a collection of formal documented procedures used to apply technical and administrative direction and surveillance to: identify and document the functional and physical characteristics of a product, result, service, or component; control any changes to such characteristics; record and report each change and its implementation status; and support the audit of the products, results, or components to verify conformance to requirements. It includes the documentation, tracking systems, and defined approval levels necessary for authorizing and controlling changes.

Configuration Identification. Identification and selection of a configuration item provides the basis for which product configuration is defined and verified, products and documents are labeled, changes are managed, and accountability is maintained.

Configuration Status Accounting. Information is recorded and reported as to when appropriate data about the configuration item should be provided. This information includes a listing of approved configuration identification, status of proposed changes to the configuration, and the implementation status of approved changes.

Configuration Verification and Audit. Configuration verification and audits ensure the composition of a project's configuration items is correct and that corresponding changes are registered, assessed, approved, tracked, and correctly implemented. This ensures the functional requirements defined in the configuration documentation have been met.

Those definitions are pretty technical. The following example shows how it would be applied and hopefully makes it easier to understand.

Car Configuration

Let's say you are developing a new car. During the design phase of the project you will have lots of drawings of the various systems and parts that will go into the car. As you elaborate and revise your ideas, you will make changes to these drawings. As one design idea changes, it may impact the layout and design of other components. You can see that the number of drawings and making sure that you have the right drawings are paramount to keeping control of the project. In order to maintain control of the drawings, you decide to label all systems with an "alpha" code. So the engine drawings start with an E, the transmission drawings with a T, the chassis drawings with a C, and so forth. Then you identify each of the individual parts in the various systems and apply a number to them. Then you put a "build number" on each drawing to indicate if it is in the concept phase, design phase, detail design phase, and so on. Each drawing will have a date and time identifier at the bottom.

Now let's assume you are in the detailed design phase of building the car. You would identify the items that need to be included in this phase of the project and the information you would want to record about those items. For example, you might want to have all the physical components listed along with their size, weight, material composition, and the supplier. By looking at the configuration control log, you should be able to identify this information and also see if and when changes to the items were submitted and the disposition of those changes. This is configuration identification.

During the production cycle of building the car, you find that a lot of changes keep coming through and you need to keep retooling part of the production line. During analysis you determine that a lot of the changes are occurring down a particular branch within the drawing (engineering) tree. Gathering this data and running the reports to show the instances within the branch are an example of configuration status accounting. This type of reporting will use metrics and statistical reports to communicate the information.

When you build your prototype car, you want to make sure that the actual parts in the car are consistent with what the drawings say is in the car. You also want to make sure the system was used to document any changes to the parts. This is configuration verification and audit.

Another aspect of the Perform Integrated Change Control process is determining the appropriate change control process. This includes deciding whether you will have one change control board or more than one, and documenting the authority levels needed to approve changes. It also includes setting up the change request forms, the change log, and the change submittal process.

Part of the change control process is determining how to account for the time necessary to understand the implications of the change. In other words, how the proposed change will impact the cost, the

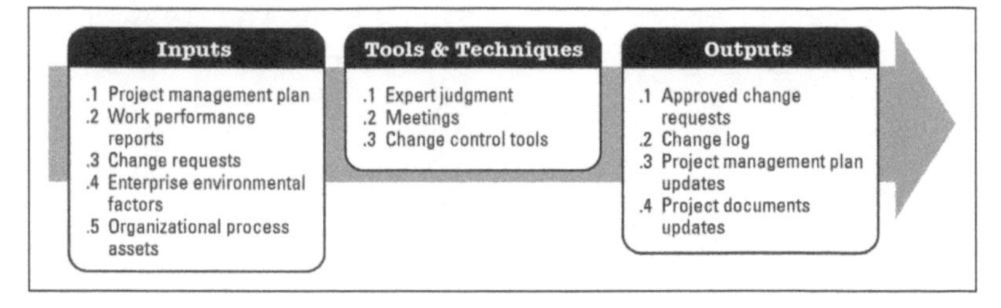

Figure 20-4
Perform Integrated Change Control: Inputs, Tools and Techniques, Outputs
Source: *PMBOK® Guide*—Fifth Edition

schedule, the quality metrics, risks, stakeholder satisfaction, and so forth. The time necessary to determine these things takes time away from project work. Sometimes the person submitting the change request isn't qualified to make these determinations and the work falls on other project resources. Failure to account for the time and cost necessary to assess the impact of the change request can have detrimental impacts on the schedule and budget.

Figure 20-4 shows the inputs, tools and techniques and outputs for the Perform Integrated Change Control process. Figure 20-5 shows a data flow diagram for the Perform Integrated Change Control process.

CHANGE REQUESTS

Change requests can come from any of the following 16 processes:

- Direct and Manage Project Work
- Monitor and Control Project Work
- Validate Scope
- Control Scope
- Control Schedule
- Control Costs
- Perform Quality Assurance
- Control Quality
- Manage Project Team
- Control Communications
- Control Risks
- Plan Procurement Management
- Conduct Procurements
- Control Procurements
- Manage Stakeholder Engagement
- Control Stakeholder Engagement

INPUTS

The project management plan includes the change management plan and all subsidiary management plans and project baselines. Depending on the nature of the change request, any subsidiary plan or baseline from the project management plan may be an input. Change requests can be related to the product or the project work. They can impact deliverables or project documents. Work performance reports provide information on the status of the project scope, schedule, and cost performance and the status of work in progress.

Any automated configuration management or change control system is considered an enterprise environmental factor. The change control system along with the policies, procedures, and templates that are organizational process assets provide the structure to implement the change control process.

TOOLS AND TECHNIQUES

Change control meetings are formal meetings where each change request is reviewed to determine the appropriate disposition. In some cases, more research on the impacts of a change request is needed. Expert judgment can be used in change control meetings to help understand the technical implications of making a change, or of not making a change. Often, expert judgment is used to help quantify the cost and schedule impact of the requested changes. Change control tools can include an automated system (or a manual system)

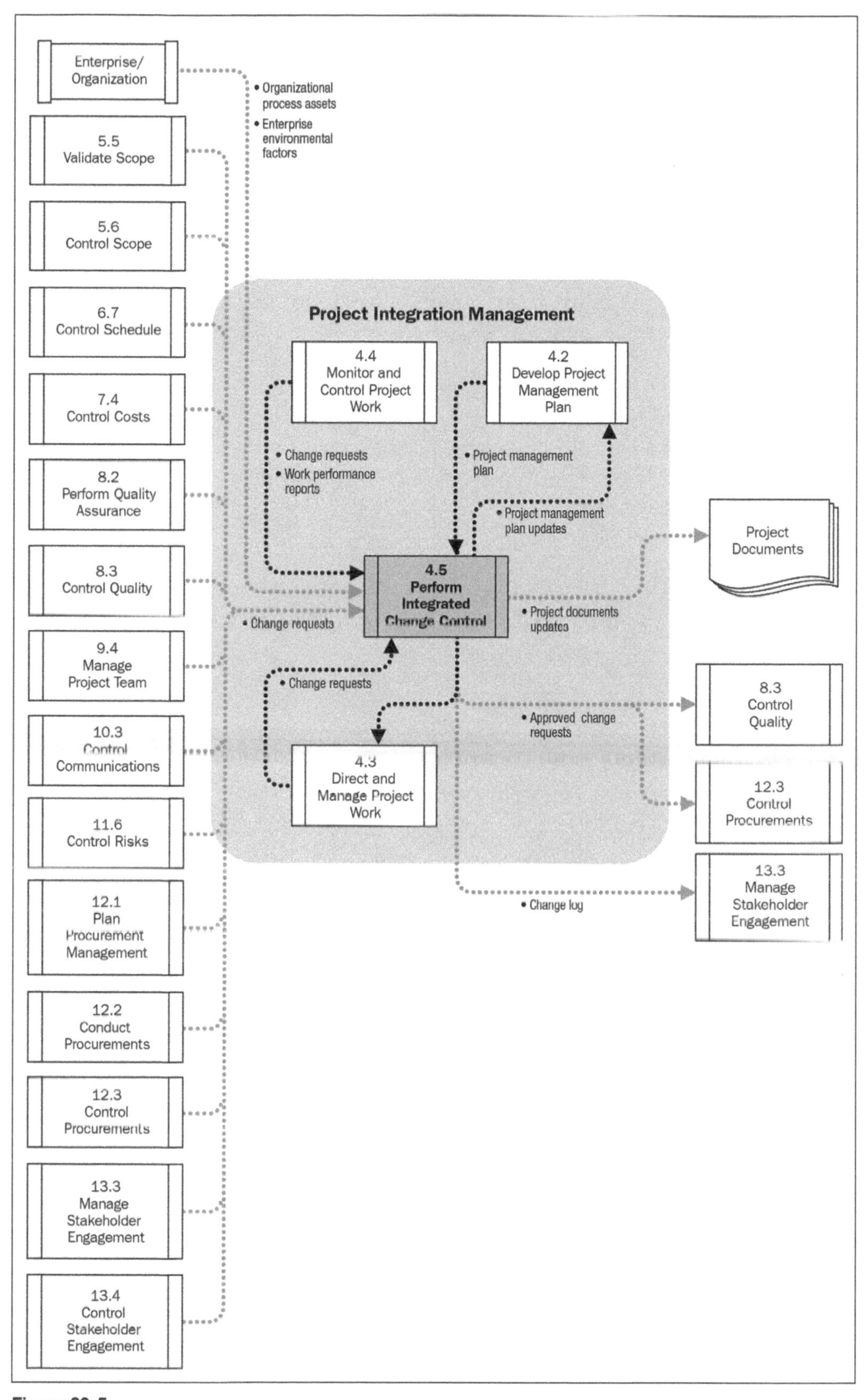

Figure 20-5

Perform Integrated Change Control Data Flow Diagram

Source: *PMBOK® Guide*—Fifth Edition

to submit changes, track changes through the system, and communicate the ultimate disposition of the change request.

OUTPUTS

Approved change requests are implemented in the Direct and Manage Project Work process and validated in the Control Quality process. Deferred changes usually require more information before a definitive decision is made. Rejected change requests are documented in the Change Log along with the reason for rejection. Regardless of the final status of the change request, the outcome needs to be communicated to the original requestor. See Appendix A for an example of a Change Request Form.

Depending on the nature of the accepted change any of the subsidiary management plans or baselines in the project management plan may be updated. Project document updates include the change log and any project document that was updated as a result of a change request.

Chapter 21

Monitoring and Controlling Scope

TOPICS COVERED

Validate Scope

Control Scope

Validate Scope

Validate Scope is the process of formalizing acceptance of the completed project deliverables. Validating project and product scope takes place with the project sponsor and/or customer. It should take place as each deliverable is completed. It signifies that the customer or sponsor approves and accepts the deliverable. By conducting scope validation throughout the project you reduce the likelihood of hearing, "Oh, that isn't what I was thinking of at all."

Figure 21-1 shows the inputs, tools and techniques and outputs for the Validate Scope process. Figure 21-2 shows a data flow diagram for the Validate Scope process.

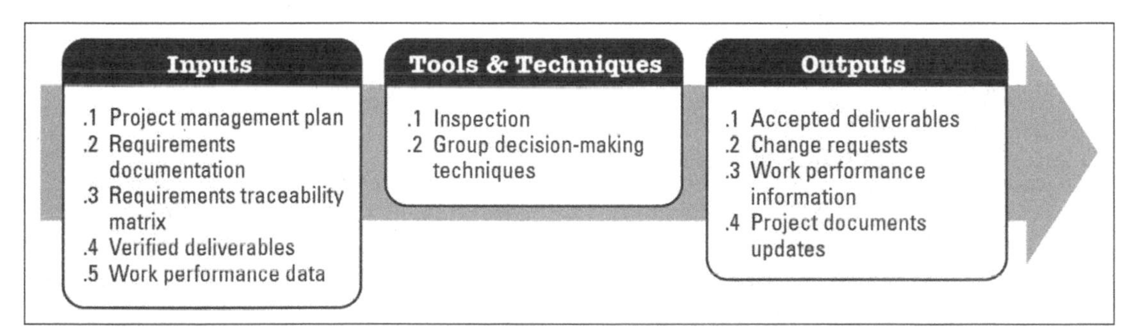

Figure 21-1
Validate Scope: Inputs, Tools and Techniques, Outputs
Source: *PMBOK® Guide*—Fifth Edition

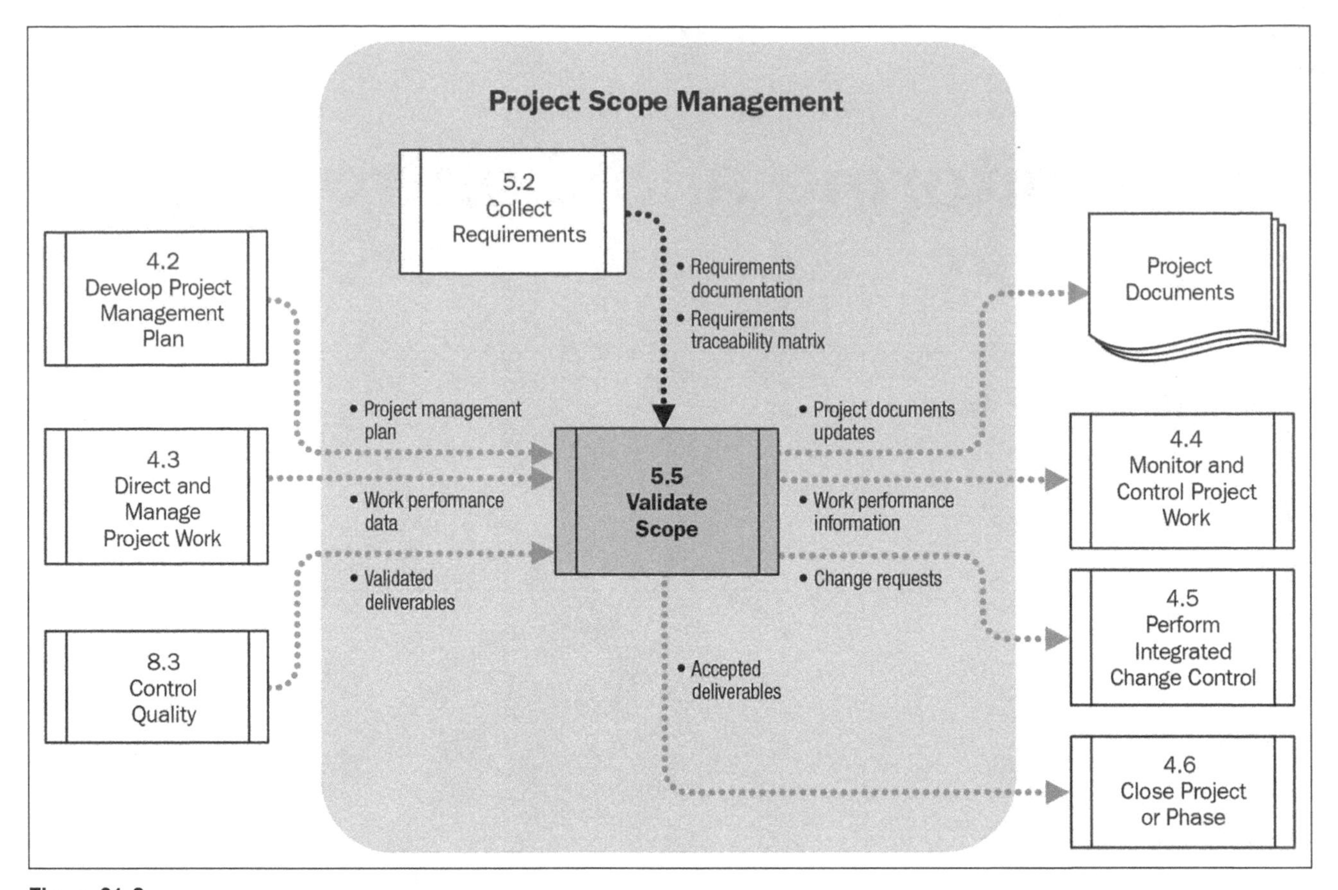

Figure 21-2
Validate Scope Data Flow Diagram
Source: *PMBOK® Guide*—Fifth Edition

INPUTS

The project management plan contains the scope management plan and the scope baseline for the project. The requirements documentation and the requirements traceability matrix contain all the required deliverables and the details associated with the deliverables. These are then compared with the validated deliverables. Validated deliverables are those deliverables that have been through the Control Quality process to ensure that they meet the quality specifications as outlined in the quality management plan and documented metrics. Work performance data includes the number and type of defects or nonconforming outputs.

VALIDATE SCOPE AND CONTROL QUALITY PROCESSES

These two processes are complementary. In some cases they can be performed at the same time. However, there is a subtle difference between them. Control Quality ensures that the deliverables are correct and built according to plan. Validate Scope is the process of attaining customer acceptance and signoff.

TOOLS AND TECHNIQUES

The primary technique used to compare the scope baseline, requirements, and deliverables is inspection. Inspection can take the form of observing a deliverable perform specified functions, measuring the performance, performing a process walkthrough, or looking at test results. Many times the acceptance criteria are documented in the scope statement and thus inspection is the process of ensuring the acceptance criteria are met. If there are multiple parties involved in scope validation you may need to employ group decision making techniques such as majority or plurality voting, consensus or unanimous decisions, or dictatorship where one person makes the final decision.

OUTPUTS

Those deliverables that meet the acceptance criteria are accepted deliverables. If a deliverable does not meet the acceptance criteria a change request is generated, usually in the form of a defect repair. The defect is logged and corrected, or in some cases the change request goes through the Perform Integrated Change Control process. Work performance information from this process includes information on which deliverables have been accepted and how many defect repairs have been logged. Project document updates can include progress tracking documents, and documents that log customer acceptance.

Control Scope

Control Scope is the process of monitoring the status of the project and product scope and managing changes to the scope baseline. You may have heard of the term "scope creep." This process is concerned with managing scope so there is no scope creep and ensuring the scope is developed according to plan.

Figure 21-3 shows the inputs, tools and techniques and outputs for the Control Scope process. Figure 21-4 shows a data flow diagram for the Control Scope process.

Scope Creep. The uncontrolled expansion to product or project scope without adjustments to time, cost, and resources.

INPUTS

The project management plan has a number of components that are used for this process. The scope baseline is compared with the work

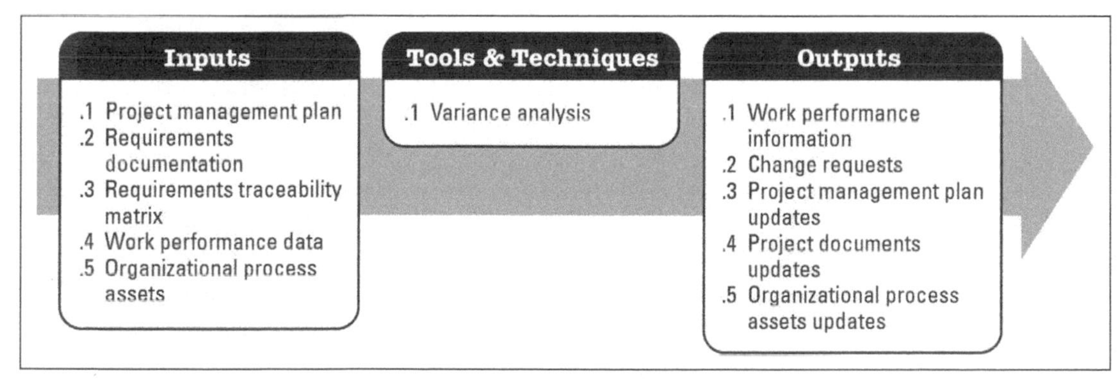

Figure 21-3
Control Scope: Inputs, Tools and Techniques, Outputs
Source: *PMBOK® Guide*—Fifth Edition

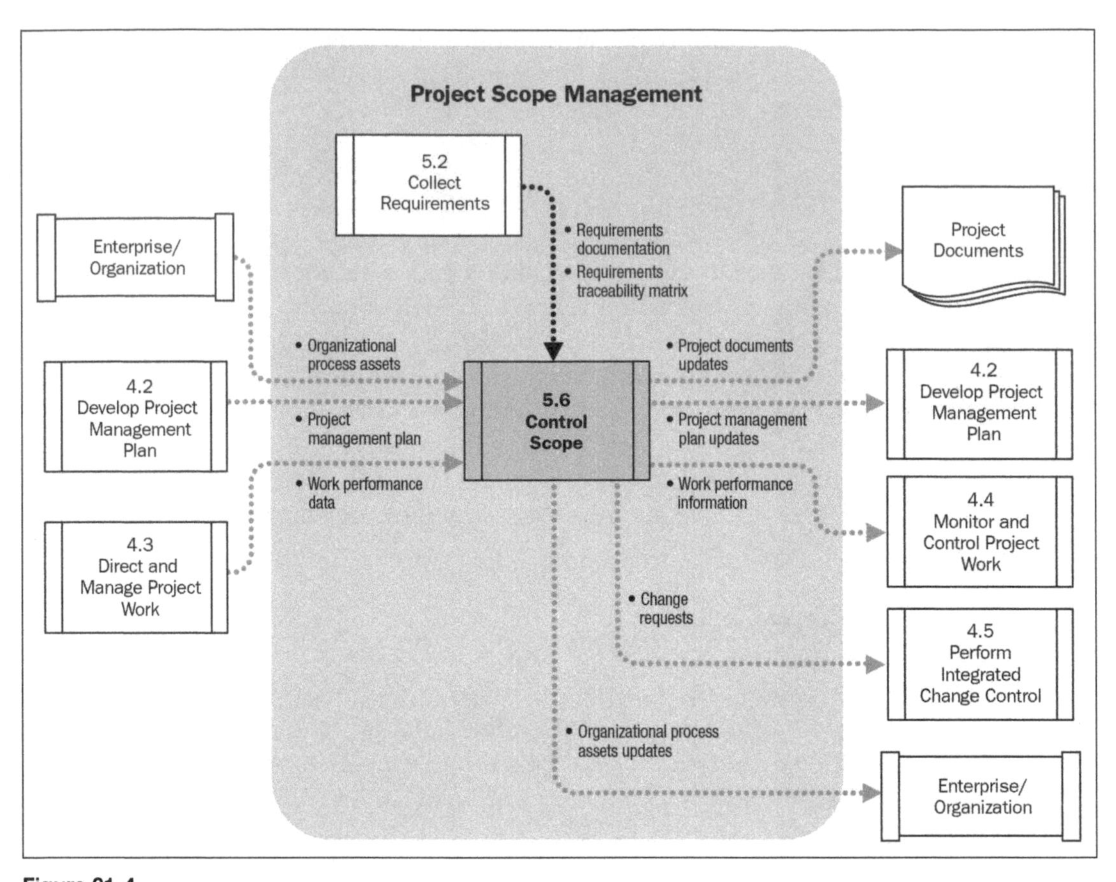

Figure 21-4
Control Scope Data Flow Diagram
Source: *PMBOK® Guide*—Fifth Edition

performance data to understand the current progress in developing deliverables or results. Some of the subsidiary management plans used to manage the scope include: the scope management plan, the requirements management plan, the change management plan, and the configuration management plan.

The scope management plan describes the process and definitions used for controlling scope. It will tie closely with the change management plan. Both of these plans will relate to the configuration management plan that identifies those deliverables under configuration control and how they will be identified and managed.

The requirements management plan describes how requirements are controlled, changed, and updated. The requirements documentation and the requirements traceability matrix provide the detailed functionality, performance, security, and other aspects of scope that need to be present to achieve project success.

Organizational process assets that might be used include change management policies. Those policies will be consistent with the change control and configuration management plans for the project.

TOOLS AND TECHNIQUES

Variance analysis determines the degree of variance from the plans to the work performance information. Variance analysis also seeks to identify the cause of the variance and determine what, if anything, should be done about it.

Variance Analysis Example

Assume you are developing a product that has to be able to operate in extreme temperature environments. Part of the product is a temperature sensor. The requirements documentation states that the sensor has to be able to sense a change in temperature in tenths of a degree increments. In the Control Quality process you measure the sensitivity of the instrument and find that it actually senses changes in hundredths of a degree increments. This is a variance. Upon investigation you find that a more sensitive material was used, but the cost and schedule were not negatively impacted. You discuss the material with the scientists and engineers to determine if there will be an impact on any other aspect of the product and they respond that there will not be any other impacts. As a result of your analysis, you determine that a change has occurred, but the only necessary action is to go through the change control process to document the new material and update the project documents to indicate the actual performance of the sensor.

OUTPUTS

Work performance information records planned versus actual technical and scope performance. Causes of any variances and the responses selected along with the logic behind the response are used to update organizational process assets. The example about the variance analysis with the sensor demonstrates the technical measurement comparison, the root cause analysis, response, and the updates to the organizational process assets and the project documents.

In some situations, the variances require a change request form be filled out and submitted via the Perform Integrated Change Control process. Changes can include project or product changes, corrective and preventive action, as well as defect repair.

If there is a documentable change to the product or project the project management plan will be updated, in particular the scope baseline, and possibly the cost and schedule baselines as well. The requirements documentation and the requirements traceability matrix are project documents that could also be updated.

Organizational process assets that are updated include information on what caused the variance, how the variance was addressed with corrective action and any lessons learned.

Chapter 22

Monitoring and Controlling the Schedule

TOPIC COVERED

Control Schedule

Control Schedule

Control Schedule is the process of monitoring the status of the project activities to update project progress and manage changes to the schedule baseline to achieve the plan. Controlling the schedule includes monitoring and controlling any and all of the project schedules. You may have several schedules you are managing on your project, for example,

- Milestone schedule
- Baseline schedule
- Target schedule
- Deliverable schedule
- Detailed testing schedule
- Network diagram

This process is associated with managing all the schedules by applying the guidelines set forth in the schedule management plan. There is a lot to monitor to make sure the schedule stays on track. You need to validate that the appropriate resources are working as scheduled, the duration estimates are valid, various activities are starting and finishing on time, and that you are maintaining vigilance on the critical path.

When the work is being done in a salaried environment people tend to work overtime to complete their tasks, but they may not report it. It won't show up as a cost variance as it would if people were paid hourly. This behavior hides the impact of inaccurate estimates and sets up a cycle of underestimating durations because the project records show that the duration was met, but it doesn't necessarily show the effort hours it took to meet the duration. The people doing the estimating don't see the effect of the overtime that the people doing the activities need to work in order to maintain the schedule.

Figure 22-1 shows the inputs, tools and techniques and outputs for the Control Schedule process. Figure 22-2 shows a data flow diagram for the Control Schedule process.

Working with the Schedule in an Agile Environment

Work done in an agile environment uses the schedule information to compare the accepted work with the estimated work completed in a sprint. Work in the sprint is estimated based on the team velocity. If the team runs into a challenge during a particular sprint, there may be a variance in the velocity for that sprint. Reprioritizing work is a matter of grooming the backlog based on actual results or a change in user story priorities.

INPUTS

The project management plan contains the schedule management plan and the baseline schedule. The schedule management plan outlines how the various schedules will be controlled. The term project schedule generally refers to the day-to-day schedule, as opposed to the baseline schedule.

The baseline schedule, project schedule, and any other schedules being used are compared to the work performance data that indicates the progress on all activities. It will indicate which activities have started, which have finished, and the status of those in progress.

Project calendars are used if there are multiple shifts, multiple organizations, or multiple locations working on the project. For

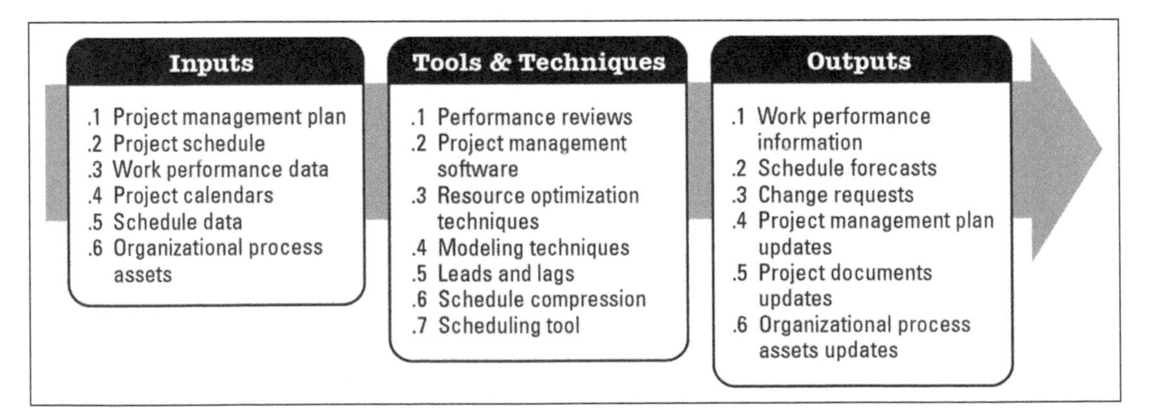

Figure 22-1
Control Schedule: Inputs, Tools and Techniques, Outputs
Source: *PMBOK® Guide*—Fifth Edition

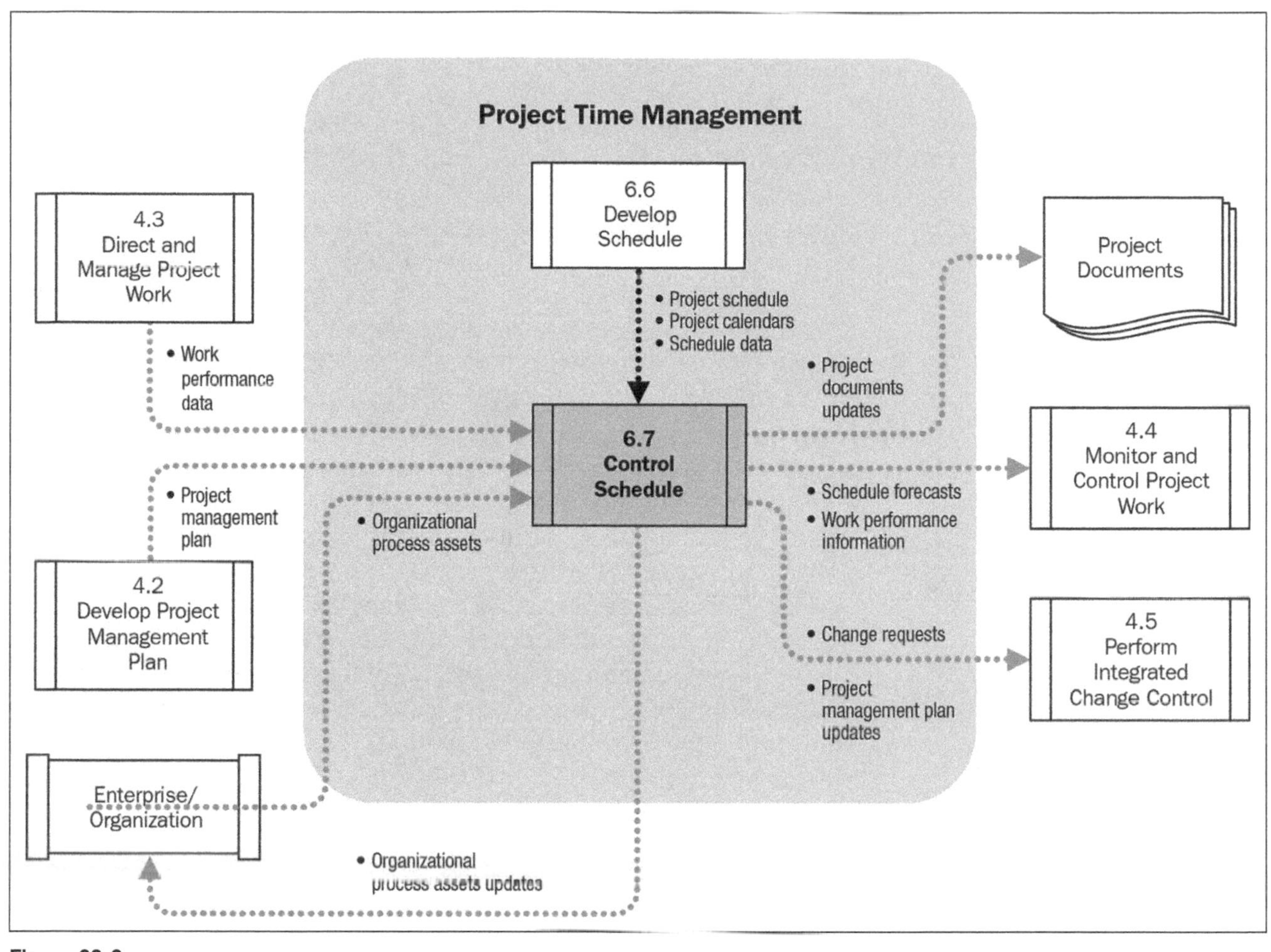

Figure 22-2
Control Schedule Data Flow Diagram
Source: *PMBOK® Guide*—Fifth Edition

example, the holidays vary in different countries and different organizations have different company work hours and policies regarding holidays and vacation time.

Schedule data includes resource histograms, alternative schedules, information about schedule reserve, and assumptions and constraints for the schedule.

Organizational process assets include policies, procedures, and guidelines on controlling the schedule. This may include guidelines for applying earned value measurements if those are being employed.

TOOLS AND TECHNIQUES

Performance reviews and variance analysis work together in this process. The performance reviews provide a detailed analysis of actual start dates, remaining duration, percent complete, and projected completion dates. This is combined with variance analysis that compares the planned to actual start dates, durations, and finish dates. The project manager uses this information to determine the degree of variance along with assessing the critical path to help determine the appropriate actions to take.

In some situations, no action is needed to correct a variance. For example, if an activity is on the critical path and finishes early, or if an activity is estimated to complete 5 days late and there are 15 days of free float for the activity.

One way of analyzing variance is using earned value metrics to compare planned value to earned value. This can help determine the schedule variance (SV) and the schedule performance index (SPI). An explanation of earned value management terms is given in Chapter 23, Control Cost.

Project management software, in particular a scheduling tool, allows you to determine the degree of variance, the impact to float, and the estimated completion dates. You can also use the software to apply resource optimization techniques (like fast tracking) and modeling techniques such as a what-if scenario analysis. This can help you determine the best and most viable responses to variances.

In response to variances, you may need to reiterate some of the steps you took when developing the schedule. This is a normal occurrence throughout the project. Specific steps include schedule compression to expedite activities that are behind, and adjusting leads and lags to find ways to fast-track and overlap activities. Once you compress the schedule you may need to employ resource optimization techniques again to ensure you don't have people assigned for 90 hours a week!

OUTPUTS

Work performance information quantifies the amount and the degree to which the project is ahead or behind schedule. As mentioned above, the schedule variance measures the amount of the variance and the schedule performance index measures the degree or ratio of schedule performance.

Schedule forecasts use the work performance information and attempt to predict future performance. In some instances the forecast is based on trend analysis, in others risk factors, adding or subtracting resources and preventive and corrective actions are taken into account.

In some situations the variances are of a nature that a change request form will need to be filled out. Change requests are usually associated with scope or resource changes and usually cause budget changes, therefore it is important to process them through the Perform Integrated Change Control process.

Elements of the project management plan that may be updated include the schedule baseline, the schedule management plan, and the cost performance baseline. Project document updates include schedules other than the baseline schedule, the resource calendars, and supporting schedule data.

Causes of any variances and the actions taken along with the logic behind those actions are used to *update organizational process assets*.

Chapter 23

Monitoring and Controlling Cost

TOPIC COVERED

Control Costs

Control Costs

Control Costs is the process of monitoring the status of the project to update the project costs and managing changes to the cost baseline. A significant part of controlling costs is comparing the status of deliverables and the cost of those deliverables. While this seems like common sense, many people just look at the budget for a particular point in time and compare the planned expenditures to what was actually spent without looking at the actual accomplishments achieved for the amount spent. If there is a variance in the amount expended to the accomplishments achieved you should investigate the cause.

Based on the current spending profile, upcoming events, and pending risks you can develop a forecast for future funding needs. If current spending or projected future spending exceeds the authorized funding amount a change request will be required.

Figure 23-1 shows the inputs, tools and techniques and outputs for the Control Costs process. Figure 23-2 shows a data flow diagram for the Control Costs process.

INPUTS

The project management plan contains the cost management plan and the cost baseline (also known as the performance measurement baseline). The cost management plan outlines how cost performance will be measured, the variance thresholds, and reporting formats. The cost performance baseline shows the time-phased budget for the project. In the Control Costs process the cost baseline is compared to work performance data. Work performance data describes which activities have started, which have finished, the status of those in

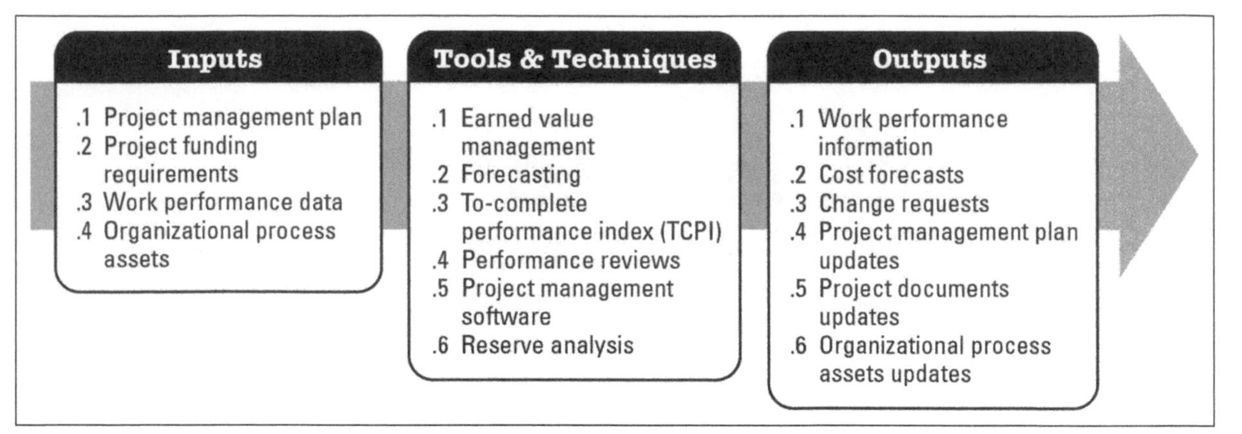

Figure 23-1
Control Costs: Inputs, Tools and Techniques, Outputs
Source: *PMBOK® Guide*—Fifth Edition

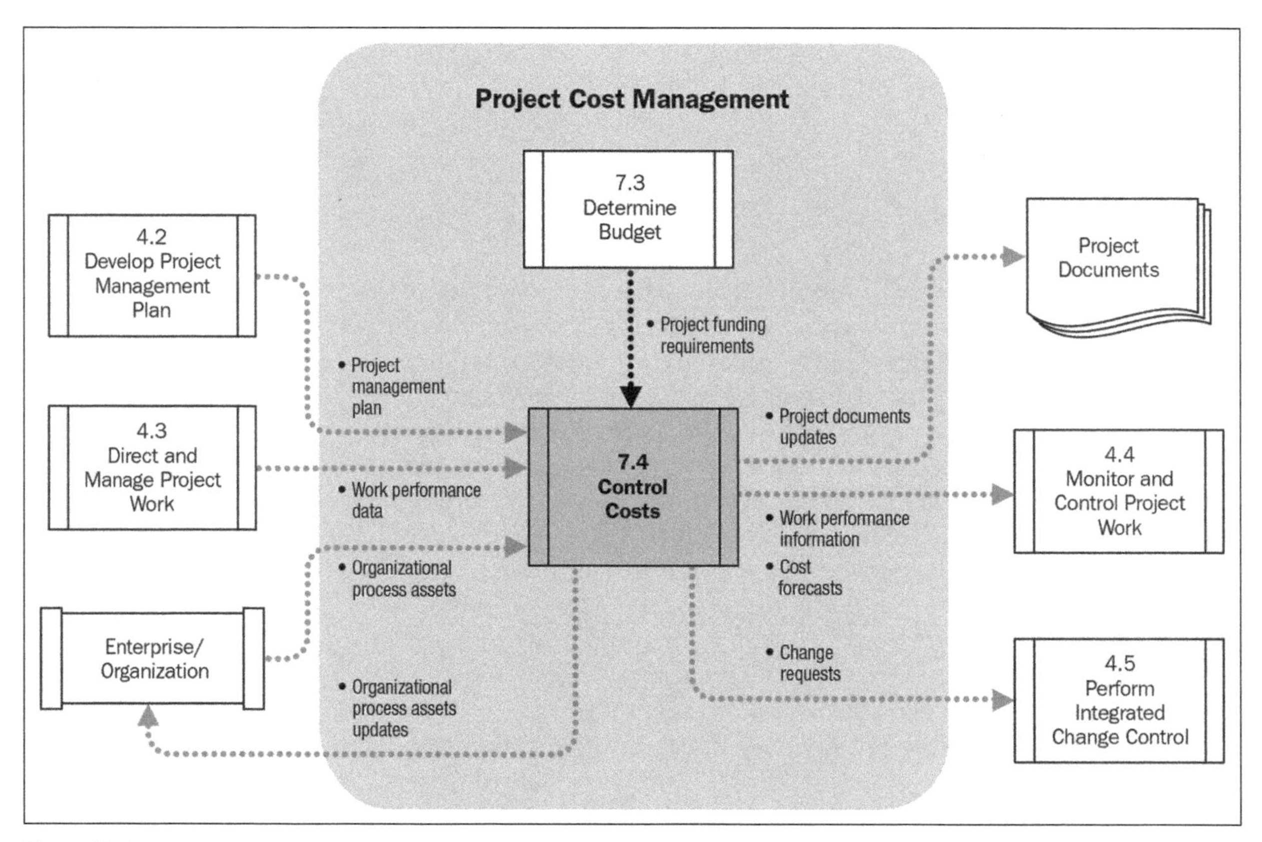

Figure 23-2
Control Costs Data Flow Diagram
Source: *PMBOK® Guide*—Fifth Edition

progress and the costs for accomplishments. The data is compared to planned accomplishments and the budgets for the planned accomplishments to determine cost efficiency and cost variances.

Project funding requirements indicate the amount of funding needed in incremental periods of time, such as quarterly, annually, and so forth. Organizational process assets include policies, procedures, and guidelines on controlling the costs. This may include guidelines for applying earned value measurements and calculations if those are being employed.

TOOLS AND TECHNIQUES

Earned value management is a technique that is used on very large projects, particularly government and defense projects. The implementation of a government "validated" earned value system can take significant investment in terms of time and money. However, the concepts of earned value are very logical and they can (and perhaps should) be applied on any project. There are some terms that are specific to earned value management that I will provide definitions for, but because the definitions can be a bit dry, I will also provide explanations and examples.

Earned Value Management. A methodology that combines scope, schedule, and resource measurements to assess project performance and progress.
Planned Value (PV). The authorized budget assigned to scheduled work.
Earned Value (EV). The measure of work performed expressed in terms of budget authorized for that work.
Actual Cost (AC). The realized costs incurred for the work performed on an activity during a specific time period.
Budget at Completion (BAC). The sum of all budgets established for the work to be performed.

Now that you have the definitions, let's look at an explanation and example of each.

The planned value is the amount of time and budget you expect to spend to accomplish work. The sum of all the work and budget is the budget at completion (BAC). Let's say you are going to redo the backyard of your house. You fully define the scope and come up with the work breakdown structure shown in Figure 23-3.

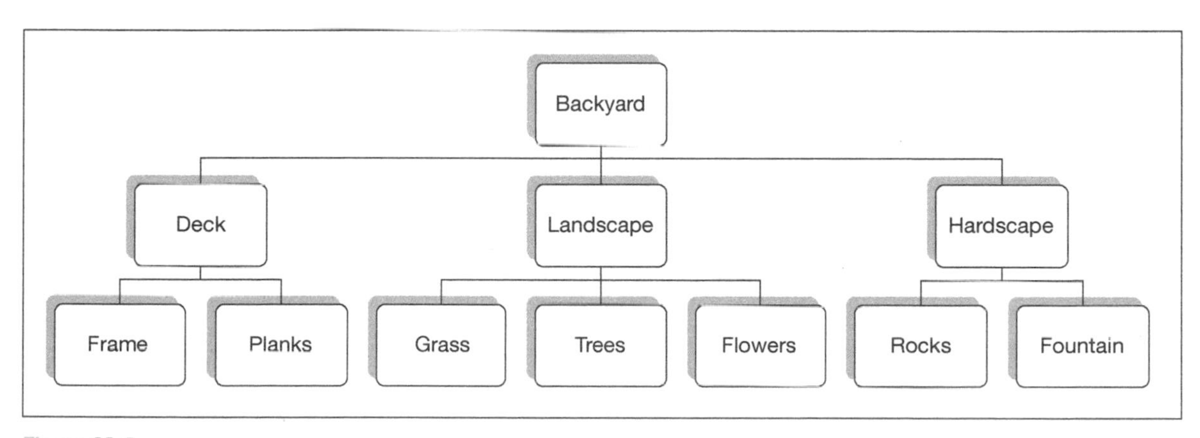

Figure 23-3
Backyard WBS

Using the processes described in the chapter on planning the schedule, you define and sequence activities, estimate resources and durations, and develop a schedule, as shown in Figure 23-4.

Then you estimate the costs for the work and develop a budget. For earned value the budget is called a performance measurement baseline (Table 23-1 and Figure 23-5).

This budget, or performance measurement baseline, provides the planned value for each element of work and shows how the scope is spread over time and budgeted. Therefore, the performance measurement baseline shows the integration of scope, schedule, and cost. This is the planned value for the project. The end point of $12,300 is the budget at completion.

The next step is to measure progress. This means determining the earned value. The earned value is the value of the work actually accomplished. So if you estimate a task to cost $1,000 and you determine you are 50 percent complete, you have earned $500.

Figure 23-4
Budget

	Estimate	Week 1	Week 2	Week 3
Frame	2500	2500		
Planks	2500		2500	
Sod	1500			1500
Trees	1800			1800
Flowers	1000			1000
Fountain	1500	1500		
Rocks	1500			1500
Weekly		4000	2500	5800
Cumulative		4000	6500	12300

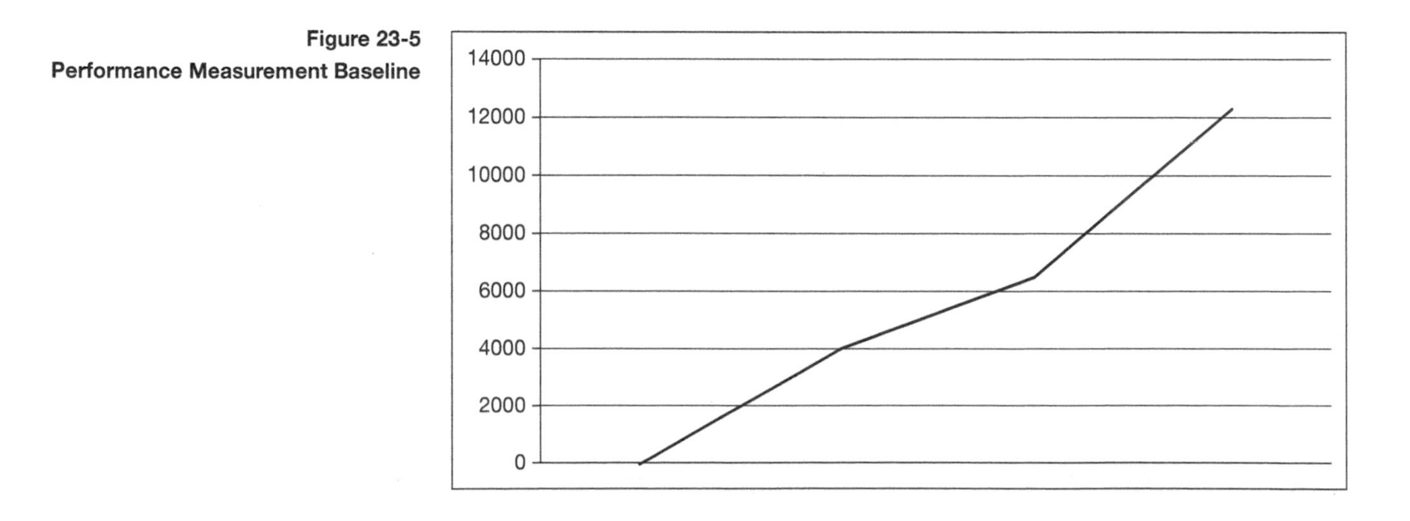

Figure 23-5
Performance Measurement Baseline

Let's look at our backyard scenario. Let's say that at the end of Week 2 we are 100 percent complete with the frame and 80 percent done with the planks, and we have ordered the fountain. We have not ordered any of the landscape or hardscape materials. Therefore, our earned value numbers are as shown in Table 23-2.

Notice that we haven't indicated our expenses for the work done, only the planned value and the earned value. The actual cost is determined by adding up the invoices to see what has been paid. In this case, we see that the frame cost $2,500, the planks cost

Table 23-1 Backyard Schedule

	Activity	Duration	Predecessor	Resource
1	**Deck**	10 days		Michael
2	Frame	4 days		Michael
3	Grade	1 day		Michael
4	Set posts	1 day	3	Michael
5	Set frame	2 days	4	Michael
6	Planks	6 days		Michael
7	Lay planks	3 days	5	Michael
8	Finish work	1 day	7	Michael
9	Stain deck	1 day	8	Michael
10	Seal deck	1 day	9	Michael
11	**Landscape**	4 days		Jose
12	Sod	3 days		Jose
13	Purchase sod	1 day	8	Larry
14	Plant sod	2 days	13	Jose
15	Trees	4 days		Jose
16	Purchase trees	1 day	13SS	Larry
17	Plant trees	3 days	16	Jorge
18	Flowers	4 days		Jose
19	Purchase flowers	1 day	13SS	Larry
20	Plant flowers	3 days	19	John
21	**Hardscape**	12 days		Mark
22	Fountain	12 days		Mark
23	Purchase fountain	1 day		Larry
24	Install fountain	1 day	23FS+2 wks	Mark
25	Rocks	3 days		Mark
26	Purchase rock	1 day	13SS	Larry
27	Fill in rockwork	2 days	26	Mark

Table 23-2 Calculation of Earned Value

Work	Planned Value	Percent Complete	Earned Value
Frame	2500	100	2500
Planks	2500	80	2000
Fountain	1500	100	1500
Total	6500		6000

$2,500, and the fountain was $1,575. Now we can add the actual costs (Table 23-3).

The next thing to look at is how the results compare to the planned results and to the costs. By looking at Table 23-3 you can intuitively see that things are not going well. It is fairly obvious that we are earning less than we planned and it is costing us more than we planned. With earned value techniques we can make some simple calculations to quantify the results.

Schedule Variance (SV). A measure of schedule performance expressed as the difference between the earned value and the planned value.
Cost Variance (CV). The amount of budget deficit or surplus at a given point in time, expressed as the difference between the earned value and the actual cost.
Schedule Performance Index (SPI). A measure of schedule efficiency expressed as the ratio of earned value to planned value.
Cost Performance Index (CPI). The measure of cost efficiency of budgeted resources expressed as the ratio of earned value to actual cost.

INDEX ETIQUETTE

For the SPI and CPI measurements, you carry the decimals out to two places.

MEASUREMENT GUIDELINES

When you are looking at variances, any negative variance indicates that your performance is not going the way you want it to. You are either accomplishing less or spending more than you planned.

When you are looking at indexes, any index less than 1.00 indicates that your performance is not going the way you want it to. A CPI of .90 indicates that for every dollar you spend you are only getting 90 cents of value.

These four simple measurements give us a lot of information about our project performance. And the information is objective, not subjective. A negative schedule variance tells us we have accomplished less work than we planned. We may not be behind on the critical path, but our accomplishment rate is less than expected. A positive schedule variance says we have accomplished more than planned. Again, we have to compare progress with our critical path to determine if we are ahead or behind schedule.

The cost variance tells us we have spent more or less than was budgeted for the work accomplished. A negative cost variance says we have accomplished less; a positive cost variance says we have accomplished more.

If we look at a cost performance index or schedule performance index we are looking to see how efficient we are. An SPI or CPI of less than 1.00 indicates we are underperforming; in other words we are behind in accomplishing work or over budget. Conversely, an SPI or CPI greater than 1.00 says we are performing well. We are accomplishing more work than planned or we are spending less to do it.

Variances allow you to determine project-specific information. When you look at indexes you can compare performance across the project or compare it to other projects in the organization.

Table 23-3 Actual Costs included in Calculations

Work	Planned Value	Percent Complete	Earned Value	Actual Cost
Frame	2500	100	2500	2500
Planks	2500	80	2000	2500
Fountain	1500	100	1500	1575
Total	6500		6000	6575

Let's see what the variances and indexes are for the backyard project. If you want, you can calculate them on your own before you look at Table 23-4.

This information tells you how far behind and how far over budget you are.

A common way of presenting information is to create a chart that shows the PV, EV, and AC plotted next to each other. (See Figure 23-6.)

Forecasting uses the information from the work accomplished to date and estimates the cost of future work given experience on the project so far, pending risks and any information about the future that could impact the project. A term that is often used to estimate future work is called an estimate to complete (ETC). By adding in the costs to date (AC) you can determine an estimate at completion (EAC).

SV AND SPI

Remember, a negative read on the SV or SPI does not mean you are behind on your schedule, it means you are behind in your accomplishments. You need to look at your critical path to determine if you are behind on your schedule.

Estimate to Complete (ETC). The expected cost to finish all the remaining project work.

Estimate at Completion (EAC). The expected total cost of completing all work expressed as the sum of the actual cost to date and the estimate to complete.

Table 23-4 Example Variances and Indexes

Work	PV	Percent Complete	EV	AC	SV	CV	SPI	CPI
Frame	2500	100	2500	2500	0	0	1	1.00
Planks	2500	80	2000	2500	500	500	0.8	0.80
Fountain	1500	100	1500	1575	0	275	1	0.95
Total	6500		6000	6575	500	575	0.92	0.91

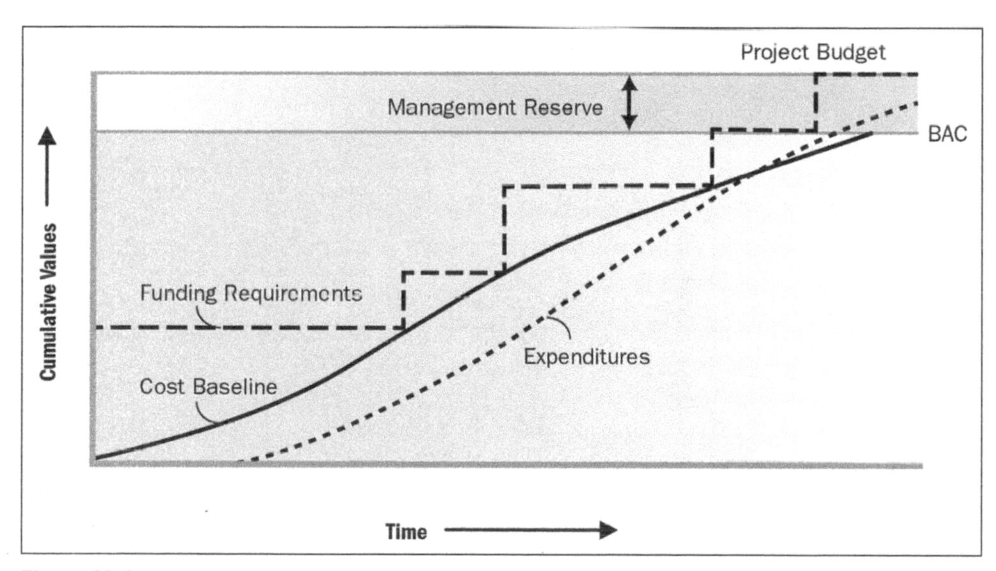

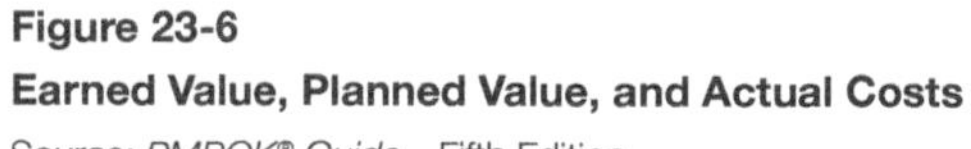
Figure 23-6
Earned Value, Planned Value, and Actual Costs

Source: *PMBOK® Guide*—Fifth Edition

The most accurate way to determine an ETC is to re-estimate the remaining work given the team's better understanding from when they developed the original estimates. However, this can be time consuming and costly for large projects. This type of estimate is called a bottom-up estimate.

There are a number of mathematical formulas you can use to develop a range of forecasted EACs. Each formula comes with a specific set of assumptions about the future work. However, these formulas should not be used in place of a bottom-up estimate if you can help it. They should provide a range in which to compare the bottom-up estimate in order to make sure the new estimate is fundamentally sound. Remember, the foundation of an EAC is the ETC − AC. Therefore, you will see this element in all the EAC calculations. Table 23-5 shows three of the more common calculations and the assumptions that go along with them.

If the EAC is significantly greater than the BAC, the team will have to find ways either to improve performance, de-scope, or get more funding.

The to-complete performance index is an interesting little equation that divides the work remaining (BAC − EV) by the funds remaining (EAC or BAC − AC). This gives the team the efficiency factor they have to achieve in order to hit the BAC or EAC (whichever factor they used in the denominator).

To-Complete Performance Index (TCPI). A measure of the cost performance that is required to be achieved with the remaining resources in order to meet a specified management goal, expressed as the ratio of the cost to finish the outstanding work to the remaining budget.

The to-complete performance index for the backyard example would be: (12,300 − 6,000)/(12,300 − 6,575) + 1.10. This means that in order to deliver the project at the original BAC, the remaining

Table 23-5 Common Calculations and Assumptions

$EAC = BAC = EV + AC$	This is used if the current cost variance is not expected to continue. It assumes that any existing variance is a one-time event. Most people do not consider this a valid way of calculating the EAC.
$EAC = (\frac{(BAC - EV)}{CPI}) + AC$ This equation has a shortcut: $\frac{BAC}{CPI}$	This is used if the current cost variance is expected to continue at the same rate. Some people consider this the best case scenario EAC.
$EAC = (\frac{(BAC - EV)}{CPI \times SPI}) + AC$	This is used if the schedule and cost are significant factors in project success, and their performance is expected to continue at the same rate. Some people consider this the most likely scenario, others the worst case scenario.

Earned Value Analysis					
Abbreviation	**Name**	**Lexicon Definition**	**How Used**	**Equation**	**Interpretation of Result**
PV	Planned Value	The authorized budget assigned to scheduled work.	The value of the work planned to be completed to a point in time, usually the data date, or project completion.		
EV	Earned Value	The measure of work performed expressed in terms of the budget authorized for that work.	The planned value of all the work completed (earned) to a point in time, usually the data date, without reference to actual costs.	EV = sum of the planned value of completed work	
AC	Actual Cost	The realized cost incurred for the work performed on an activity during a specific time period.	The actual cost of all the work completed to a point in time, usually the data date.		
BAC	Budget at Completion	The sum of all budgets established for the work to be performed.	The value of total planned work, the project cost baseline.		
CV	Cost Variance	The amount of budget deficit or surplus at a given point in time, expressed as the difference between the earned value and the actual cost.	The difference between the value of work completed to a point in time, usually the data date, and the actual costs to the same point in time.	CV = EV - AC	Positive = Under planned cost Neutral = On planned cost Negative = Over planned cost
SV	Schedule Variance	The amount by which the project is ahead or behind the planned delivery date, at a given point in time, expressed as the difference between the earned value and the planned value.	The difference between the work completed to a point in time, usually the data date, and the work planned to be completed to the same point in time.	SV = EV - PV	Positive = Ahead of Schedule Neutral = On schedule Negative = Behind Schedule
VAC	Variance at Completion	A projection of the amount of budget deficit or surplus, expressed as the difference between the budget at completion and the estimate at completion.	The estimated difference in cost at the completion of the project.	VAC = BAC - EAC	Positive = Under planned cost Neutral = On planned cost Negative = Over planned cost
CPI	Cost Performance Index	A measure of the cost efficiency of budgeted resources expressed as the ratio of earned value to actual cost.	A CPI of 1.0 means the project is exactly on budget, that the work actually done so far is exactly the same as the cost so far. Other values show the percentage of how much costs are over or under the budgeted amount for work accomplished.	CPI = EV/AC	Greater than 1.0 = Under planned cost Exactly 1.0 = On planned cost Less than 1.0 = Over planned cost
SPI	Schedule Performance Index	A measure of schedule efficiency expressed as the ratio of earned value to planned value.	An SPI of 1.0 means that the project is exactly on schedule, that the work actually done so far is exactly the same as the work planned to be done so far. Other values show the percentage of how much costs are over or under the budgeted amount for work planned.	SPI = EV/PV	Greater than 1.0 = Ahead of schedule Exactly 1.0 = On schedule Less than 1.0 = Behind schedule
EAC	Estimate At Completion	The expected total cost of completing all work expressed as the sum of the actual cost to date and the estimate to complete.	If the CPI is expected to be the same for the remainder of the project, EAC can be calculated using:	EAC = BAC/CPI	
			If future work will be accomplished at the planned rate, use:	EAC = AC + BAC - EV	
			If the initial plan is no longer valid, use:	EAC = AC + Bottom-up ETC	
			If both the CPI and SPI influence the remaining work, use:	EAC = AC + [(BAC - EV)/(CPI x SPI)]	
ETC	Estimate to Complete	The expected cost to finish all the remaining project work.	Assuming work is proceeding on plan, the cost of completing the remaining authorized work can be calculated using:	ETC = EAC - AC	
			Reestimate the remaining work from the bottom up.	ETC = Reestimate	
TCPI	To Complete Performance Index	A measure of the cost performance that must be achieved with the remaining resources in order to meet a specified management goal, expressed as the ratio of the cost to finish the outstanding work to the budget available.	The efficiency that must be maintained in order to complete on plan.	TCPI = (BAC-EV)/(BAC-AC)	Greater than 1.0 = Harder to complete Exactly 1.0 = Same to complete Less than 1.0 = Easier to complete
			The efficiency that must be maintained in order to complete the current EAC.	TCPI = (BAC - EV)/(EAC - AC)	Greater than 1.0 = Harder to complete Exactly 1.0 = Same to complete Less than 1.0 = Easier to complete

Figure 23-7

Earned Value Calculations Summary Table

Source: *PMBOK® Guide*—Fifth Edition

work must be 10 percent more efficient than the original estimate. Note that the existing work with a CPI of .92 is 8 percent less efficient than the estimate. So, this is really saying that the rest of the work will need to improve by 18 percent, not a likely scenario. But by having advance warning, we can choose to order less costly trees, or not as many. Or we can decide to forego some of the flowers. This advance warning lets us choose how we want to address the cost overrun.

Trend analysis compares previous performance reviews to the current status to determine if the trend is improving or deteriorating.

Project management software is used to calculate variances, indexes, forecasts, trends, and other useful scenarios and trends.

Reserve analysis is used to monitor the depletion rate of the cost reserves for the project. If the project is 50 percent complete and 70 percent of the reserves have been allocated, the team may end up asking for additional reserves. Conversely, if there are excess reserves, those funds may be released for other projects or business operations.

OUTPUTS

Work performance information takes the data from a variance analysis and puts them into calculable form. Variance analysis compares the work performance data with the cost baseline to determine the variance. The variance thresholds established in the cost management plan are referenced to determine if the variance is acceptable, or if corrective or preventive actions need to be deployed. If the variance is trending toward a threshold then some preventive actions will be an output. If it has surpassed the threshold, corrective actions are appropriate.

The work performance information is where you will see the earned value numbers such as the CPI and SPI. Cost forecasts project the funds needed for the next phase, reporting period, or the balance of the project. If the team is using earned value, the ETC and EAC are calculated here. The Earned Value Calculations Summary Table shown in Figure 23-7 shows an overview of some of the most common earned value calculations.

Causes of any variances and the actions taken along with the logic behind those actions are used to update organizational process assets.

In some situations the variances are of a nature that a change request form will need to be filled out. Change requests are usually associated with scope changes or cost overruns. They are processed through the Perform Integrated Change Control process. Preventive and corrective actions may be considered a change request if they impact elements of the project management plan or project documents.

Elements of the project management plan that may be updated include the cost baseline and the cost management plan. Project document updates include cost estimates, the basis of estimates, and the funding requirements.

Chapter 24

Monitoring and Controlling Quality

TOPIC COVERED

Control Quality

Control Quality

Control Quality is the process of monitoring and recording results of executing the quality activities to assess performance and recommend necessary changes. As you recall, quality is concerned with both the project and the product. In this process you will review the project results and compare them to the project metrics, such as budget performance and variance, and schedule performance and variance. You will also assess the degree to which the deliverables meet the quality metrics and requirements. As mentioned previously, product quality metrics are specific to the particular product and organization.

The Control Quality process has some specific concepts associated with measuring compliance that you should be familiar with. Let's take a look at those.

Quality Control Terminology

Inspection. Examining or measuring to verify whether an activity, component, product, result, or service conforms to specified requirements.

Attribute Sampling. Method of measuring quality that consists of noting the presence (or absence) of some characteristic (attribute) in each of the units under consideration. After each unit is inspected, the decision is made to accept a lot, reject it, or to inspect another unit.

Tolerance. The quantified description of acceptable variation for a quality requirement.

Control Limits. The area composed of three standard deviations on either side of the centerline, or mean, of a normal distribution of data plotted on a control chart that reflects the expected variation in the data.

Table 24-1 Examples of Quality Control Terminology

Prevention. Establishing processes to make sure errors or defects don't occur.	If we are building a new heating component we employ the information from past successful components, use our most senior designers, and establish a backup temporary heating device that will start in the event the primary device does not start, or fails mid-cycle.
Inspection	We can test every heating component against the requirements before it is approved.
Attribute sampling	If the metric is that the oven must heat to 525, you can use a thermometer to determine if the temperature reaches 525. An easier definition of attribute sampling is determining if the result conforms or not.
Variable sampling. Determining the degree of conformity.	If a target is to heat to 525 and the oven temperature is 510, then it has achieved 97 percent of the target.
Tolerance	If the target is 525 you may have a ± 10 tolerance. In other words, any measurement from 515 to 535 is acceptable.
Control limits	If you are measuring how long it takes the oven to reach 525 and you have an average time of 12 minutes and a range of 8 to 16 minutes, you may determine that any oven you test that heats faster or slower than that is not behaving as expected. There is something that is causing it to be out of the control limit and you should determine the variable that is causing the behavior. An easier way of thinking of control limits is as the thresholds that determine whether a process is in or out of control.

Some of those definitions are pretty complex. Table 24-1 provides some examples to help make the terms easier to understand in the context of quality control. The table also includes definitions for *prevention* and *variable sampling,* which are not included in the *PMBOK® Guide*—Fifth Edition Glossary, but are used when you are controlling quality.

In the Plan Quality Management process we used some of the same techniques we will be looking at for this process, however, in that planning process we were establishing the control limits, tolerances and determining the sampling populations. In this process we will be applying those tools to measure the correctness of deliverables to determine if we are within in the control limits and tolerances. As with all the control processes we will compare the plan to the results to determine if we need to take any corrective or preventive actions, repair any defects, or make any changes.

Figure 24-1 shows the inputs, tools and techniques and outputs for the Control Quality process. Figure 24-2 shows a data flow diagram for the Control Quality process.

INPUTS

The project management plan contains the quality management plan which outlines how the team will meet the quality metrics, standards, and requirements for the project and product. In some cases a quality checklist is used to make sure the team performs a series of steps needed to meet the quality requirements.

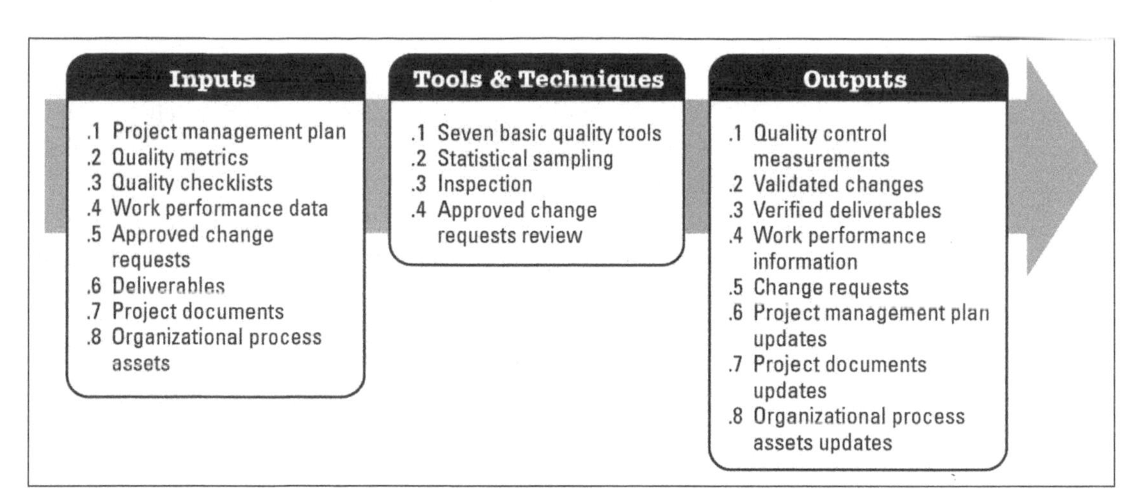

Figure 24-1
Control Quality: Inputs, Tools and Techniques, Outputs
Source: *PMBOK® Guide*—Fifth Edition

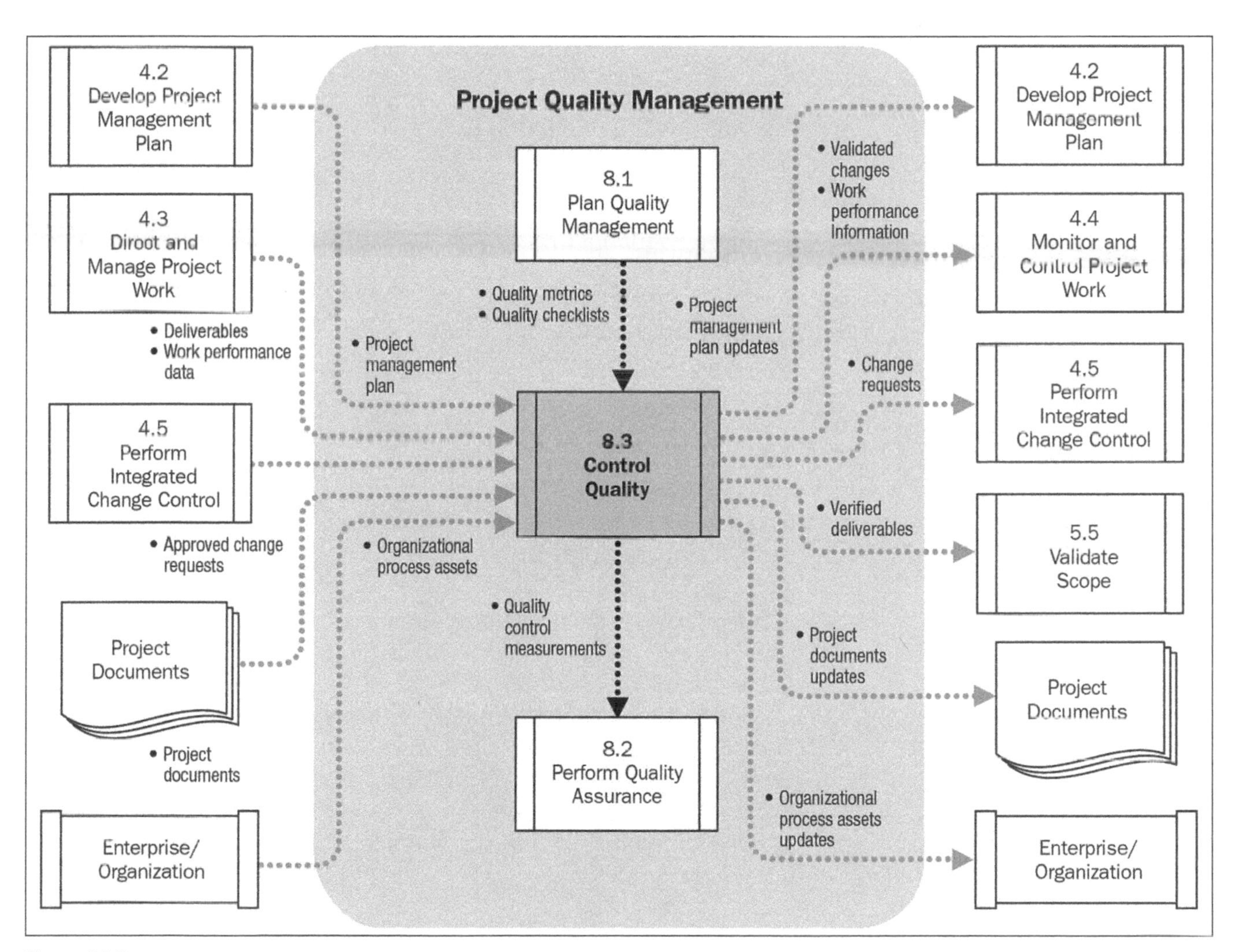

Figure 24-2
Control Quality Data Flow Diagram
Source: *PMBOK® Guide*—Fifth Edition

Work performance data are compared to the established tolerances and control limits to determine if the project performance for scope, schedule, and cost are performing acceptably. The deliverables are compared to the quality metrics to determine if they meet the specified acceptance criteria.

One of the outputs of the Perform Integrated Change Control process is approved change requests. For those changes that were approved, the implementation of the change needs to be validated to ensure the changes are implemented in a timely fashion and appropriately.

Project documents, such as audit reports from Perform Quality Assurance, corrective action plans, process documentation and results from deploying any of the quality management tools and techniques are used in the Control Quality process.

The organizational process assets include project and product performance standards, quality management policies, procedures, forms and templates, and any defect management procedures. These items provide a system and guidance in how to apply the quality management plan in a manner consistent with the organization's infrastructure.

TOOLS AND TECHNIQUES

The seven basic quality tools that were described in the Plan Quality Management process are implemented here to determine if the deliverables meet the metrics and quality requirements and specifications. As a reminder, those seven basic quality tools are:

- Cause and effect diagram
- Flowcharts
- Checksheets
- Pareto charts
- Histograms
- Control charts
- Scatter diagrams

In the Plan Quality Management process we looked at statistical sampling to determine the number of items we should test and the timing of the tests. In the Control Quality process we perform the actual testing of items to ensure they comply with requirements. Once we select the items, we will perform an inspection. Inspection can take the obvious form of looking at the item to see if it conforms to a specification, however, not all requirements and standards can be determined just by looking at something. Sometimes we need to measure the outcome (such as size, weight, or proportion). Other times we may need to conduct tests to determine if the item performs as expected. If we are inspecting a process or a deliverable composed of several pieces and parts, an inspection can be called a walkthrough, a review, or other terms that are industry specific.

The approved change requests from the Perform Integrated Change Control process are reviewed to ensure they were implemented properly.

OUTPUTS

All the testing and tools and techniques applied in this process result in quality control measurements. These are the quantifiable results and outcomes of the quality control activities. If the results are satisfactory, they are documented and the deliverables are verified as being correct. They will then go through the Validate Scope process. If the results are not satisfactory a change request is initiated. The change request may entail a defect repair for the deliverable, or it may entail preventive or corrective action for the process involved in creating the deliverable. The approved change requests are considered validated changes using this same approach.

Work performance information contains contextualized information about how well the quality processes are performing, defect rates and sources, causes of rework and rejections, and information on any processes that need modifications or improvements.

The quality management plan and process improvement plan are components of the project management plan that may be updated. Quality standards or metrics are project documents that may be updated.

The completed checklists, inspection results, and other corresponding documentation are considered organizational process asset updates. Lessons learned about the cause of defects and the resulting repair or corrective actions should also be recorded and added to the organizational process assets.

Chapter 25

Monitoring and Controlling Communications

TOPIC COVERED

Control Communications

Control Communications

Control Communications is the process of monitoring and controlling communications throughout the entire project lifecycle to ensure the information needs of project stakeholders are met. Essentially the Control Communications process seeks to ensure that the right people are getting the correct information in a timely manner. If they aren't, the change process is initiated through this process.

Figure 25-1 shows the inputs, tools and techniques and outputs for the Control Communications process. Figure 25-2 shows a data flow diagram for the Control Communications process.

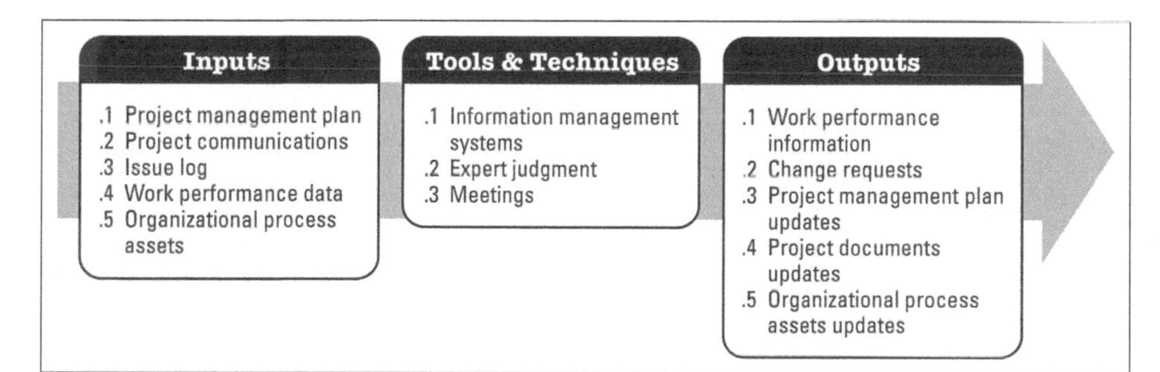

Figure 25–1

Control Communications: Inputs, Tools and Techniques, Outputs

Source: *PMBOK® Guide*—Fifth Edition

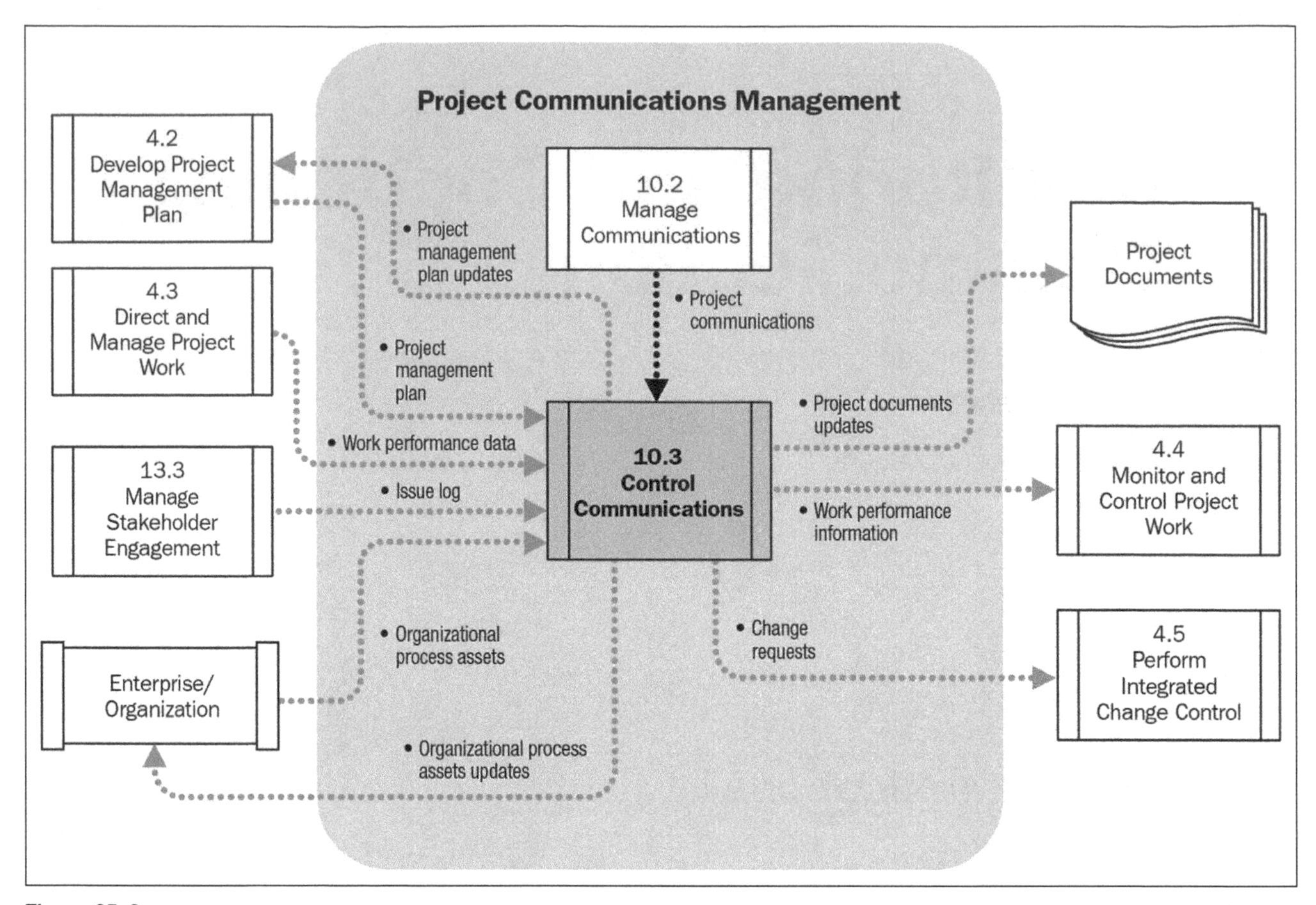

Figure 25–2
Control Communications Data Flow Diagram
Source: *PMBOK® Guide*—Fifth Edition

INPUTS

The project management plan contains the communications management plan and stakeholder management plan which identify the stakeholders who need project communications, when they will receive the information, and the format of the information.

Issue logs may contain documentation on issues associated with communication and the parties accountable for resolving the issues. The work performance data comes from the Direct and Manage Project Work process. Organizational process assets, such as policies, procedures, templates, technology, and security requirements influence all aspects of communication.

TOOLS AND TECHNIQUES

The organization's reporting systems collect, consolidate, and distribute information. Expert judgment can be employed to determine whether the communications processes are working as intended. If not, one or more meetings may be required to determine the issue and generate options to resolve the issue.

OUTPUTS

Work performance information contains contextualized information about how well the project is performing. If performance is not as expected a change request may be submitted to align the communications processes with the needs of the project.

The project management plan components that may get updated as part of this process include the communications management plan, the stakeholder management plan and the human resource management plan.

Issue logs, performance reports, and forecasts are project documents that may be updated. If aspects of the communication process are not working, templates, procedures, and reporting formats are organizational process assets that can be updated.

Chapter 26

Monitoring and Controlling Risks

TOPIC COVERED

Control Risks

Control Risks

Control Risks is the process of implementing risk response plans, tracking identified risks, monitoring residual risks, identifying new risks, and evaluating risk process effectiveness throughout the project. Control Risk includes implementing risk response plans and monitoring the effectiveness of the specific responses as well as the risk management process as a whole. As the project progresses, the team will continually revisit all the risk management planning processes to identify, analyze, and respond to risks and opportunities throughout the project.

Information on the project performance is reviewed throughout the project to identify trends that can lead to negative project results, assumptions that are not valid, and the use of schedule and cost contingency reserve.

If results are not favorable the project team develops or employs previously identified strategies, takes corrective or preventive actions, and acts on contingency and fall-back plans as necessary.

Figure 26-1 shows the inputs, tools and techniques and outputs for the Control Risks process. Figure 26-2 shows a data flow diagram for the Control Risks process.

INPUTS

The project management plan contains the risk management plan as well as the project baselines. The risk management plan provides guidance on how often risks will be reassessed, the tools and techniques that will be used to monitor risk, and how risk reserve will be utilized.

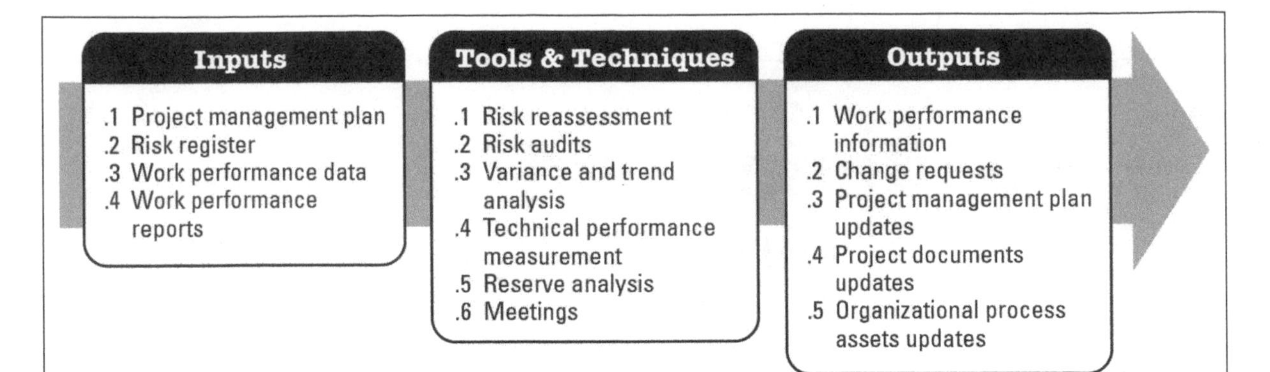

Figure 26-1
Control Risks: Inputs, Tools and Techniques, Outputs
Source: *PMBOK® Guide*—Fifth Edition

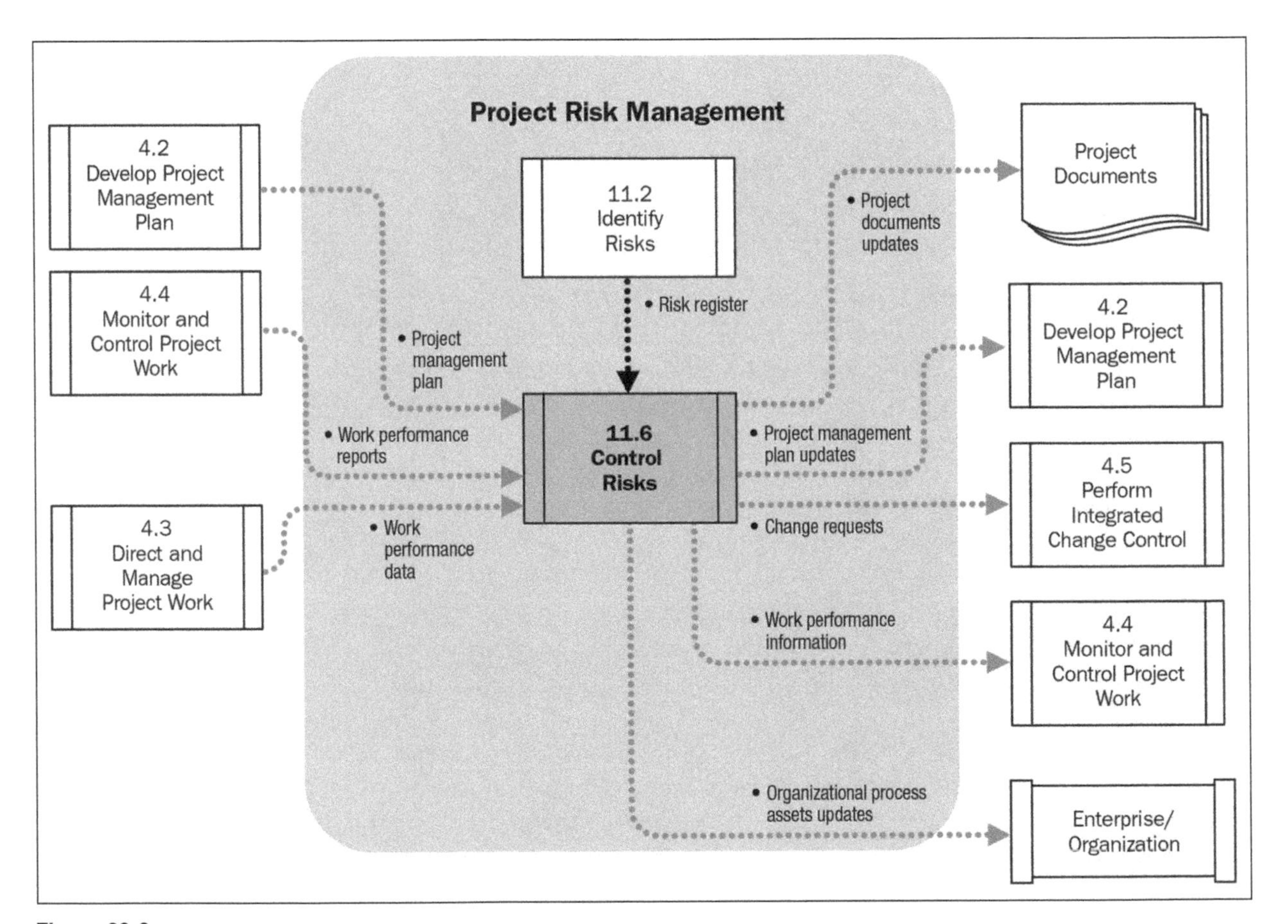

Figure 26-2
Control Risks Data Flow Diagram
Source: *PMBOK® Guide*—Fifth Edition

The Risk Register contains all the identified risks, an analysis of their probability and impact, and risk responses. It also identifies the person responsible for the risk and risk triggers.

The baselines are used along with the work performance data and work performance reports to determine if project performance is at risk. The work performance data will indicate status of work in progress and the incurred costs. Work performance reports integrate information from variance analysis, earned value, and forecasts.

TOOLS AND TECHNIQUES

Risk reassessment is the process of reviewing the existing information on the Risk Register and updating it. Common questions you can ask include:

- Are there any new risks?
- Have any existing risks changed in probability or impact?
- Can any risks be closed?
- Have any trigger events occurred?
- How effective have my risk response strategies been?

A risk audit is used to review the risk management process to determine its effectiveness. If there are numerous events occurring that are having a negative impact on the project, then you can assume that something in either the Identify Risks process or the Plan Risk Responses processes is not adequate. The risk audit can also determine if the risk process is being used as planned. Are risk meetings being conducted? Are the probability and impact matrixes being used appropriately for the project? Are risk owners following through? Finally, the risk audit also tracks the effectiveness of risk triggers and risk responses to determine if the triggers were appropriate and if the responses managed the risk the way the team expected they would.

Technical performance measurement looks at the technical achievement of the deliverables and determines if they are progressing according to plan. For example, if a process is supposed to achieve a certain volume of throughput and there are interim checkpoints of specified volume achievement, this process would compare the actual achievement with the planned technical achievement.

Variance and trend analysis looks at cost, schedule, and technical achievement trends to determine if any of the project objectives are at risk based on current and predicted future performance.

Reserve analysis is conducted to determine if the existing reserve is adequate for the project given the expected remaining duration, the progress to date, and the forecasted estimate at completion. Some organizations have specific metrics that indicate how much reserve should be left at various points in the project. These metrics help the project management team determine how they are doing compared to the benchmarks established by the organization.

Status meetings can be used to implement any of the above techniques. It is a good practice to address risks at each status meeting, even if it is just as a standing agenda item at weekly meetings to inquire if there are any new risks.

RISK MANAGEMENT PLANNING

Since the planning processes continue throughout the project it is not really necessary to include them in the monitoring and control processes. However, because it is important to continually identify, analyze, and develop responses for risks, the risk reassessment activities are specifically identified to remind project managers that risk management occurs throughout the project, not just in the beginning.

OUTPUTS

Work performance information is used to support decision making and alternatives analysis for the project. For example, if the team is looking at several corrective actions, taking the work performance information from the Control Risk process can provide useful information about the burn rate of contingency, risks associated with forecasts, and the current risks on the project.

Based on the outcomes of the variance and trend analysis it may be necessary to generate a change request in the form of preventive or corrective action to keep the project performance on track.

Project management plan updates include any of the project management subsidiary plans or baselines. These may change based on the risk reassessment or the implementation of contingency actions or implementing risk responses. The project documents that can be updated include the risk register, assumption logs, and technical documentation.

Organizational process assets are updated to provide information for future projects, in particular the risk register, the probability and impact matrices, and the risk response plans.

Chapter 27

Monitoring and Controlling Procurements

TOPIC COVERED

Control Procurements

Control Procurements

Control Procurements is the process of managing procurement relationships, monitoring contract performance, and making changes and corrections as appropriate. This process requires much coordination and integration with the project and between vendors on the project. Controlling procurements involves:

- Reviewing performance and comparing it to the agreed upon plan and contractual provisions
- Implementing preventive and corrective actions as appropriate
- Managing changes and revisions to the work and the agreements
- Ensuring appropriate payments to the vendors
- Coordinating work between vendors and the project

Because contracts are legal documents there is usually someone from the contracts or procurement department on the team who handles all the contractual issues and obligations.

Figure 27-1 shows the inputs, tools and techniques and outputs for the Control Procurements process. Figure 27-2 shows a data flow diagram for the Control Procurements process.

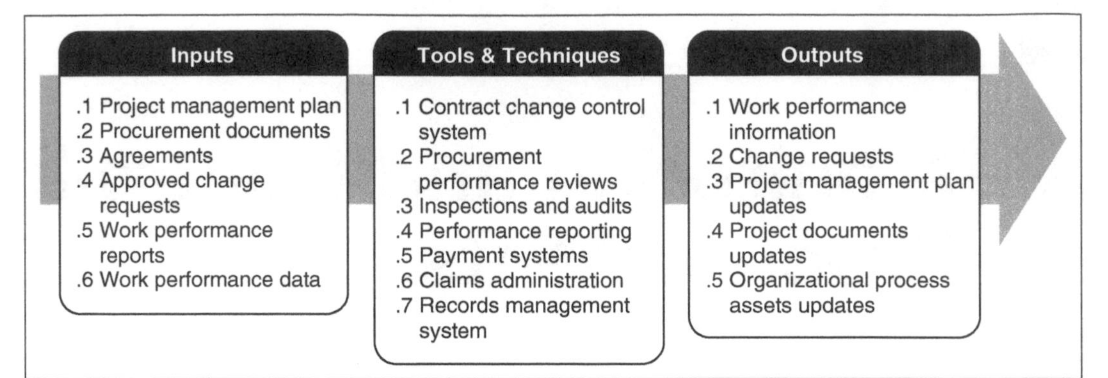

Figure 27-1
Control Procurements: Inputs, Tools and Techniques, Outputs
Source: *PMBOK® Guide*—Fifth Edition

INPUTS

The project management plan contains the procurement management plan which provides direction on:

- Managing multiple suppliers
- Coordinating procurements with other project aspects, such as scheduling and performance reporting
- Metrics that will be used to manage contracts

The procurement documents include the statement of work. The agreement establishes the legal foundation of the relationship. It can include the following information that is useful when managing the contract:

- Technical information on deliverables
- Schedule baseline
- Performance reporting requirements
- Roles and responsibilities
- Payment terms
- Acceptance criteria

Approved change requests have been through the Perform Integrated Change Control process. Change requests can include changes to the statement of work, deliverables, due dates, cost estimates, terms and conditions, or services and results.

Work performance reports and work performance data provide detailed information on vendor status. Much like the work performance reports and work performance data on the internally produced deliverables, they identify the work that has started, the progress to date, the work that has completed, committed costs, forecasts, technical achievement to date, and other information as required.

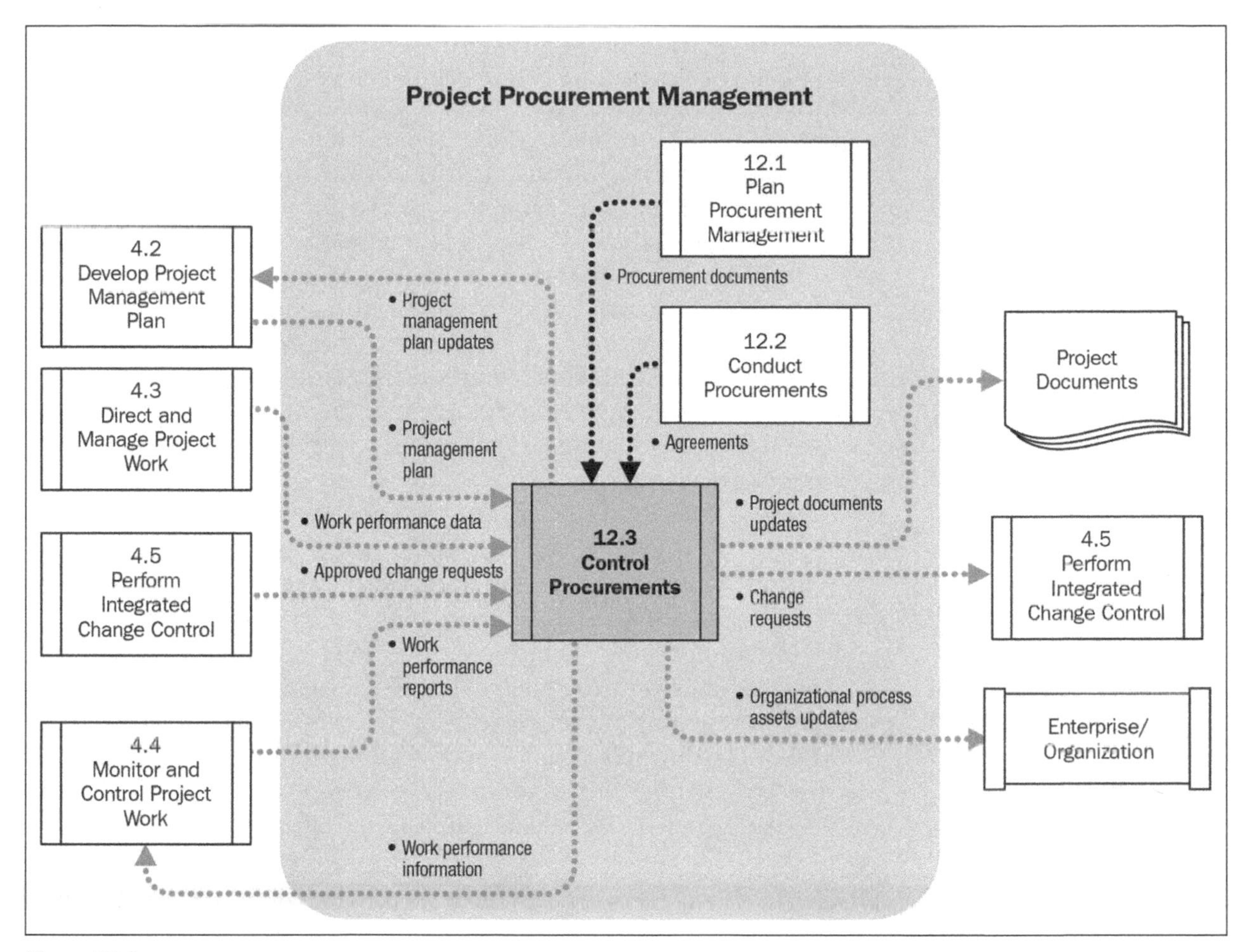

Figure 27-2
Control Procurements Data Flow Diagram
Source: *PMBOK® Guide*—Fifth Edition

TOOLS AND TECHNIQUES

There are a number of systems that are used to manage suppliers:

- *Contract change control system.* The contract change control system is part of and conforms to the integrated change control system. It also contains information on approval levels for authorizing contract changes, the paperwork needed for contract changes, and tracking systems.
- *Payment system.* The payment system includes the necessary paperwork to submit charges, the process to validate that the amount being charged is consistent with the work actually done, and the accounts payable function in the organization.
- *Records management system.* Contracts often have a large amount of paperwork and documentation. A records management system provides a structure to organize all the contract documents and ensure correspondence is controlled, safe, and readily accessible.

Performance reporting for vendors is much the same as it is for internal work. The baselines are compared to the work performance information to determine technical, schedule, and cost performance. A procurement performance review is a more in-depth, structured review of the contract status. These reviews can take place at the client site to permit the buyer to observe the status of in-progress work. They can take all day on large complex projects. Where there are variances from plan, the buyer and seller discuss the cause of the variance and the corrective actions the appropriate party will take. The parties also discuss risks and issues that have arisen since the last review, along with reviewing existing risks and issues and any changes in status since the prior review. Many times these reviews include the project manager, technical leads, project businesspeople, and the contracts person.

An inspection and audit may be required in the contract. The inspection or audit is not the same as a performance review. The inspection or audit is more focused on compliance to process, procedure, and adherence to contractual stipulations. In other words, a performance review determines the seller's progress, while the audit determines how they achieved that progress.

From time to time on projects, there will be disputes. Working through these disputes is called claims administration. A claim can be a contested change (the buyer says it is in scope; the seller says it is new scope), or a disagreement on compensation for a change. The contract usually spells out the dispute resolution steps that should be taken to work through claims. Most contracts require the use of alternative dispute resolution (ADR) rather than going through the court system. ADR can take the form of mediation or arbitration. Mediation uses a third party to work with the two parties to try to reach some kind of agreement. The agreement is voluntary. Arbitration uses a professional mediator and the parties can agree that the outcome is binding. The advantage of ADR is that it is not subject to the rules, time, and costs that formal litigation imposes.

OUTPUTS

Work performance information provides the status on the vendor's work, similar to the work performance reports for in-house work. Information on scope, schedule, cost, and quality are used to determine if the vendor is performing consistent with the agreement. The work performance information can also provide insight as to whether the organization should use the vendor on future work.

Change requests can take the form of changes to the contract, or because of the vendor performance, the changes can impact the project management plan including baseline schedules, budgets, or the scope baseline. All change requests go through the Perform Integrated Change Control process.

Project management plan updates can include changes to the procurement management plan, or the schedule and budget. As mentioned above, those changes will have to go through the Perform Integrated Change Control process.

One thing procurements do well is generate paper! All the records, communication, invoices, change requests, status reports, and so forth are project documents that are updated, or organizational process assets that are updated. The organizational assets can also include correspondence with the vendor, evaluation of the vendor from the audit and inspection process, performance reports, and results of the reviews. Assets can also be updated with evaluations of the vendor performance that can be used for future procurements.

Chapter 28

Monitoring and Controlling Stakeholder Engagement

TOPIC COVERED

Control Stakeholder Engagement

Control Stakeholder Engagement

Control Stakeholder Engagement is the process of monitoring overall project stakeholder relationships and adjusting strategies and plans for engaging stakeholders. This process compares the stakeholder management plan with the results from engaging stakeholders. If stakeholder engagement is not occurring as desired a change request is initiated. You will notice that the artifacts used to control stakeholder engagement are similar to those used to control communications. These two processes are very tightly linked. They may occur simultaneously without much distinction between the two. However, there are times when the focus is explicitly on how to most effectively engage and manage stakeholders, or how to most effectively communicate.

Figure 28-1 shows the inputs, tools and techniques and outputs for the Control Stakeholder Engagement process. Figure 28-2 shows a data flow diagram for the Control Stakeholder Engagement process.

INPUTS

The project management plan contains the stakeholder management plan, the life cycle, the communications management plan, and the human resource management plan. All of this information is used to assess the effectiveness of stakeholder engagement.

Issue logs identify any issues and the status of issue resolution. The work performance data provides information on the project results. Other project documents that provide insight into the effectiveness of stakeholder engagement include the schedule, change logs, the stakeholder register, and various communications.

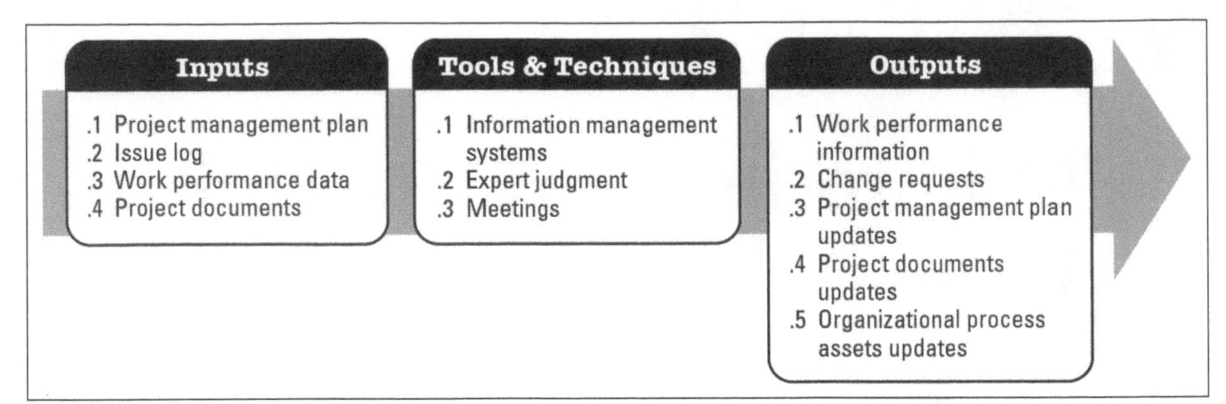

Figure 28-1
Control Stakeholder Engagement Inputs, Tools and Techniques, Outputs
Source: *PMBOK® Guide*—Fifth Edition

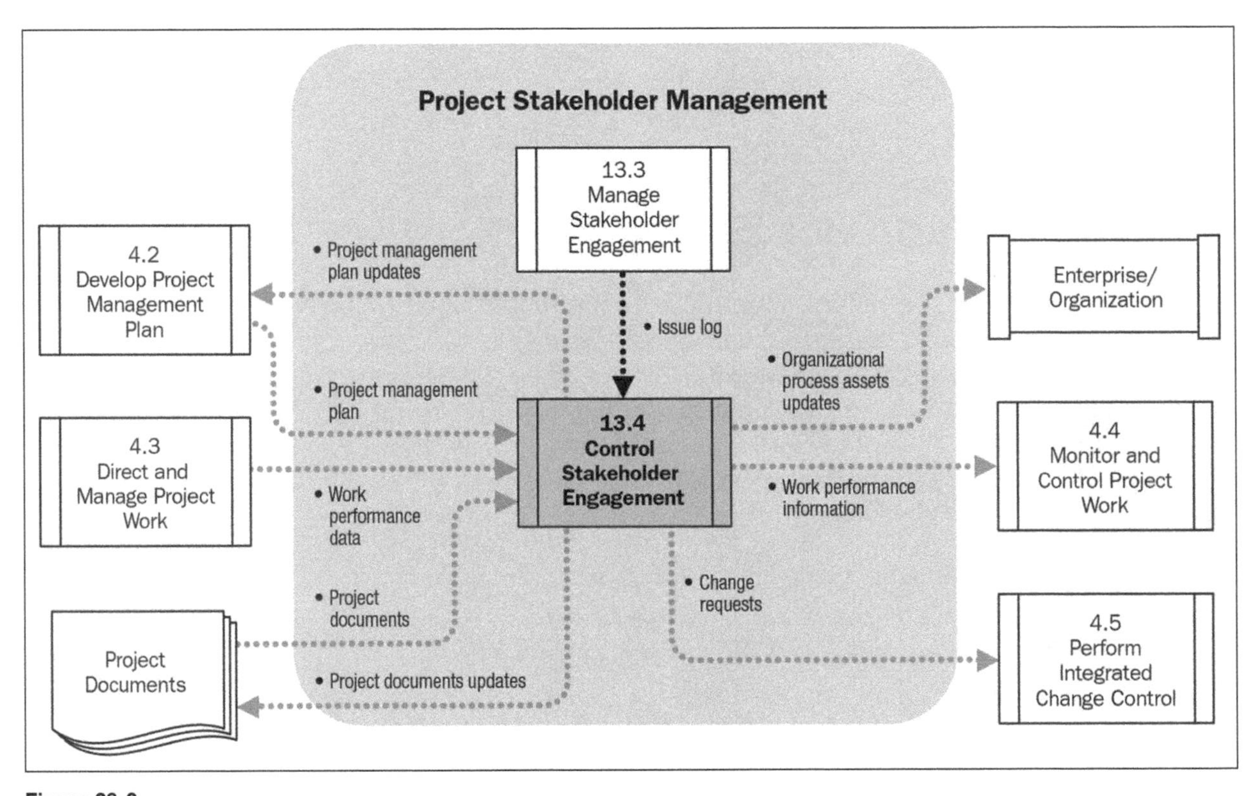

Figure 28-2
Control Stakeholder Engagement Data Flow Diagram
Source: *PMBOK® Guide*—Fifth Edition

TOOLS AND TECHNIQUES

The organization's information management systems collect, consolidate, and distribute information about the project performance. Expert judgment can be employed to determine whether the stakeholder management processes are working as intended. If not, one or more meetings may be required to determine the issue and generate options to resolve any stakeholder engagement issues.

OUTPUTS

Work performance information contains contextualized information about how well the project is performing. If performance is not as expected a change request may be submitted to correct deviation from the plan or prevent deviation from the plan.

Any aspect of the project management plan may be updated based on input from stakeholders. For example, if communication with stakeholders is not effective, the communications management plan will be updated. If there are a lot of issues concerning scope creep vs. progressive elaboration the scope management plan may need to be updated.

Issue logs and the stakeholder register are project documents that may be updated. Project reports, records, presentations, feedback, and lessons learned are examples of updates to the organizational process assets.

Chapter 29

Closing the Project

TOPICS COVERED

Closing Process Group

Close Project or Phase

Close Procurements

Closing Process Group

The Closing Process Group consists of those processes performed to finalize all activities across all the Project Management Process Groups to formally close the project or phase. The most common activities include:

- Obtaining formal acceptance of all deliverables
- Comparing final project results to the project objectives; product requirements; and the scope, schedule, and cost baselines
- Conducting a lessons learned session, or series of sessions, documenting the results, and distributing them as indicated in the communication management plan
- Releasing project resources
- Transitioning the final product
- Organizing and archiving all project documentation
- Closing out all contracts
- Acknowledging the team and celebrating success

Close Project or Phase

Close Project or Phase is the process of finalizing all activities across all of the Project Management Process Groups to formally complete the project or phase. This process is invoked at the end of each phase and at the end of the project. If the project is terminated prior to completion, this process will make sure all the documentation is collected and archived.

During the process all the exit criteria needed to close out one phase of the project life cycle and move to the next are addressed. If there is a transfer of deliverables or products at the end of the phase or the project the transition activities are also addressed.

Figure 29-1 shows the inputs, tools and techniques and outputs for the Close Project or Phase process. Figure 29-2 shows a data flow diagram for the Close Project or Phase process.

INPUTS

The project management plan describes the activities necessary to complete each phase, such as a list of documents that must be signed

Figure 29-1
Close Project or Phase: Inputs, Tools and Techniques, Outputs
Source: *PMBOK® Guide*—Fifth Edition

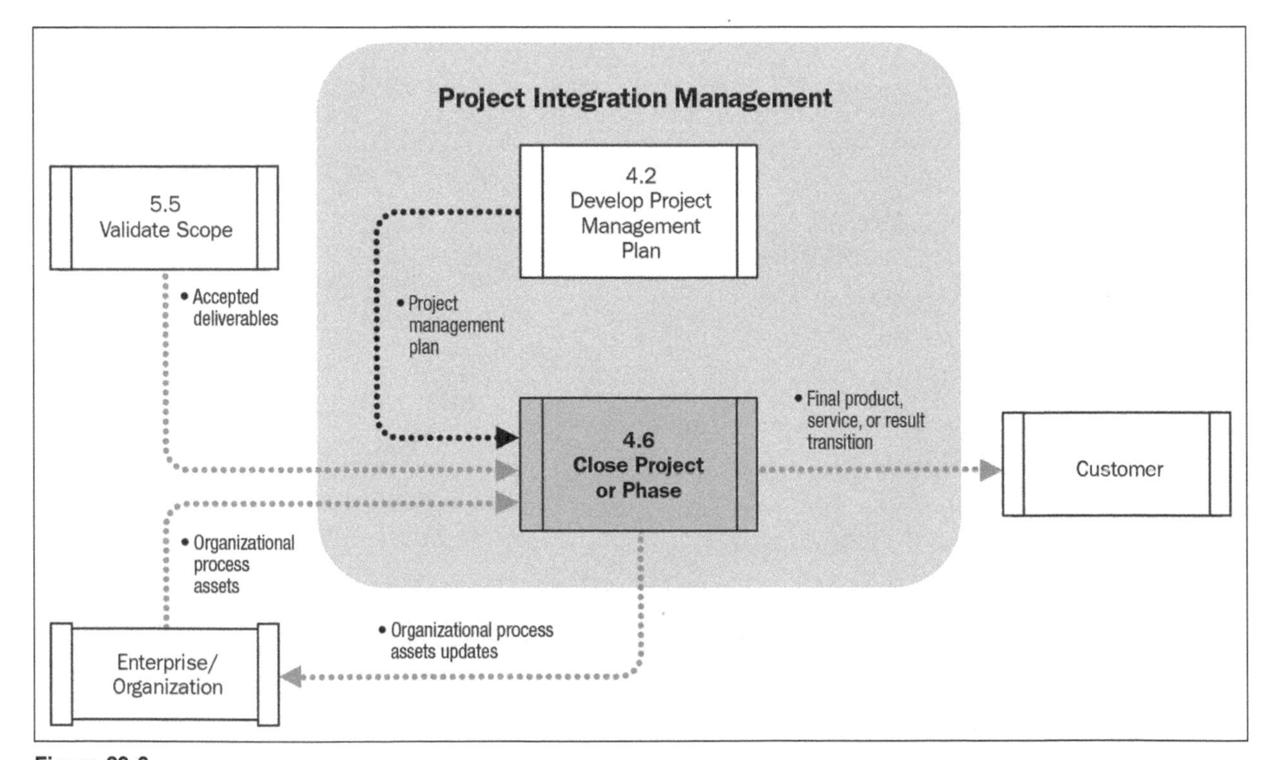

Figure 29-2
Close Project or Phase Data Flow Diagram
Source: *PMBOK® Guide*—Fifth Edition

off at a phase gate review, and the activities necessary to complete the project. Project completion is measured against the project management plan and the work is measured against the baselines to determine that all the work is done according to schedule and within budget—or hopefully somewhere close to budget!

Accepted deliverables come from the Validate Scope process. When there are multiple deliverables in the project they should each be accepted as completed prior to accepting the final product in the Close Project or Phase process. By the time the project is closing, the documentation from the prior phase closures and the documentation that shows accepted deliverables is compiled and presented for final project and product acceptance and signoff.

Guidelines, procedures, and forms used to close out the project are the organizational process assets that are most often used.

TOOLS AND TECHNIQUES

Technical expert judgment is used to attain product signoff. Project management expertise, financial expertise, and sometimes the project management office are involved to ensure a clean and complete project close-out. When all or part of the project is done under contract, the procurement and legal departments may provide expertise to make sure project contractual obligations are complete.

Analytical techniques are used in writing up the final project report and compiling lessons learned. For example, the team may use a regression analysis to determine the root cause of any realized risks, contentious issues, or other project challenges.

As part of the product or service transition there will likely be several meetings to ensure the person or organization receiving the deliverables is fully briefed and ready to support and maintain the deliverables. Meetings are also used to elicit lessons learned from project team members and other stakeholders. A lessons-learned meeting or series of meetings can analyze performance and document what went well and what could be improved in the future. It may focus on specific project processes, such as estimating or risk management, or it may focus on internal team dynamics.

OUTPUTS

The final product, service, or result is transitioned either to the customer or operations. At this point there may still be some open action items. The project manager assigns these to someone for follow-up. Transition can include on-the-job training, transition meetings, or sometimes onsite support for a specified period of time.

The organizational process assets that are updated include lessons learned, project documentation, and the transition documentation. When writing up the final project report that will become an organizational process asset, it is a good idea to compare the project results with the objectives and success criteria identified in the charter and document how the objectives were met.

See Appendix A for an example of a Lessons Learned form and a Project Close-Out form.

Close Procurements

Close Procurements is the process of completing each project procurement. Contracts can be terminated or closed due to completion, for cause, or for convenience. Most contracts are closed based on completion—all the work is done per the contractual agreement. When contracts are terminated for cause, it is because one party has breached the contract or is about to breach the contract. For example, they are filing for bankruptcy or they do not have the capacity or resources to complete the contract. Termination for convenience is when the buyer terminates the contract because they no longer have the need for, or the desire for the product, service, or result.

Figure 29-3 shows the inputs, tools and techniques and outputs for the Close Procurements process. Figure 29-4 shows a data flow diagram for the Close Procurements process.

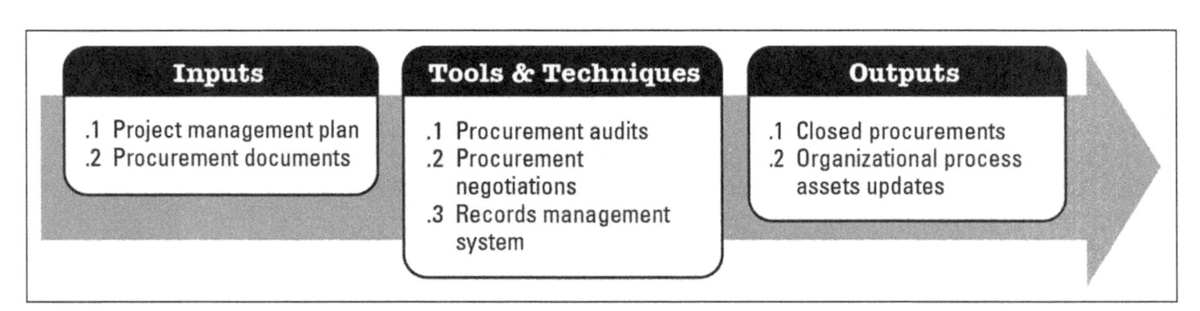

Figure 29-3
Close Procurements: Inputs, Tools and Techniques, Outputs
Source: *PMBOK® Guide*—Fifth Edition

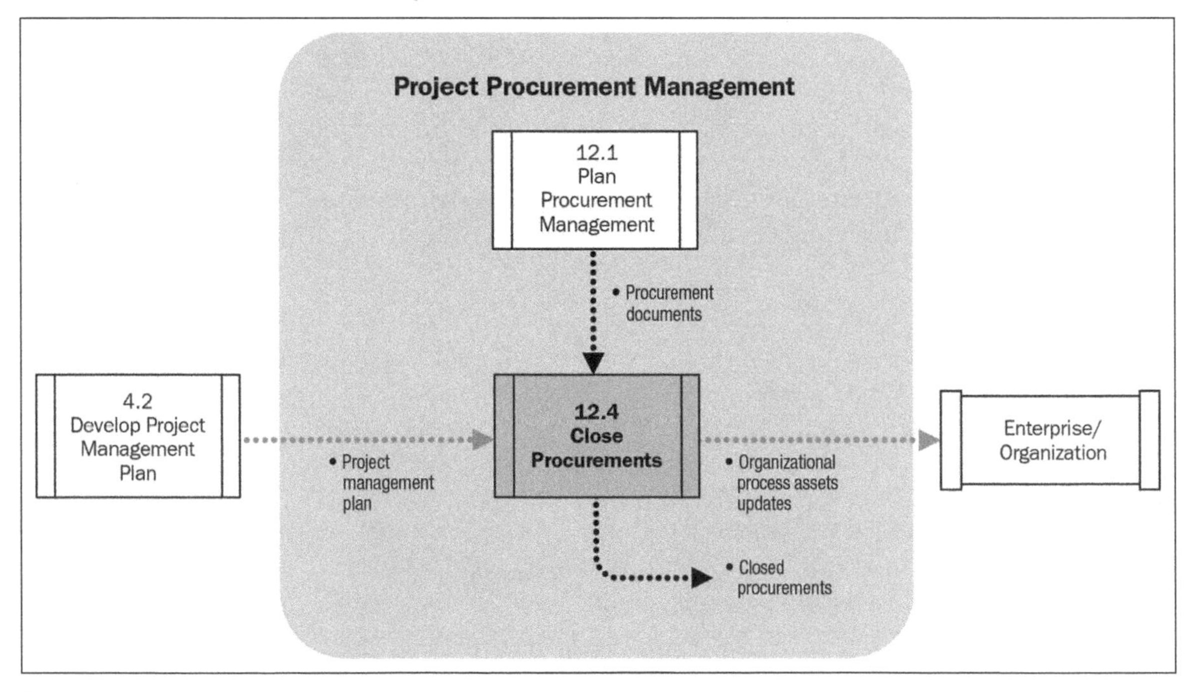

Figure 29-4
Close Procurements Data Flow Diagram
Source: *PMBOK® Guide*—Fifth Edition

INPUTS

The project management plan contains the procurement management plan which outlines the steps necessary to close out each procurement. This can include a checklist of items that need to be completed, forms for which signoff is needed, and equipment that needs to be returned.

Procurement documents include all correspondence, performance reports, invoices, inspection results, and contract changes that have been accumulated throughout the life of the contract. In short, it includes anything written that pertains to the contract. This will need to be indexed and archived.

TOOLS AND TECHNIQUES

A procurement audit is conducted to document the procurement process on a project or on a specific contract. It can record both positive and negative aspects of the procurement. See Appendix A for an example of a Procurement Audit form.

Procurement negotiations are used to resolve any open claims or disputes between the buyer and seller. Generally, mediation or arbitration is preferred to litigation because of timeliness and cost concerns. The records management system organizes and compiles all the procurement documentation.

OUTPUTS

The agreement defines who has the authority to close procurements. Usually the project manager will notify the person with contractual authority that the statement of work has been fulfilled and this will start the formal closure paperwork.

Organizational process assets that are updated include all the procurement documentation, the acceptance of all deliverables, the results of the procurement audit, and any lessons learned from the procurement.

Appendix

Process:	Appendix Form:
Collect Requirements Output	Figure A-1: Requirements Documentation 262
Collect Requirements Output	Figure A-2: Requirements Traceability Matrix 263
Define Scope Output	Figure A-3: Assumption and Constraint Log 264
Create WBS Output	Figure A-4: WBS Dictionary 265
Define Activities Output	Figure A-5: Activity Attributes 266
Estimate Activity Resources Output	Figure A-6: Resource Breakdown Structure 267
Estimate Activity Durations Output	Figure A-7: Activity Duration Estimates 268
Estimate Costs Output	Figure A-8: Cost Estimating Worksheet 269
Plan Human Resource Management T&T	Figure A-9: Responsibility Assignment Matrix 270
Plan Communications Management Output	Figure A-10: Communications Management Plan 271
Identify Risks (Output)	Figure A-11: Risk Register 272
Identify Risks (Output)	Figure A-12: Risk Data Sheet 273
Plan Procurement Management Output	Figure A-13: Procurement Management Plan 274
Develop Project Team Tool and Technique	Figure A-14: Team Operating Agreement 277
Perform Integrated Change Control Output	Figure A-15: Change Request 279
Close Project or Phase Output	Figure A-16: Lessons Learned 282
Close Project or Phase Output	Figure A-17: Project Close-Out 284
Close Procurements Tool and Technique	Figure A-18: Procurement Audit 286

REQUIREMENTS DOCUMENTATION

Project Title: ______________________ Date Prepared: ______________________

ID	Requirement	Stakeholder	Category	Priority	Acceptance Criteria	Validation Method

Page 1 of 1

Figure A-1
Requirements Documentation

REQUIREMENTS TRACEABILITY MATRIX

Project Title: ______________________ Date Prepared: ______________________

Requirement Information					Relationship Traceability			
ID	Requirement	Priority	Category	Source	Objective	WBS Deliverable	Metric	Validation

Figure A-2
Requirements Traceability Matrix

ASSUMPTION AND CONSTRAINT LOG

Project Title: ______________________ Date Prepared: ______________________

ID	Category	Assumption/Constraint	Responsible Party	Due Date	Actions	Status	Comments

Page 1 of 1

Figure A-3
Assumption and Constraint Log

WBS DICTIONARY

Project Title: ______________________ Date Prepared: ______________________

Work Package Name:	Code of Account:
Description of Work:	Assumptions and Constraints:
Milestones: 1. 2. 3.	Due Dates:

ID	Activity	Resource	Labor			Material			Total Cost
			Hours	Rate	Total	Units	Cost	Total	

Quality Requirements:

Acceptance Criteria:

Technical Information:

Agreement Information:

Page 1 of 1

Figure A-4
WBS Dictionary

ACTIVITY ATTRIBUTES

Project Title: ______________________ Date Prepared: ______________________

ID:	Activity:				
Description of Work:					
Predecessors	Relationship	Lead or Lag	Successor	Relationship	Lead or Lag
Number and Type of Resources Required:	Skill Requirements:		Other Required Resources:		
Type of Effort:					
Location of Performance:					
Imposed Dates or Other Constraints:					
Assumptions:					

Page 1 of 1

Figure A-5
Activity Attributes

RESOURCE BREAKDOWN STRUCTURE

Project Title: ______________________ Date Prepared: ______________________

2. Project
    2.1. People
        2.1.1. Quantity of Role 1
            2.1.1.1. Quantity of Level 1
            2.1.1.2. Quantity of Level 2
            2.1.1.3. Quantity of Level 3
        2.1.2. Quantity of Role 2
    2.2. Equipment
        2.2.1. Quantity of Type 1
        2.2.2. Quantity of Type 2
    2.3. Materials
        2.3.1. Quantity of Material 1
            2.3.1.1. Quantity of Grade 1
            2.3.1.2. Quantity of Grade 2
    2.4. Supplies
        2.4.1. Quantity of Supply 1
        2.4.2. Quantity of Supply 2
    2.5. Locations
        2.5.1. Location 1
        2.5.2. Location 2

Figure A-6
Resource Breakdown Structure

Page 1 of 1

ACTIVITY DURATION ESTIMATES

Project Title: ______________________ Date Prepared: ______________________

WBS ID	Activity Description	Effort Hours	Duration Estimate

Figure A-7
Activity Duration Estimates

Page 1 of 1

COST ESTIMATING WORKSHEET

Project Title: ______________________ Date Prepared: ______________________

Parametric Estimates				
WBS ID	Cost Variable	Cost per Unit	Number of Units	Cost Estimate

Analogous Estimates					
WBS ID	Previous Activity	Previous Cost	Current Activity	Multiplier	Cost Estimate

Three Point Estimates					
WBS ID	Optimistic Cost	Most Likely Cost	Pessimistic Cost	Weighting Equation	Expected Cost Estimate

Figure A-8
Cost Estimating Worksheet

RESPONSIBILITY ASSIGNMENT MATRIX

Project Title: ______________________ **Date Prepared:** ______________________

	Person 1	Person 2	Person 3	Person 4	Etc.
Work package 1	R	C	A		
Work package 2		A		I	R
Work package 3		R	R	A	
Work package 4	A	R	I	C	
Work package 5	C	R	R		A
Work package 6	R		A	I	
Etc.	C	A		R	R

R = Responsible: The person performing the work.

A = Accountable: The person who is answerable to the project manager that the work is done on time, meets requirements, and is acceptable.

C = Consult: The person who has information necessary to complete the work.

I = Inform: This person should be notified when the work is complete.

Figure A-9
Responsibility Assignment Matrix

COMMUNICATIONS MANAGEMENT PLAN

Project Title: ______________________ Date Prepared: ______________________

Stakeholder	Information	Method	Timing or Frequency	Sender

Assumptions	Constraints

Glossary of Terms or Acronyms

Attach relevant communication diagrams or flowcharts.

Figure A-10
Communications Management Plan

Page 1 of 1

RISK REGISTER

Project Title: ______________________ Date Prepared: ______________________

Risk ID	Risk Statement	Probability	Impact		Score	Response
			Scope	Quality	Schedule	Cost

Revised Probability	Revised Impact		Revised Score	Responsible Party	Actions	Status	Comments
	Scope		Quality	Schedule	Cost		

Page 1 of 1

Figure A-11
Risk Register

RISK DATA SHEET

Project Title: ______________________ Date Prepared: ______________________

Risk ID	Risk Description						
Status	Risk Cause						
Probability	Impact				Score	Responses	
	Scope	Quality	Schedule	Cost			
Revised Probability	Revised Impact				Revised Score	Responsible Party	Actions
	Scope	Quality	Schedule	Cost			
Secondary Risks							
Residual Risk							
Contingency Plan					Contingency Funds		
					Contingency Time		
Fallback Plans							
Comments							

Figure A-12
Risk Data Sheet

PROCUREMENT MANAGEMENT PLAN

Project Title: ______________________ Date Prepared: ______________________

Procurement Authority

Roles and Responsibilities:

Project Manager	Procurement Department
1.	1.
2.	2.
3.	3.
4.	4.
5.	5.

Standard Procurement Documents

1.
2.
3.
4.
5.

Contract Type

Figure A-13
Procurement Management Plan

PROCUREMENT MANAGEMENT PLAN

Bonding and Insurance Requirements

Selection Criteria

Weight	Criteria

Procurement Assumptions and Constraints

Figure A-13
Procurement Management Plan (*continued*)

PROCUREMENT MANAGEMENT PLAN

Integration Requirements

WBS	
Schedule	
Documentation	
Risk	
Performance Reporting	

Performance Metrics

Domain	Metric Measurement

Figure A-13
Procurement Management Plan (*continued*)

TEAM OPERATING AGREEMENT

Project Title: ______________________ Date Prepared: ______________________

Team Values and Principles

1.
2.
3.
4.
5.

Meeting Guidelines

1.
2.
3.
4.
5.

Communication Guidelines

1.
2.
3.
4.
5.

Decision-Making Process

Figure A-14
Team Operating Agreement

TEAM OPERATING AGREEMENT

Conflict Management Approach

Other Agreements

Signature: Date:

Figure A-14
Team Operating Agreement (*continued*)

Page 2 of 2

CHANGE REQUEST

Project Title: ______________________ Date Prepared: ______________

Person Requesting Change: ______________ Change Number: ______________

Category of Change:

☐ Scope ☐ Quality ☐ Requirements

☐ Cost ☐ Schedule ☐ Documents

Detailed Description of Proposed Change

Justification for Proposed Change

Impacts of Change

Scope	☐ Increase	☐ Decrease	☐ Modify
Description:			
Grade:	☐ Increase	☐ Decrease	☐ Modify
Description:			

Figure A-15
Change Request

CHANGE REQUEST

Requirements	☐ Increase	☐ Decrease	☐ Modify
Description:			
Cost	☐ Increase	☐ Decrease	☐ Modify
Description:			
Schedule	☐ Increase	☐ Decrease	☐ Modify
Description:			
Stakeholder Impact	☐ High risk	☐ Medium risk	☐ Low risk
Description: Project Documents			

Comments

Figure A-15
Change Request (*continued*)

CHANGE REQUEST

Disposition ☐ Approve ☐ Defer ☐ Reject

Justification

Change Control Board Signatures

Name	Role	Signature

Date: ______________________

Figure A-15
Change Request (*continued*)

LESSONS LEARNED

Project Title: ______________________ Date Prepared: ______________________

Project Performance Analysis

	What Worked Well	What Can Be Improved
Requirements definition and management		
Scope definition and management		
Schedule development and control		
Cost estimating and control		
Quality planning and control		
Human resource availability, team development, and performance		
Communication management		
Stakeholder management		
Reporting		
Risk management		
Procurement planning and management		
Process improvement information		
Product-specific information		
Other		

Page 1 of 2

Figure A-16
Lessons Learned

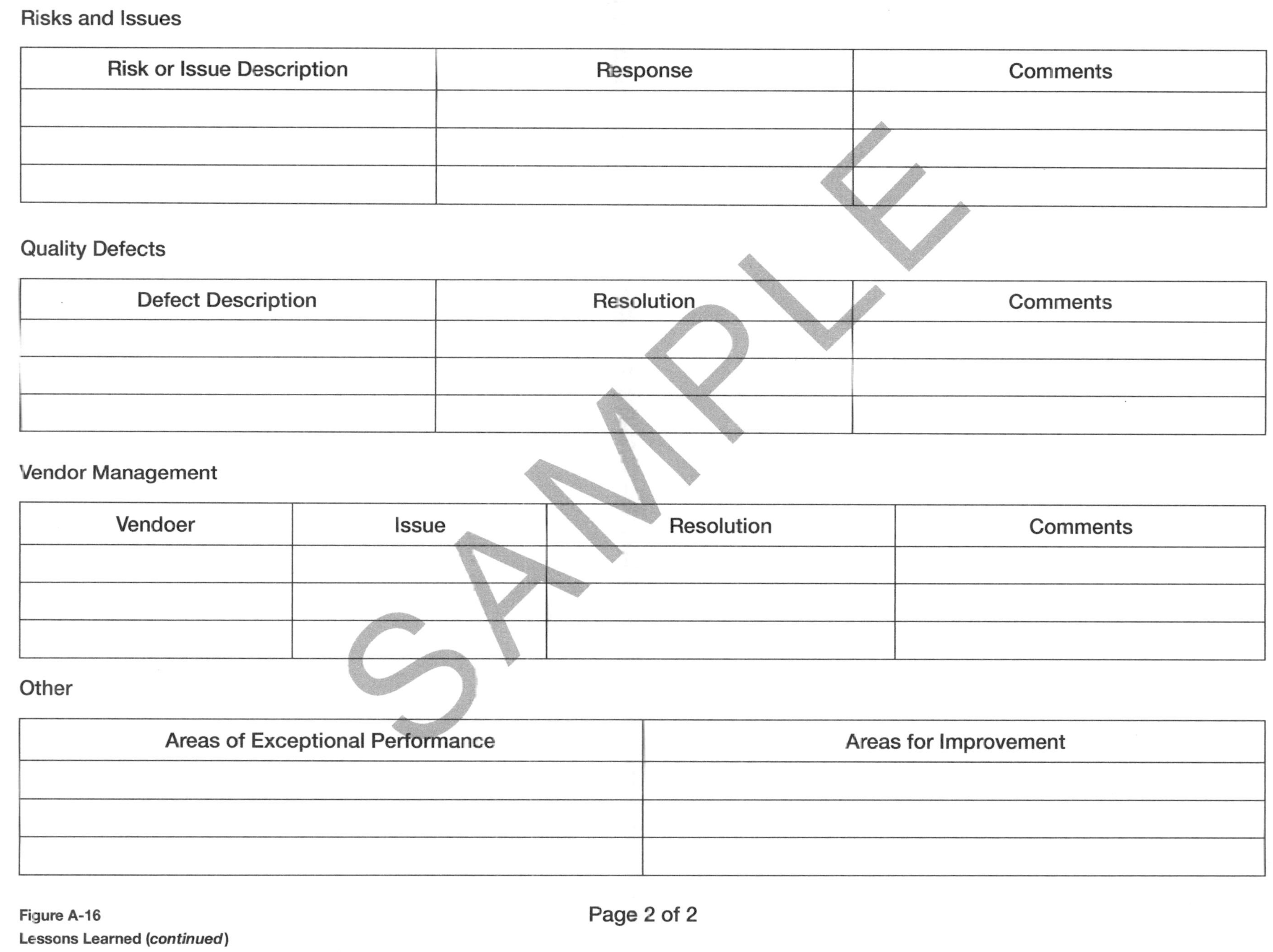

LESSONS LEARNED

Risks and Issues

Risk or Issue Description	Response	Comments

Quality Defects

Defect Description	Resolution	Comments

Vendor Management

Vendoer	Issue	Resolution	Comments

Other

Areas of Exceptional Performance	Areas for Improvement

Page 2 of 2

Figure A-16
Lessons Learned (*continued*)

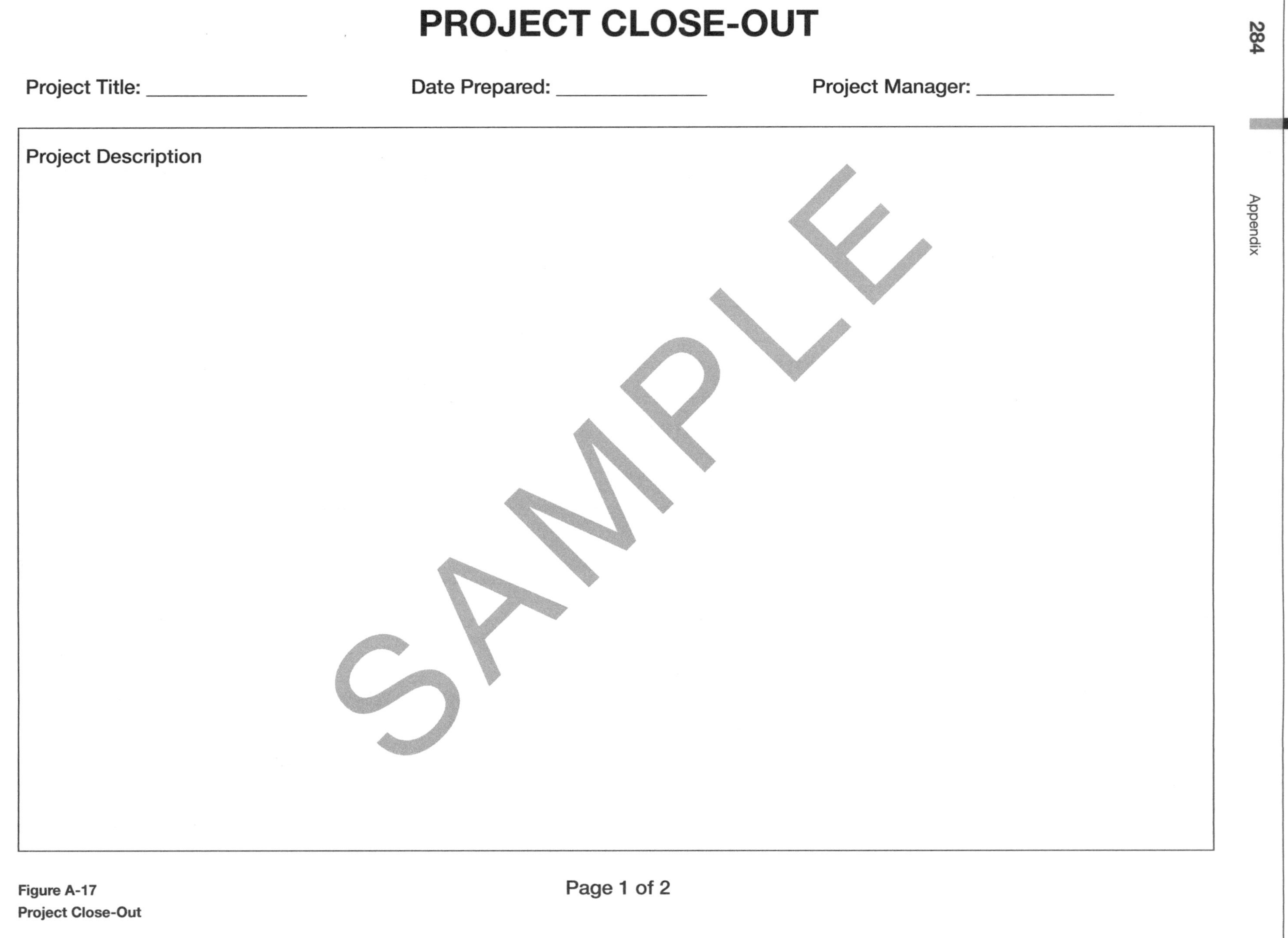

PROJECT CLOSE-OUT

Project Title: ______________ Date Prepared: ______________ Project Manager: ______________

Project Description

Page 1 of 2

Figure A-17
Project Close-Out

PROJECT CLOSE-OUT

Performance Summary

	Project Objectives	Completion Criteria	How Met
Scope			
Quality			
Time			
Cost			

Figure A-17
Project Close-Out (*continued*)

Page 2 of 2

PROCUREMENT AUDIT

Project Title: ____________________ Date Prepared: ____________________

Project Auditor: ____________________ Audit Date: ____________________

Vendor Performance Audit

What Worked Well	
Scope	
Quality	
Schedule	
Cost	
Other	
What Can Be Improved	
Scope	
Quality	
Schedule	
Cost	
Other	

Procurement Management Process Audit

Process	Followed	Tools and Techniques Used
Plan Procurements		
Conduct Procurements		
Administer Procurements		
Close Procurements		

Figure A-18
Procurement Audit

PROCUREMENT AUDIT

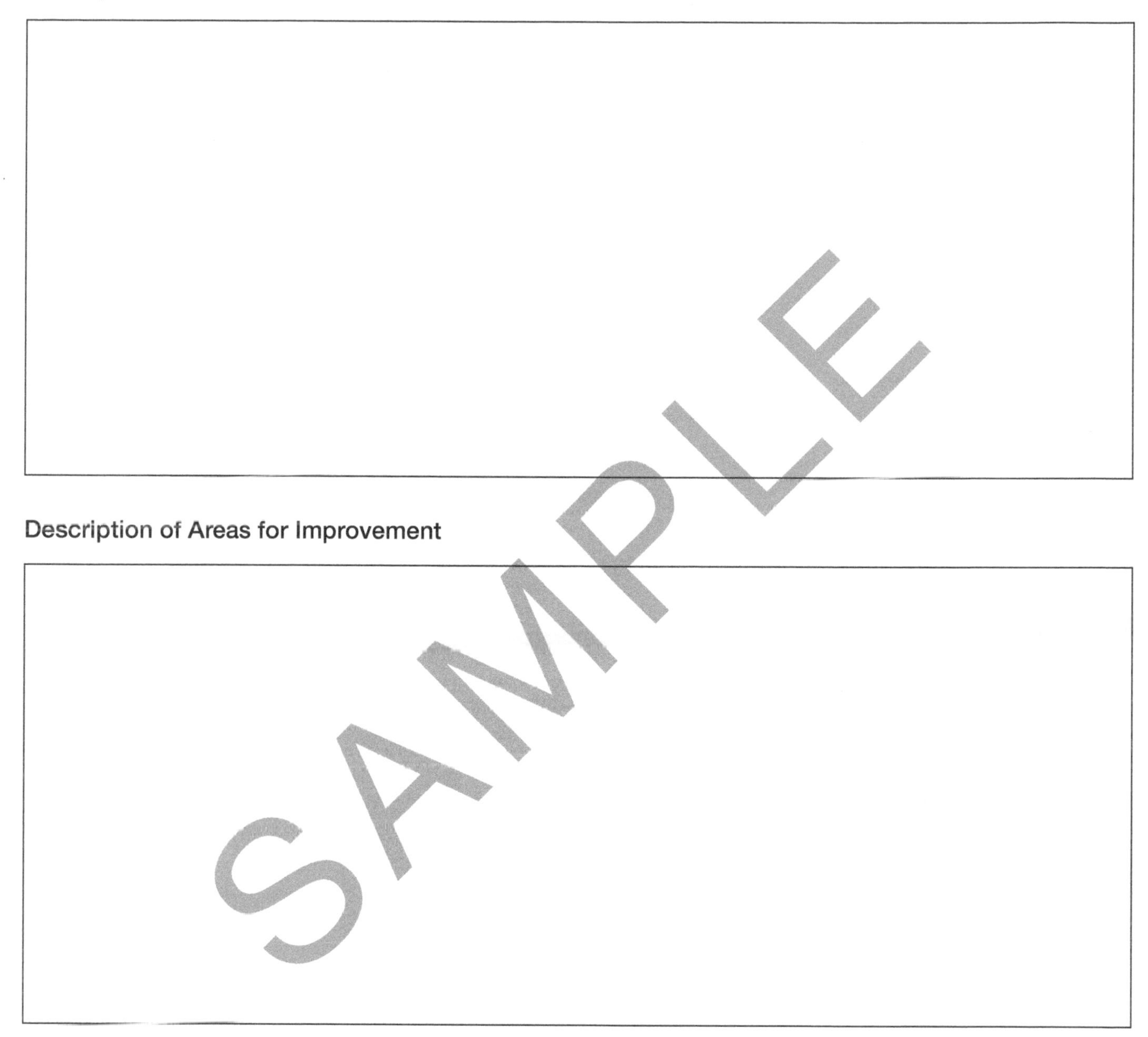

Description of Good Practices to Share

Description of Areas for Improvement

Figure A-18
Procurement Audit (*continued*)

Index

A

Accept (negative risk event/threat strategy), 143, 144–145
Accept (positive risks/opportunities strategies), 145, 146
Acceptance criteria, 90, 246
Accuracy, 100–101
  definition, 100
Acquire Project Team, 173–176
  acquisition, usage, 175
  data flow diagram, 174f
  enterprise environmental factors, 175
  flowchart, 174f
  inputs, 174–175
  multicriteria decision analysis, 175–176
  organizational process assets, 175
  outputs, 176
  resources, 176
  tools/techniques, 175–176
Activity cost estimate
  categorization, 93
  determination, 130, 132
Activity on Node (AON) diagramming, 61
Activity resource requirements, development, 113
Actor, definition, 41
Actual Cost (AC)
  definition, 223
  example, 227f
  inclusion, 226t
Affinity diagram
  example, 38f
  usage, 38
Agreements, 148
  input consideration, 17
Analogous cost estimates, 91
  example, 92
Analogous duration estimates, 91
Analogous estimating, information (usage), 91
Analogous time estimate, 70
Appraisal
  conformance cost element, 103
  costs, 104
Assumptions analysis, 133
Attribute sampling, quality control terminology, 231
Authority, definition, 115
Avoid (negative risk event/threat strategy), 143, 144

B

Backward pass, example, 78f
Backyard WBS, example, 223f
Baseline
  creation, 83
  schedule, 217
Basis of estimates, 93
  example, 94
Benchmarking, 41, 169
  usage, 107
Bidder conferences, 194
Bottom-up estimates, development, 92
Bottom-up estimating, usage, 92
Bottom-up re-forecast, 203
Brainstorming
  sessions, 45
  usage, 38, 200

Budget
example, 224f
metric, 108
Budget at Completion (BAC)
calculations, 228
definition, 223
Buyers, names, 148

C

Capability Maturity Models (CMMs), 108
Cause and effect diagram, 169
flowchart, 105f
quality tool, 234
usage, 105
Change control processes, 30
Change Log
contents, 199
usage, 198
Change management/control, 28
Change Request, form contents, 165
Change requests, 208, 248
approval, 210
Checklist analysis, 133
Checksheets
quality tool, 234
usage, 105–106
Childcare center activities
affinity diagram, example, 38f
context diagram, 41f
mind map, example, 39f
Close Procurements, 258–259
data flow diagram, 258f
documents, usage, 259
flowchart, 258f
inputs, 259
negotiations, 259
organizational process assets, update, 259
outputs, 259
tools/techniques, 259
Close Project or Phase, 255–257
accepted deliverables, 257
analytical techniques, 257
data flow diagram, 256f
flowchart, 256f
inputs, 256–257
organizational process assets, update, 257
outputs, 257
technical expert judgment, usage, 257
tools/techniques, 257
Closing Process Group, 3, 255
Collaboration, conflict management technique, 184–185
Collect requirements, 35–43
data flow diagram, 36f
flowchart, 36f
inputs, 36–37
outputs, 42–43
process, 45
project/product requirements, 35
tools/techniques, 37–41
types, 37f
Communication management plan, 159
Communications
elements, 118
equation, 120–121
monitoring/control, 237
Communications management
execution, 187
plan, 28
Compromise, conflict management technique, 185
Conduct Procurements, 191–195
bidder conferences, 194
data flow diagram, 193f
flowchart, 192f
inputs, 191–193
Invitation for Bid (IFB), 191
make-or-buy decisions, 193
negotiated contract elements, 194
organizational process assets, 193
outputs, 194–195
Request for Proposal (RFP), 191
Request for Quotation (RFQ), 191–192
resource calendars, 195
subject matter expertise, impact, 194
tools/techniques, 193–194
Configuration identification, 206
Configuration management
processes, 30
system, 206
Configuration management/control, 28
Configuration status accounting, 206
Configuration verification/audit, 206
Conflict management techniques, 183t–184t
example, 184–185

Conflict resolution, 199
Conflict sources, 183
Conformance, cost, 103
Context diagrams, usage, 41
Contingent response strategies, usage, 146
Contract change control system, 247
Contractors (identification), procurement documents (usage), 20
Contract templates, 151
Contract terminology, 152
Control account, definition, 50
Control charts, 169
  example, 107
  quality tool, 234
  usage, 106
Control Communications, 237–239
  data flow diagram, 238f
  flowchart, 237f
  inputs, 238
  issue logs, documentation, 238
  outputs, 239
  tools/techniques, 238
  work performance information, contextualized information, 239
Control Costs, 221–230
  calculations/assumptions, 228t
  data flow diagram, 222f
  flowchart, 222f
  forecasting, impact, 227
  indexes, 266
  inputs, 221–223
  measurement guidelines, 226
  outputs, 229
  project funding requirements, 222–223
  spending profile, impact, 221
  tools/techniques, 223–229
  variances, example, 227t
Control limits, quality control terminology, 231
Control Procurements, 245–249
  change requests, 248
  contract administration, 246
  contract change control system, 247
  data flow diagram, 247f
  inputs, 246
  outputs, 248–249
  payment system, 247
  records management system, 247
  tools/techniques, 247–248
Control Quality, 231–235
  data flow diagram, 233f
  flowchart, 233f
  inputs, 232, 234
  organizational process assets, 234
  outputs, 235
  tools/techniques, 234–235
  work performance data, 234
Control Risks, 241–244
  baselines, usage, 242–243
  data flow diagram, 242f
  flowchart, 242f
  inputs, 241–243
  outputs, 244
  Risk Register, 242
  status meetings, usage, 243
  tools/techniques, 243
  trend analysis, 243
  variance analysis, 243
Control Schedule, 217–220
  agile environment, impact, 218
  data flow diagram, 219f
  flowchart, 218f
  forecasts, 220
  inputs, 218–219
  organizational process assets, 219
  outputs, 220
  performance reviews, 219–220
  project calendars, usage, 218–219
  schedule data, 219
  tools/techniques, 219–220
  variance analysis, 219–220
  variances, response, 220
Control Scope, 213–215
  data flow diagram, 214f
  flowchart, 213f
  inputs, 213–214
  organizational process assets, 214, 215
  outputs, 215
  requirements management plan, 214
  tools/techniques, 214–215
  variance analysis, example, 215
Control Stakeholder Engagement, 251–253
  data flow diagram, 252f
  flow chart, 252f
  inputs, 251
  issue logs, 251, 253
  outputs, 253

Control Stakeholder Engagement (*continued*)
 tools/techniques, 252
 work performance information, usage, 253
Control thresholds, 87
Control tools, storyboard, 171
Core competencies, 153
Corporate knowledge base, 11–12
Corrective action, definition, 161
Cost
 aggregation, 96
 baseline, 28, 96–97
 definition, 152
 management plan, 28, 88, 130
 monitoring/control, 221
Cost-benefit analysis
 quality cost, combination, 103
 techniques, 17
Cost estimate
 contents, 90
 detail, 93
Cost estimating
 policies/procedures/guidelines, 90–91
 sensitivity, 88
Cost of quality
 analysis, 93
 example, 104–105
Cost Performance Index (CPI), definition, 226
Cost-plus-award-fee (CPAF) contracts, 152
Cost-plus-fixed-fee (CPFF) contracts, 152
Cost-plus-incentive-fee (CPIF) contracts, 152
Cost-reimbursable contracts, 152
Cost Variance (CV), definition, 226
Crashing, 82
Create WBS, 46–51
 activity-orient elements, 49–50
 data flow diagram, 47f
 deliverables, 49
 enterprise environmental factors, 48
 flowchart, 47f
 inputs, 48
 organizational process assets, 48
 outputs, 49–51
 product deliverables, 48
 project document updates, 51
 project management elements, 49–50
 project scope description, 48
 requirements documentation, 48
 scope baseline, 49
 scope statement contents, 46
 tools/techniques, 48–49
Critical chain method, example, 79f
Critical chain methodology
 duration estimates, padding (absence), 79
 estimates, usage, 76
Critical path
 calculation, 76
 first iteration, 80
 float identification, 78f
 method, 75
Cultural awareness, 186
Customer requests, 17
Customer satisfaction
 impact, 99
 metric, 108

D

Data-gathering techniques, 140
Data-representation techniques, 140
Decision tree
 diagram, 141f
 usage, 142
Decode, definition, 120
Decomposition, definition, 48
Defect frequency metric, 108
Defect repair, definition, 161
Define Activities, 56–59
 activity attributes, 59
 data flow diagram, 57f
 flowchart, 57f
 inputs, 57–58
 outputs, 59
 tools/techniques, 58
Define Scope, 43–46
 analysis, types, 45
 assumptions, 46
 data flow diagram, 44f
 flowchart, 43f
 inputs, 43–44
 organizational process assets, 44
 outputs, 45–46
 project deliverables, 45
 project exclusions/constraints, impact, 46
 questions, 43
 requirements documentation, 44
 tools/techniques, 45

Deliverables, technical information, 246
Deliverable schedule, 217
Delphi technique, 71–72, 93
design, 133
usage, 38
Dependencies
determination, 62
types, 62–63
Design of experiments, 107–108, 169
Detailed testing schedule, 217
Determine Budget, 94–97
agreements, information, 95
cost baseline, 96, 97
data flow diagram, 95f
flowchart, 94f
historical relationships, 96
inputs, 94–95
outputs, 96–97
project schedule, usage, 95
resource calendars, usage, 95
tools/techniques, 96
Develop Project Charter, 15–18
data flow diagram, 16f
inputs/tools/techniques/outputs, flowchart, 15f
Develop Project Management Plan, 26–30
data flow diagram, 27f
enterprise environmental factors, 28
flowchart, 26f
industry standards/regulations, 28
inputs, 28–29
organizational process assets, policies, 28–29
outputs, 29–30
planning processes, outputs, 28
tools/techniques, 29
Develop Project Team, 176–181
co-location, 180
data flow diagram, 177f
flowchart, 177f
ground rules, 180
hygiene/motivation factors, 178t
inputs, 178
interpersonal skills, 178
leadership ability, 178
motivation, impact, 178
outputs, 181
personnel assessment tools, usage, 181
recognition/rewards, 180
team building, stages, 179–180
Theory X/Theory Y, 179, 179t
tools/techniques, 178–181
training, 180
Develop Schedule, 73–83
baseline, creation, 83
baselining, 83
compression, 82
crashing, 82
data, 83
data flow diagram, 74f
fixed duration, 82
flowchart, 73f
inputs, 74–75
outputs, 82–83
tools/techniques, 75–81
Direct and Manage Project Execution, 26
Direct and Manage Project Work, 162–165
change requests, 165
data flow diagram, 163f
expert judgment, 164
flowchart, 163f
inputs, 162, 164
outputs, 164–165
project management information system, 164
tools/techniques, 164
Discretionary dependencies, 62
Documentation reviews, usage, 133
Documents
analysis, requirements gathering, 41
reviews, 132
Duration
conversion, 73
definition, 72

E

Early dates/late dates, mathematical difference, 75
Early finish date (EF), definition, 76
Early start date (ES), definition, 76
Earned Value (EV)
calculation, 225t
calculations summary table, 230f
definition, 223
example, 227f
Earned value management, definition, 223
Effort
definition, 72
duration conversion, 73

Encode, definition, 120
Enhance (positive risks/opportunities strategies), 145
Enterprise environment factors (EEFs), 9–10
  consideration, 17, 20
  organizational process assets, contrast, 12
  updates, 186
Estimate Activity Durations, 68–73
  data flow diagram, 69f
  effort, duration conversion, 73
  enterprise environmental factors, 70
  flowchart, 68f
  group decision making techniques, 71–72
  inputs, 69–70
  outputs, 72
  project scope statement, assumptions/constraints, 69
  resource calendars, usage, 69
  three-point time estimate, 71
  tools/techniques, 70–72
Estimate Activity Resources, 63–67
  data flow diagram, 65f
  flowchart, 65f
  inputs, 64–66
  outputs, 67
  resource estimating iterations, 66f
  tools/techniques, 66–67
Estimate at completion (EAC), 203
  calculations, 228
  forecast, 228
Estimate at Completion (EAC), definition, 227
Estimate Costs, 88–94
  data flow diagram, 89f
  enterprise environmental factors, 90
  flowchart, 89f
  inputs, 88–91
  outputs, 93–94
  risk register, risks, 90
  tools/techniques, 91–93
Estimates
  analogous time estimate, 70
  basis, 94
  parametric time estimate, 71
  refining, 68
  usage, 67
Estimate to Complete (ETC), definition, 227
Executing Process Group, 2, 161–162
Expected cost, 92
Expected monetary value, 140
Experiments, design. *See* Design of experiments
Expert judgment, 45
  usage, 137
Exploit (positive risks/opportunities strategies), 145
External dependencies, 62
External failure, 103
  costs, 104–105

F

Facilitated workshops, 37–38
Fast tracking, 82
Fee, definition, 152
Financial investment/return, factors, 17
Finish dates, adjustment, 80
Finish-to-finish (FF), 61
Finish-to-start (FS), 61
Firm-fixed-price (FFP) contracts, 151
Fixed duration, 82
Fixed-priced contracts, 151–152
Fixed-price-incentive-fee (FPIF) contracts, 151–152
Flowcharts, 169
  quality tool, 234
  usage, 105
Focus groups, interviews (comparison), 37
Forcing, conflict management technique, 185
Forward pass, example, 77f
Free float
  definition, 76
  mathematical difference, 75
Funding limit reconciliation, usage, 96
Future value (FV), 17

G

Grade
  definition, 100
  usage, 100
Group decision making techniques, 71–72, 93

H

Herzberg, Frederick (motivation theory), 178
Histograms (bar charts), 106, 169
  example, 106f
  reference, 116

Human resource management
execution, 173
plan, 174
Human resource plan, 90
development, expert judgment (usage), 115
project management plan component, 115
Human resources management plan, 28, 132
Human resources plan, information, 159
Hygiene/motivation factors, 178t

I

Identify Risks, 129–134
cost management plan, 130, 132
data flow diagram, 131f
documentation reviews, usage, 133
enterprise environmental factors, 132
flowchart, 130f
inputs, 130–132
outputs, 134
schedule management plan, 132
tools/techniques, 133–134
Identify Stakeholders, 18–21
data flow diagram, 19f
flowchart, 19f
inputs, 20
outputs, 21
tools/techniques, 20–21
Impact matrix
numerical ratings, usage, 136
probability, relationship, 129f
Indexes, etiquette, 226
Information sensitivity, 153
Information technology (IT) project, phases (example), 8
Initial resource utilization, 81f
Initiating process group, 2, 13
Input, 2, 16–17
Inspection, quality control terminology, 231
Intellectual property, 153
Interactive communication, 199
Internal dependencies, 62–63
Internal failure, 103
costs, 104
Internal organizational need, 16
Internal rate of return (IRR), 17
Interviews, focus groups (comparison), 37
Invitation for Bid (IFB), 154
Issue log, 251
documentation, 238
usage, 182–183

J

Joint Application Development (JAD)
sessions, 45
workshops, 38

K

Knowledge areas mapping, 5f
Knowledge workers, 38

L

Lag
application, 82
definition, 63
Late finish date (LF), definition, 76
Lateral thinking, 45
Late start date (LS), definition, 76
Lead
application, 82
definition, 63
Linkages, 87
London Organizing Committee for the Olympic Games (LOCOG), portfolio, 8

M

Make-or-buy analysis, 154
Manage Communications, 187–189
data flow diagram, 188f
flowchart, 188f
inputs, 187, 189
management, 189
organizational process assets, 189
outputs, 189
performance reporting, 189
tools/techniques, 189
work performance reports, 187
Management reserve, unplanned in-scope work, 96
Manage Project Team, 181–186
conflict management techniques, 184–185
conflict sources, 183
cultural awareness, 186

Manage Project Team (*continued*)
data flow diagram, 182f
decision-making techniques, 185–186
enterprise environmental factors, updates, 186
flowchart, 181f
human resource management, 186
inputs, 182–183
interpersonal skills, 185
issue log, usage, 182–183
organizational process assets, 183
outputs, 186
project staff assignments, 182
team performance assessments, 182
tools/techniques, 183–186
Manage Stakeholder Engagement, 197–200
active listening, 199
change, resistance (overcoming), 199–200
Change Log
conflict resolution, 199
data flow diagram, 198f
flowchart, 198f
inputs, 197–199
interactive communication, 199
organizational behavior, modification, 200
organizational process assets, updates, 200
outputs, 200
project management plan, updates, 200
project needs (satisfaction), agreements (negotiation), 200
project objectives, consensus (facilitation), 200
project support, people (influencing), 200
pull communication, 199
push communication, 199
tools/techniques, 199–200
trust, building, 199
Mandatory dependencies, 62
Market demand, 16
Marketplace conditions, enterprise environmental factors, 10
Medium, definition, 120
Meetings, usage, 108
Metrics
ID, usage, 109
sample, 108
Milestone
definition, 59
list, usage, 59
schedule, 217
Mind map, example, 39f
Mind mapping
ideas, usage, 39
usage, 38
Mitigate (negative risk event/threat strategy), 143, 144
Monitor and Control Project Work, 26, 202–205
activities, 202–203
analytical techniques, 205
changes, approval, 203
data flow diagram, 204f
flowchart, 203f
inputs, 203, 205
outputs, 205
performance problems, 205
performance report information, 205
schedule and cost forecasts project, 203
tools/techniques, 205
Monitoring and Controlling Process Group, 201
Monitoring and controlling process group, 3
Monitoring and Controlling Scope, 211
Monte Carlo analysis, usage, 80
Motivation, theory (Herzberg), 178
Multicriteria decision analysis
example, 176t
ratedmulticriteria decision analysis, 40t
usage, 38, 39
weighting criteria, 40t
Multiphase projects, charter (usage), 15

N

Negative risk events/threats, strategies, 143
Negotiated contract elements, 194
Net present value (NPV), 17
Network diagram, 64f, 217
durations, inclusion, 77f
Noise, definition, 120
Nominal group technique, usage, 38
Nonconformance, cost, 103

O

Observation, usage, 40
Olympic Delivery Authority (ODA), management facilitation (example), 8
Opportunity
definition, 123
identification, 130, 135

Organizational breakdown structure (OBS), definition, 114
Organizational charts, usage, 114
Organizational culture, enterprise environmental factors, 10
Organizational process assets (OPAs), 10–12
enterprise environmental factors, contrast, 12
helpfulness, 17, 20
update, 220
Organizational processes/procedures, 11
Output, definition, 2

P

Parametric cost estimates, example, 92
Parametric estimates, usage, 70–71
Parametric estimating, mathematical relationship (usage), 91
Parametric time estimate, 71
Pareto charts (bar charts), 169
example, 106f
quality tool, 234
usage, 106
Payback period, 17
Payment system, 247
Performance
reporting requirements, 246
Performance measurement
baseline, 224, 224f
guidelines, 29
rules, 87
Performance report information, 205
Performance reporting, 189
Performing Quality Assurance, 167–171
data flow diagram, 168f
flowchart, 168f
inputs, 168–169
outputs, 171
quality management and control tools, 170f
tools/techniques, 169–171
Perform Integrated Change Control, 205–210
analytical techniques, 205
change control meetings, formality, 208, 210
change control system, consideration, 206
change requests, 208
configuration,example, 207
data flow diagram, 209f
flowchart, 208f
inputs, 208
outputs, 210
process, 206
tools/techniques, 208, 210
Perform Qualitative Risk Analysis, 134–138
data flow diagram, 136f
enterprise environmental factors, 136
expert judgment, usage, 137
flowchart, 135f
inputs, 135–136
opportunities, analysis, 135
outputs, 138
project documents, usage, 138
risk assessment, 137
risk urgency assessment, usage, 137
scope baseline, 135–136
tools/techniques, 136–137
Perform Quantitative Risk Analysis, 138–142
cost management plan, 139
data flow diagram, 139f
data-gathering techniques, 140
data-representation techniques, 140
decision tree, 141f, 142
expected monetary value, 140
expert judgment, usefulness, 142
flowchart, 138f
inputs, 139
organizational process assets, 139
outputs, 142
probability distribution, 140–141
schedule management plan, 139
tools/techniques, 140–142
Personnel assessment tools, usage, 181
Plan Communications, 117–121
data flow diagram, 119f
enterprise environmental factors, 118–119
flowchart, 118f
inputs, 118–119
organizational process assets, 119
outputs, 121
technology, usage, 119–120
tools/techniques, 119–121
Plan Cost Management, 85–88
accuracy level, 87
contents, 87–88
control thresholds, 87
cost management plan, 88
data flow diagram, 86f

Plan Cost Management (*continued*)
flowchart, 86f
linkages, 87
measure units, 87
outputs, 87–88
performance measurement, rules, 88
reporting, 87
tools/techniques, 87
Plan-do-check-act (PDCA) process, 99
Plan Human Resource Management, 112–116
data flow diagram, 113f
enterprise environmental factors, impact, 113–114
flowchart, 112f
inputs, 112–114
outputs, 115–116
tools/techniques, 114–115
Planned Value (PV)
definition, 223
example, 227f
Planning loops, 24–25
Planning processes
outputs, 28
overlap/interaction, 24
Planning process group, 2, 23
high-level information, usage, 3
Plan Procurement Management, 148–155
contract templates, 151
cost-reimbursable contracts, 152
data flow diagram, 149f
enterprise environmental factors, importance, 150–151
fixed-priced contracts, 151–152
flowchart, 149f
inputs, 150–153
make-or-buy analysis, 154
make-or-buy decision, 153
market research, usage, 153–154
organizational process assets, 150–151
outputs, 154–155
request for proposal (RFP), 154
schedule, 150–151
selection criteria, 154–155
time-material contracts, 152–153
tools/techniques, 153–154
Plan Quality, 101–109
data flow diagram, 102f
flowchart, 101f
organizational process assets, 103
outputs, 108–109
tools/techniques, 103–108
Plan Quality Management process, 234
Plan Risk Management, 124–129
contents, 127
data flow diagram, 125f
enterprise environmental factors, 126
flowchart, 125f
inputs, 125–126
organizational process assets, 126
outputs, 126–129
plan, components, 126
planning meetings, 126
probability, 127
risk, identification, 127
risk breakdown structure, example, 128f
tools/techniques, 126
Plan Risk Responses, 142–146
contingent response strategies, usage, 146
data flow diagram, 143f
expert judgment, usage, 146
flowchart, 143f
inputs, 142–143
outputs, 146
tools/techniques, 143–146
Plan Schedule Management, 54–56
data flow diagram, 55f
flowchart, 54f
inputs, 54–55
outputs, 56
tools/techniques, 55–56
Plan scope management, 32–35
data flow diagram, 33f
enterprise environmental factors, 34
flowchart, 32f
inputs, 32–34
outputs, 34–35
tools/techniques, 34
Plan Stakeholder Management, 157–160
data flow diagram, 158f
engagement levels, 159
enterprise environmental factors, 159
flowchart, 158f
inputs, 158–159
outputs, 160
process, 197
tools/techniques, 159–160

Portfolio, 7
 definition, 7
Position descriptions, roles/responsibilities, 114
Positive risks/opportunities strategies, 145
Precedence diagramming method (PDM), 61
Precision, 100–101
 definition, 100
Predecessor Activity, definition, 61
Prevention
 conformance cost element, 103
 costs, 104
Preventive action, definition, 161
Price, definition, 152
Probability, 127
 defining, 128–129
 distribution, 140–141
 impact matrix, relationship, 129f
 matrixes, numerical ratings (usage), 136
Problem solving, conflict management technique, 184
Process improvement plan, usage, 109
Procurement management, 147
 plan, 28, 155, 191
Procurements
 documents, 20, 154
 monitoring/control, 245
 paper generation, 249
 statement of work (SOW), 154
Procurement statement of work, 192
Product
 breakdown structure, 45
 description/deliverables, 90
 quality processes/tools, 99
 scope, definition, 31
Program, 7
Program Evaluation & Review Technique (PERT) estimate, usage, 92
Progressive elaboration
 definition, 9, 23
 example, 24
Project, 7
 baselines, performance (maintenance), 14
 calendars, usage, 218–219
 closure, 255
 cost management, 4
 definition, 7
 deliverables, 45
 exclusions/constraints, impact, 46
 files, definition, 12
 funding requirements, impact, 96
 human resource management, 4
 interfaces, 29
 knowledge areas mapping, list, 5f
 monitoring/controlling, 201
 needs (satisfaction), agreements (negotiation), 200
 objectives, documentation, 14
 performance, information, 241
 portfolio, example, 8
 reviews, 29
 sponsor role, 13–14
 staff assignments, 176
 status, reporting, 14
 undertaking, reasons, 16–17
 work, conducting, 162
Project charter
 high-level information, 37
 project document, 15
Project Communications Management, 6, 117
Project Cost Management, 85
Project documents, 132
 components, 169
 project management plan, contrast, 30f
 updates, 97, 109, 121, 138
Project Human Resource Management, 111–112
Project integration management, 4, 25–26
 knowledge area, processes, 26
Project life cycle, 8–9, 29
 definition, 8
Project management
 elements, 49–50
 information system, enterprise environmental factors, 10
 knowledge areas, 4–6
 software, usage, 229
 team, definition, 112
Project Management Office (PMO), usage, 13
Project management plan, 176
 components, updates, 239
 development, 24, 26
 elements, 161
 procurement management plan, usage, 246
 project documents, contrast, 30f
 quality management plan, relationship, 108
 refinement loop, 25f
 risk management plan, 241

Project management plan (*continued*)
subsidiary management plans, 203
updates, 200, 220, 244
Project management process groups, 2–3
flowchart, 3f
list, 5f
Project manager
definition, 112
role, responsibilities, 14–15
tailoring, 9
Project Manager's Book of Forms, The, 2
Project phase, 8
ending, usage, 8–9
Project Procurement Management, 147
Project procurement management, 4, 6
Project quality management, 4, 99–101
terms, 100
Project Risk Management, 123–124
Project risk management, 6
Project schedule
display, 82–83
usage, 95
Project scope
definition, 31
management, 4, 31–32
organization/definition, work breakdown structure (usage), 49
statement, 45, 69
Project-specific needs, 116
Project stakeholder
communication, 14
management, 4, 6
Project team, tailoring, 9
Project Time Management, 53
Project time management, 4
Prototypes
usage/examples, 40–41
Pull communication, 199
Push communication, 199

Q

Quality
audit, 169
checklist, development, 109
cost. *See* Cost of quality.
definition, 100
metrics, 108, 109, 146, 169
monitoring/control, 231
planning tools, dependence, 108
tools, 234
usage, 100
Quality control
measurements, 169
terminology, 231
terminology, examples, 232t
Quality Function Deployment (QFD), 108
workshops, 38
Quality management
methodologies, enterprise environmental factors, 108
plan, 28, 132, 168, 235
tools, storyboard, 170
Quantitative analysis technique, 140
Quantitative opportunity analysis, 139
Quantitative/qualitative data collection, 40
Quantitative risk analysis, purposes, 138

R

Rated multicriteria decision analysis, 40t
Records management system, 247
Refinement loop.*See* Project management plan; Subsidiary management plan
Reporting templates, definition, 11
Request for proposal (RFP), 154
Request for quotation (RFQ), 154
Requirements documentation
information, 42, 48
measurements/specifications, 102
plan, 28
project requirements, impact, 44
Requirements management plan, 28
definition, 34
Requirements traceability matrix
benefit, 42
usage, 42
Reserve analysis, 93, 243
conducting, 96
usage, 229
Resource breakdown structure, definition, 114
Resources
balancing scenario, 80–81
calendars, usage, 69, 95, 195
estimating iterations, 66f

initial resource utilization, 81f
leveling, 81f
smoothing, 80
Responsibility, definition, 115
Responsibility Accountable Consulted and Informed (RACI) chart, usage, 115
Responsibility assignment matrix (RAM)
definition, 114
updating, 109
Return on investment (ROI), 17
Risk
assessment, 137
audit, 243
brainstorming, 133
definition, 123
monitoring/control, 241
statement, 134
tolerance, definition, 123
urgency assessment, usage, 137
Risk breakdown structure, example, 128f
Risk management, 14
activities, conducting, 124
plan, 28, 130
planning, 243
procedures, definition, 11
process, stakeholder participation, 128
project initiation/cessation, 124
responsibility, 136
Risk Register
definition, 134
information, review, 243
risks, 70, 91
usage, 102
Role, definition, 115
Rolling wave planning, 23
definition, 24
Root cause analysis, 133
Run charts, 169

S

Scatter diagrams, 169
plot, 107
quality tool, 234
Schedule
baseline, 28, 101, 246
components, 54
data, 219
monitoring/controlling, 217
network analysis, 75
Schedule management plan, 28
definition, 56
usage, 132
Schedule Performance Index (SPI), 220
definition, 226
Schedule Variance (SV), 220
definition, 226
SPI, relationship, 227
Scheduling tool, usage, 75
Scope baseline, 28
documents, 90
information, 132
scope statement, relationship, 101
work breakdown structure (WBS), relationship, 49
Scope management plan, 28
definition, 34
process, 214
project management plan component, 34
Sellers, names, 148
Sequence Activities, 59–63
data flow diagram, 60f
flowchart, 60f
inputs, 59–61
network diagram, 64f
outputs, 63
process, 82
tools/techniques, 61–63
Share (positive risks/opportunities strategies), 145
Six Sigma, 99, 108
Skill sets, availability, 153
Smoothing, conflict management technique, 185
Social needs, 17
Software Development Life Cycle, JAD sessions (presence), 38
Sponsor, definition, 112
Staffing histograms, usage, 116
Staffing management plan, 116
Stakeholders
analysis, conducting, 20
definition, 18
engagement, 159f, 251
identification, 18–21, 37
management plan, 28, 36–37
power/interest grid, 20f
register, 21, 102, 118–119, 158

Start dates, adjustment, 80
Start-to-finish (SF), 62
Start-to-start (SS), 62
Statement of work (SOW), 154
  procurement statement of work, 192
Statistical sampling, 169
  determination, 108
Status meetings, usage, 243
Strengths, Weaknesses, Opportunities, and Threats (SWOT) analysis, 133–134
  definition, 134
Subject matter expertise, impact, 194
Subsidiary management plan
  development, 33
  refinement loop, 33f
Successor Activity, definition, 61
Supplier management, systems (usage), 247
Systems analysis/engineering, 45

T

Tailoring, 9
  decisions, 29
  guidelines, 29
  usage, 45
Tally sheets, usage, 105–106
Target schedule, 217
Team building, stages, 179–180
Team members, networking, 115
Team performance assessments
  information, provision, 182
  usage, 181
Technical expert judgment, usage, 257
Technical performance measurement, usage, 243
Technique, 18
  definition, 2
Technology, advancement, 17
Theory X, 179, 179t
Theory Y, 179, 179t
Three-point communication, 120
Three-point cost estimate, example, 93
Three-point estimate, modification, 71, 92
Three-point estimating, usage, 71, 92
Three-point time estimate, 71
Time and material (T&M) contracts, 152–153
Time constraints, 153
Time estimates
  analogous time estimate, 70
  parametric time estimates, 71
Tolerance, quality control terminology, 231
Tool, 18
  definition, 2
Total float, definition, 76
Total Quality Management (TQM), 99
Transfer (negative risk event/threat strategy), 143, 144
Trend analysis, 243
  usage, 229
Tuckman, Bruce, 179

V

Validate Scope, 211–213
  acceptance, formalization, 211
  data flow diagram, 212f
  deliverables, 213
  flowchart, 211f
  inputs, 212
  outputs, 213
  tools/techniques, 212–213
Validate Scope and Control quality processes, 212
Value engineering/analysis, 45
Variable (impact), project time (basis), 124f
Variance analysis, 243
  example, 215
  impact, 214–215
Variance thresholds, 28, 29
Vendors
  bid analysis, 93
  identification, procurement documents (usage), 20
  performance reporting, 248
  status, work performance information, 248
  work, coordination, 245

W

What-if scenario analysis, 80
Withdrawal, conflict management technique, 185
Workperformance
  data, 202f, 234
  reports, 187, 202f
  technical description, 90
  transfer, 144
Work authorization, 29

Work breakdown structure (WBS), 31, 85. *See also* Create WBS, Backyard WBS
  decomposition, work packages (usage), 50f
  development, challenge, 49
  dictionary, 51, 90
  implementation, example, 48
  input, 46
  templates, organizational process assets, 48
Work package, 90
  definition, 50
  usage, 50
Work performance information, 202f
  contextualized information, 239
  data, 229
  usage, 244
Work project statement, 16